J.-F. Bonneville   F. Cattin   J.-L. Dietemann

Computed Tomography of the
Pituitary Gland

With a Chapter on
Magnetic Resonance Imaging of the Sellar
and Juxtasellar Region
By M. Mu Huo Teng and K. Sartor

With 275 Figures in 598 Separate Illustrations

Springer-Verlag Berlin Heidelberg NewYork Tokyo

Professeur JEAN-FRANÇOIS BONNEVILLE
Département de Neuroradiologie
Centre Hospitalier et Universitaire de Besançon
F-25030 Besançon

Dr. FRANÇOISE CATTIN
Département de Neuroradiologie
Centre Hospitalier et Universitaire de Besançon
F-25030 Besançon

Dr. JEAN-LOUIS DIETEMANN
Hospices Civils de Strasbourg, Centre Hospitalier Régional
Service de Radiologie, 1, Place de l'Hôpital
F-67091 Strasbourg

The cover picture shows a microprolactinoma as demonstrated by dynamic CT scan (the "tuft sign"). (See figure 5.15a, page 80)

ISBN 978-3-642-70377-5   ISBN 978-3-642-70375-1 (eBook)
DOI 10.1007/978-3-642-70375-1

Library of Congress Cataloging-in-Publication Data. Bonneville, Jean-François, 1943– .
Computed tomography of the pituitary gland. Includes bibliographies and index. 1. Pituitary
body – Diseases – Diagnosis. 2. Tomography. 3. Pituitary body – Cancer – Diagnosis. 4.
Magnetic resonance imaging. I. Cattin, F. II. Dietemann, J.L. (Jean Louis), 1951– . III.
Title. [DNLM: 1. Pituitary Gland – radiography. 2. Tomography, X-Ray Computed.
WK 500 B717c] RC658.2.B66 1986   616.4'7'07572   86-6534

Reproduction of the figures: Gustav Dreher GmbH, Stuttgart

2127/3130-543210

"The surgery of the hypophysis at the present time is practically in the stone age of its development. The time will come, before long, perhaps, when the biochemists will have shown us how to cure most of the common functional adenomas of this gland."

Harvey Cushing, 1926

Preface

The present volume is the results of 6 years' work by our team, during which time 2300 CT scans of the pituitary region were carried out. This was made possible by the close collaboration between physicians and technicians in our neuroradiological department, as well as by numerous corresponding physicians. We wish to express our gratitude for their confidence and our sincere thanks to our colleagues at Besançon, Dijon, Grenoble, Lyon, Montpellier, and Strasbourg. Furthermore, we especially wish to thank the patients who willingly accepted the difficult requirements of these studies. We are grateful to the technicians at the Neuroradiology Department of the Centre Hospitalier et Universitaire de Besançon, who have perfected the methodology so as to meet the ever increasing imperatives for precise anatomical mapping of the pituitary gland and the surrounding region; without their efforts, this book would never have been possible. Finally, we wish to express our thanks to the medical photographer of our group, as well as the secretarial staff for their contribution to the successful production of this work. We thank Laboratoires Guerbet and General Electric for their excellent assistance, and Springer-Verlag for their care and competence in the production of this book.

In writing *Computed Tomography of the Pituitary Gland*, we have sought to develop morphological study of the pituitary gland to a degree of reliability comparable to that of laboratory findings in endocrine disorders. This should provide the physician and surgeon with precise guidance for treatment of disorders of the pituitary region.

Besançon, 1986
J.-F. BONNEVILLE
F. CATTIN
J.-L. DIETEMANN

Contents

Chapter 1
Technical Aspects

Diagnosis of pituitary disorders requires optimal imaging quality, obtainable only with the most rigorous possible technique. There is also no substitute for experience. Without wishing to seem unduly chauvinistic, we would compare it with French cooking: while a good recipe and the best ingredients are imperative, a touch of art is also required.

The machine is of primary importance: a high resolution CT with a lateral digital localizer (scoutview) allowing scanning of thin sections (1 to 1.5 mm thickness). Furthermore, we find that dynamic scanning is essential for diagnosis of purely intrasellar lesions. For pituitary scanning to be successful, the team must be well trained, and know exactly what it is looking for. Finally, the patients require a good deal of psychological preparation; they must be aware that the examination is of the greatest importance for their health, and quality of results will be largely dependent upon their complete cooperation (Fig. 1.1).

Projections

The choice of projections is fundamental. If the patient's physical condition allows, direct coronal sections are to be preferred, since they provide more precise information than axial sections, even when the latter are completed by reformatting. It is clear that direct coronal sections are to be preferred to axial sections for all lesions of the sellar region. For intrasellar pathologies, we consider that direct coronal sections are absolutely imperative (Fig. 1.2). Choice of the thickness of the section is primarily a function of the tumor volume.

Axial Sections

Axial sections are made with the patient in supine position, and are thus generally easy to carry out except in cases of major gibbosity. Axial sections can be done first in the elderly or where the prone position required for direct coronal sections is precluded by the patient's condition: e.g., obesity, pregnancy, respiratory failure, agitation. Axial sections are generally acceptable where the lesion volume is great, i.e., where there is suprasellar expansion or the tumor originates outside the sella turcica. Nevertheless, in the latter case, coronal sections are preferable for clear differentiation of the tumor from the pituitary gland. In terms of positioning of the patient's head, it is most important to avoid inclusion of the sellar region and the petrous bones in the same section so as to prevent linear artifacts which render reformatting unreliable. This is best ensured by using a lateral scoutview. In general, the gantry is inclined as little as possible to facilitate reformatting, and the patient's chin is raised approximately OM-10° (Fig. 1.3).

Thickness of the sections should be appropriate for the lesion size, which can sometimes be estimated from X-ray films. In any case, quality of coronal and sagittal reconstructions is improved when sections are thin, contiguous, or overlapping, and when the patient is strictly immobile. This is particularly important for intrasellar lesions when direct coronal sections are precluded.

For suprasellar lesions, a relatively large area should be scanned, from the sphenoid sinus to the floor of the third ventricle. An acceptable compromise to avoid an excessive number of sections for the largest tumors is to scan a 5-mm thick section every 3 mm.

In some cases, such as the following, axial sections *must* be made in addition to coronal sections:

1. Where there is suspicion of a very anterior or very posterior pituitary lesion, possibly noted on an X-ray film (Fig. 5.10)

2. Where the lesion seen in coronal sections is ambiguous, poorly contrasted, or poorly defined; in a few cases, coronal reformatted images of thin axial sections with a relatively large number of pixels yield results better than those with thin direct coronal sections

3. Where there is ambiguity between a posterior pituitary microadenoma and the posterior lobe of the pituitary gland, an axial dynamic scan can show early and transient enhancement of the vascular space of the posterior pituitary lobe (see Chaps. 2 and 3).

In general, one must always bear in mind the fact that partial volume effect is inevitable for axial sections contiguous with the sellar floor or the chiasmatic cistern. For this reason, wherever possible we prefer direct coronal sections for exploration of the pituitary region.

The Reconstructions

Coronal and sagittal reformatting is virtually always carried out using axial sections. In our opinion, coronal reformatting can only serve to supplement or confirm results of direct coronal sections. Sagittal reformatting is useful to determine the exact topography and relationships of a lesion expanding outside the confines of the sella or with a suprasellar point of origin. Sagittal sections can sometimes confirm the existence of an intrasellar lesion larger than 5 mm in diameter; for the small intrasellar lesions (2–3 mm in diameter), we have only very limited confidence in reformatted images. Reformatting requires complete immobilization of the patient as well as very thin sections, sometimes with overlapping.

In all cases, several reconstructions will be carried out with various numbers of pixels. In general, a choice of 3 or 5 pixels yields satisfactory results. The majority of currently available systems allow oblique reformatting, of some interest for sellar tumors with intraorbital expansion or for study of the optic canal, as well as three-dimensional reconstruction (Fig. 1.4).

Coronal Sections

An increase in understanding of pituitary pathological phenomena has arisen from the use of direct coronal scans. Coronal sections require competent and fast performance by the medical team, as well as good cooperation by the patient. They are preferably carried out with the patient in prone position, with the head in hyperextension. In particular for dynamic scanning, a wide-gauge needle (preferably 16 gauge) is inserted into a vein of the forearm or the back of the hand before the patient is positioned. The entire procedure is carefully explained before the beginning of the study, and since extension of the head is uncomfortable, it is essential to proceed rapidly after the patient is in place. A lateral scoutview allows localization of the sella turcica, and the axis of sections is chosen less as a function of sellar morphology than to avoid inclusion of metallic dental prostheses in the section (Fig. 1.5). With an increasing degree of hyperextension of the head, the tilting of the gantry from the vertical axis can be decreased. In acromegaly, prognathism frequently prevents sufficient hyperextension and coronal sections thus pass through the maxillary sinus and the symphysis of the chin rather than the sphenoid sinus; quality of coronal sections is diminished in such patients (Fig. 8.2).

Coronal sections are taken from the planum sphenoidale to the interpeduncular cistern, so as to widely bracket the sellar contents. Thickness of sections is generally between 1 and 1.5 mm; nevertheless, with major enlargement of the sella turcica, thickness of sections can be increased to 3 mm.

The Dynamic CT

Dynamic CT is always carried out with direct coronal sections except when seeking to visualize the posterior pituitary lobe. This procedure is discussed in detail in Chap. 3. The ideal section level passes through the center of the sella turcica; the 1.5 mm section includes neither the anterior clinoid processes nor the dorsum sellae. Six to eight sections are made at the same level as quickly as possible after a bolus injection of contrast medium. The examination finishes with a further section 30 s later, followed by a series of coronal sections covering the sella turcica from front to back.

Study of Bone Structures

The sellar floor is always studied on the direct coronal sections; the anterior wall and dorsum sellae are better visualized in axial section. Study of osseous structures of the sellar region can be carried out with bone window, or with retrospective review using the General Electric review software package, for instance. The CT scan with reconstruction algorithm for bone detail allows a more exact study of bone structures, while tending to underestimate cortical thinning (Fig. 1.6).

Contrast Media

Pituitary scanning is generally carried out after intravenous injection of a water-soluble iodinated contrast medium.

We consider that in the majority of cases, complete examination without injection of contrast media is useless and can be omitted, thus shortening duration of the examination. Nevertheless, a native scan may be carried out where injection of contrast media is contraindicated for medical reasons. Such examinations can allow adequate estimation of the volume of the pituitary, sometimes even showing abnormalities in density of intrasellar structures (see Chaps. 4 and 5). Scans without use of contrast media can also allow diagnosis of an empty sella turcica (see Chap. 13). A native scan is also carried out during the localization phase preceding the dynamic scan. Finally, CT scans may be carried out without contrast media for identification of intrasellar or suprasellar calcifications, or where recent hemorrhage in the pituitary region is suspected.

Since there is no blood-brain barrier protecting the pituitary gland, contrast media diffuse freely from the vascular space to the interstitial space. The intensity of enhancement after injection of contrast medium is thus primarily a function of the interstitial space. Following injection, the posterior pituitary lobe appears less dense than the anterior lobe due to the relatively smaller volume of the interstitial compartment (see Chap. 3).

Contrast media used for pituitary CT scans should have a high iodine concentration, good diffusion into the interstitial compartment, be well tolerated by the venous system even after very rapid bolus injection, and provoke a minimum of nausea. For pituitary scanning, we have chosen a hexiodinated contrast medium containing 32% iodine (Hexabrix, Laboratoire Guerbet, Aulnay-Sous-Bois, France). For dynamic CT, 60 cc warm contrast medium is injected in 6–8 s.

Where dynamic CT is not performed, contrast medium can be injected more slowly, e.g., over 30 s. Coronal and axial sections are taken immediately. If axial sections are required after the coronal scan, 100 cc of the same contrast medium are injected so as to maintain plasma iodine levels. Accidents due to intolerance to contrast media are rare; treatment is the same as for other iodine hypersensitivity incidents.

Indications for CT cisternography are increasingly uncommon since development of high resolution scanners. In our experience, of 2300 pituitary scans over 6 years, this technique was required in only about 20 cases. CT cisternography permits opacification of the subarachnoid spaces by intrathecal injection of 5 ml water-soluble contrast medium, containing 170 mg iodine/ml, following which the patient is kept head downwards for a few minutes. This technique allows very good resolution of the superior pole of the pituitary or limits of sellar tumors, as well as the visualization of neighboring vascular and nervous structures. While the excellent image quality is undeniable, it generally does not justify the discomfort of lumbar puncture and cisternal injection. Furthermore, CT cisternography is of no help in the study of intrasellar lesions. The few remaining indications include differential diagnosis between empty sella turcica and cystic or necrotic pituitary adenomas where CT imaging is insufficient (Fig. 1.7), for precise study of the optic chiasm (Fig. 1.8); or for study of rhinorrhea.

Artifacts

Elimination or identification of artifacts is necessary for correct interpretation of pituitary images (Fig. 1.9). The hardest artifacts to avoid are those caused by involuntary movements of the patient, particularly for direct coronal sections. We consider that good psychological preparation of the patient is far more effective than rigid contention of the head; we use no sedatives. Where patients are non-cooperative, where injection gives rise to pain or nausea, or where dyspnea causes involuntary movements and artifacts, direct coronal scanning could be stopped and axial sections scanned in supine position.

Nevertheless, other types of artifacts can occur in axial or direct coronal sections. For axial sections, it is necessary to avoid streak artifacts due to scanning of the superior part of the petrous bones (Fig. 1.10) or less frequently, of the orbits. These artifacts, narrow hypodense bands, are caused by beam hardening and abrupt change in the attenuation of adjacent structures. They are generally readily recognized in axial sections; however, in coronal or sagittal reformatting they can show up as rounded hypodensities which can be mistaken for pituitary adenomas (Fig. 1.11). This should be avoided by scanning at an axis where the sella turcica is projected above the petrous bones (OM-10° or OM-20°).

For coronal scans, the scoutview virtually always allows us to avoid artifacts due to dental fillings, provided good hyperextension of the head is possible. If hyperextension is insufficient, one can attempt to place the patient in supine position with maximal hyperextension; however, this position is generally less well tolerated and more difficult to maintain than the classical prone position and, thus, we use it only very rarely.

A linear cross-like artifact can sometimes be seen immediately in front of the dorsum sellae (Fig. 1.12). Other streak artifacts can appear prolonging the septa of the sphenoid sinus (Fig. 1.13).

More troublesome is the hypodense artifact, frequently seen in anterior sections, caused by partial volume effect attributable to the air of the sphenoid sinus; this hypodense area, just below the curvilinear shadow of the tuberculum sellae, can be mistaken for an anterior pituitary microadenoma (Fig. 1.14) ("tuberculum sign") (Bonneville).

Scan Details

With the GE CT/T 8800 scanner
  1. Conventional CT scan:
     320 mA
     120 KVp
     pulse width code 2 (2.7 ms)
     576 pulses/scan
     scan time: 9.6 s
  2. Dynamic CT scan:
     500 mA
     120 KVp
     pulse width code 1 (1.3 ms)
     scan time: 9.6 s
     interscan delay time: 1.4 s

With the GE CT/T 9800 scanner
  1. Conventional CT scan
     200 mA
     120 KVp
     576 pulses/scan
     scan time: 3 s
  2. Dynamic CT scan
     120 mA
     120 KVp
     576 pulses/scan
     scan time: 2 s
     interscan delay time: 2–6 s

Radiation Exposure

For Earnest, high-resolution axial CT scanning of adjacent 1.5 mm sections produces an average entrance dose of 3.6 rad (0.036 Gy). We have calculated that the maximum surface dose

Table 1. Protocols for pituitary scans

Symptoms	Plain films	CT procedure
Hyperprolactinemia	Normal sella turcica	Dynamic CT + 1.5 mm coronal
	Possible posterior microadenoma	+ Axial 1.5 mm + reconstructions
Hyperprolactinemia and pregnancy (>4 months)		Native coronal scan: 2–3 slices 1.5 mm
Follow-up after medical treatment		Dynamic CT + coronal 1.5 mm
Acromegaly	Enlarged sella turcica	Dynamic CT (carotid siphons) + coronal 3 or 5 mm. If impossible, 1.5 mm axial + reconstructions
	Normal sella turcica	Dynamic CT + coronal 1.5 mm
Cushing's disease	Sella turcica generally normal or overall demineralized	Dynamic CT + coronal 1.5 mm + axial 1.5 mm + reconstructions. If impossible axial 1.5 mm with overlap + reconstructions
Pituitary insufficiency	Enlarged sella turcica	Coronal 3 or 5 mm with or without dynamic CT or axial + reconstructions
	Normal sella turcica	Coronal 1.5 mm with or without dynamic CT
Pituitary adenomas after surgery	(Surgical clips?)	Dynamic CT (intracavernous residual tumor) + coronal or axial with contrast infusion and reconstructions
		Occasionally, CT metrizamide cisternography
Nontraumatic diabetes insipidus (BBS, histiocytosis, tuberculosis, tumors of the pituitary stalk, etc.)	Sella turcica generally normal	Coronal 1.5 mm (pituitary stalk) + axial 1.5 mm (posterior lobe) + reconstructions (suprasellar tumor?)
		Occasionally, CT metrizamide cisternography
Delayed growth	Pathological sella turcica (craniopharyngioma, etc.)	Dynamic CT + coronal + axial sections with reconstructions
	Normal sella turcica	Head axial 10 mm + axial 1.5 mm (hypothalamus, 3rd ventricle)
Premature puberty (harmartoma)	Sella turcica generally normal	Axial 1.5 mm + reconstructions
Abnormal visual field	Generally pathological sella turcica	Coronal 1.5–3 or 5 mm depending upon sellar volume, or axial 3–5 mm + reconstructions
"Radiologic indications" (no clinical symptomatology; laboratory values normal: generally variants of the normal or empty sella)	Sella "suspect" (?)	Coronal 1.5 mm

for eight scans is approximately 20 rad (0.2 Gy), while in women patients the corresponding dose to the ovaries is only about 10 mR (2.58 μC/kg).

Exposure to the lens of the eye is lower with coronal sections than with axial sections.

Examination Protocols

Techniques for study of the pituitary region vary as a function of the putative diagnosis, as well as of clinical, laboratory, and X-ray data. Radiologic studies before the CT should be kept to a minimum. We generally limit these to anteroposterior and lateral views focused on the sella turcica, so as to determine whether it is enlarged or deformed, whether there are intra- or suprasellar calcifications, or whether the floor is tilted or demineralized. The quality of the bone structure of the sellar walls can be evaluated by magnified lateral view, if possible with an ultrafine focus or electron gun.

We no longer carry out tomography, since this is both expensive and involves a high radiation dose, while almost never providing decisive information conditioning the CT examination.

Table 1 shows our protocols for pituitary scans as a function of the more common clinical and laboratory profiles of pathologies of the sellar or parasellar region. These protocols are shown only as a general guideline, and are, of course, adapted for each individual case.

References

Aubert B (1981) Mesures des doses en radiodiagnostic et tomodensitométrie. J Radiol 62:587–589

Bonneville JF, Poulignot D, Cattin F, Couturier M, Mollet E, Dietemann JL (1982) Apport des méthodes nouvelles dans l'exploration morphologique des tumeurs hypophysaires. Ann Endocrinol (Paris) 43:303–308

Bonneville JF, Poulignot D, Coche G, Portha C, Cattin F, Bacha M (1982) Radiological techniques in the diagnosis of microprolactinoma. In: Molinatti GM (ed). A clinical problem: microprolactinoma. Diagnosis and treatment. Excerpta Medica, Amsterdam-Oxford-Princeton, p 57

Bonneville JF, Cattin F, Dietemann JL (1985) CT of pituitary microadenomas (Letter). AJNR 6:650

Daniels D, Haughton V, Williams A, Gager W, Berns TF (1980) CT of the optic chiasm. Radiology 137:123–127

Davis KR, Zito JL, Hesselink JR, Taveras JM, Kjellberg RN (1979) Metrizamide sagittal tomography: Adjunct to CT cisternography of the sellar region. AJR 134:1205–1208

Drayer BP, Kattah J, Rosenbaum A, Kennerdell J, Maroon J (1979) Diagnostic approaches to pituitary adenomas. Neurology 29:161–169

Drayer BP, Rosenbaum AE, Kennerdell JS, Robinson AG, Bank WO, Deeb ZL (1977) CT diagnosis of suprasellar masses by intrathecal enhancement. Radiology 123:339–344

Drayer BP, Rosenbaum AE, Riegel DB, Bank WO, Deeb ZL (1977) Metrizamide CT cisternography: pediatric applications. Radiology 124:349–357

Earnest FIV, McCullough EC, Frank DA (1981) Fact or artifact: an analysis of artifact in high-resolution CT scanning of the sella. Radiology 140:109–113

Enzmann DR, Norman D, Newton TH (1977) Computer tomography in cisternography with metrizamide. Acta Radiologica (suppl 355):294–298

Eresue J, Drouillard J, Philippe JC, Guibert JL, Poux P, Tavernier J (1982) L'exploration des adénomes hypophysaires par scanographie à haute résolution et angioscanographie. Ann Radiol 25:509–517

Etling N, Gehin-Pouque F, Vielh JP, Gautray JP (1979) The iodine content of amniotic fluid and placental transfer of iodinated drugs. Obstet Gynecol 53:376–380

Fargason RD, Jacques S, Rand RW, Shelden CH, McCann GD, Linn P (1981) Visualization and three-dimensional reconstruction of pituitary microadenomas from CT data: a technical report. Surg Neurology 15:450–454

Gardeur D, Nachanakian A, Kulesza E, Metzger J (1979) La tomodensitométrie dans les adénomes hypophysaires. Ann Radiol 22:489–499

Geehr RB, Allen WE, Rothman SGL (1978) Pluridirectional tomography in the evaluation of pituitary tumors. AJR 130:105–109

Ghoshhajra K (1981) High-resolution metrizamide CT cisternography in sellar and suprasellar abnormalities. J Neurosurg 54:232–239

Glover GH, Pely NJ (1980) Nonlinear partial volume artifacts in X-ray CT. Med Phys 7:238–248

Gross CE, Binet EF, Esguerra JV (1979) Metrizamide cisternography in the evolution of pituitary adenomas and the empty sella syndrome. J Neurosurg 50:472–476

Guibert-Tranier F, Elie G, Guibert JL, Piton J, Caille JM (1980) Selles turciques vides. Diagnostic TDM. J Neuroradiology 7:105–119

Hall K, McAllister VL (1980) Metrizamide cisternography in pituitary and juxtapituitary lesions. Radiology 134:101–108

Haverling M, Johanson H, Ahren L (1978) Approximate sagittal CT of the sellar and suprasellar regions. Acta Radiol (Diagn Stockh) 19:918–920

Hayman LA, Evans RA, Hinck VC (1979) Rapid high

dose (RHD) contrast CT of perisellar vessels. Radiology 131:121–123

Hoffman JC Jr, Tindall GT (1980) Diagnosis of empty sella syndrome using Amipaque cisternography combined with CT. J Neurosurg 52:99–102

Kline LB, Acker JD, Post JD (1982) CT evaluation of the cavernous sinus. Ophthalmology 89:374–385

Kricheff II (1979) The radiologic diagnosis of pituitary adenoma. Radiology 131:263–265

Kuuliala I, Katevuo K, Ketonen L (1981) Metrizamide cisternography with hypocycloid and CT in sellar and suprasellar lesions. Clin Radiol 32:403–408

Lemaître G, Linquette M, Fossati P, Cappoen JR (1982) Détection des adénomes hypophysaires sécrétants par tomodensitométrie. Ann Med Int 133:33–34

Lipman JK, Marshal W (1982) Practical errors in measurement of the pituitary at CT (letter). AJNR 3:87

Malena S, Babuti D, Borrelli P, Fariello G, Carnevale A, Parri C, Gugliantini P (1981) Present value of conventional X-ray examination in pituitary microadenomas in childhood: relation with CT. Ann Radiol (Paris) 24:109–110

Manelfe C, Bonafé A, Morel C, Sancier A, Treil J (1978) Reconstructions frontales et sagittales en tomodensitométrie crânienne. J Neuroradiology 5:175–186

Manelfe C, Giraud B, Espagno J, Kascol A (1978) Cisternographie computérisée au metrizamide. Rev Neurol (Paris) 134:471–484

Mass S, Norman D, Newton TH (1978) Coronal computed tomography: indications and accurancy. AJR 131:875–879

McCullough EC, Payne JT (1978) Patient dosage in CT. Radiology 129:457–463

Newton DR, Witz S, Norman D, Newton TH (1983) Economic impact of CT scanning on the evaluation of pituitary adenomas. AJNR 4:57–60

Osborn AG, Anderson RE (1978) Direct sagittal computed tomography scans of the face and paranasal sinuses. Radiology 129:81–87

Parsons C, Hodson N (1979) CT of paranasal sinus tumors. Radiology 132:641–645

Penley MW, Pribram HFW (1980) Diagnosis of empty sella with small amount of air at CT. Surg Neurol 14:296–301

Roberson GH, Tadmor R, Taveras JM, Kleefield J, Ellis G (1977) CT in Metrizamide cisternography. Importance of coronal and axial views. J Comput Assist Tomogr 1:241–245

Rogers RT (1969) Radiation dose to the skin in diagnostic radiography. Br J Radiol 42:511

Salcman M (1979) Correlation of absorption coefficients with intracranial fluid protein concentrations and specific gravities. Neurosurgery 5:16–20

Sheldon P, Molyneux A (1978) Metrizamide cisternography and CT for the investigation of pituitary regions. Neuroradiology 17:83–87

Shope TB, Morgan TJ, Showalter CK, Pentlow KS, Rothenberg LN, White DR, Speller RD (1982) Radiation dosimetry survey of CT systems from ten manufactures. Br J Radiol 55:60–69

Strother CM, Sackett JF, Appen RE (1977) Anatomic considerations for CT of the optic chiasma. Arch Neurol 34:713–714

Taylor S (1982) High resolution computed tomography of the sella. Radiologic clinics of North America 20:207–236

Tubiana M (1979) Problèmes posés par l'irradiation des femmes enceintes. Effects des radiations ionisantes sur l'embryon et le foetus. Bull Cancer 66:155–164

Wiggli U, Benz UF (1978) Normal CT anatomy of the suprasellar subarachnoid space. Radiology 128:65–70

Wolfman NT, Boehnke M (1978) The use of coronal sections in evaluating lesions of the sellar and parasellar regions. J Comput Assist Tomogr 2:308–313

Zull DN, Falko JM (1981) Metrizamide cisternography in the investigation of the empty sella syndrome. Arch Intern Med 141:487–489

Zwicker RD, Zamenhof RG, Wolpert SM (1982) Comparative dosimetry of high-detail CT using the Siemens Somatom 2 and complex motion tomography for examination of the sella turcica. AJNR 3:354–355

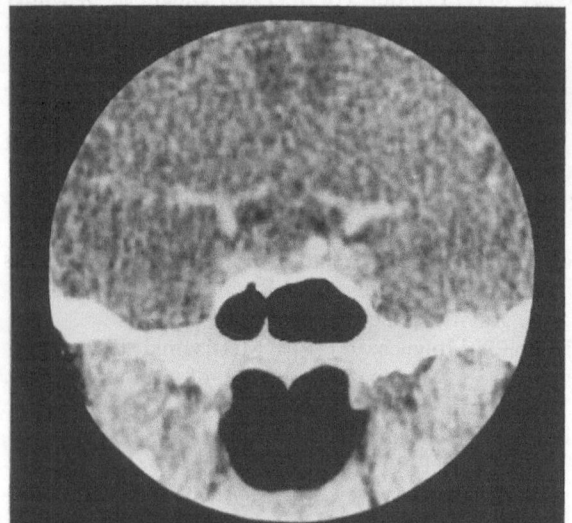

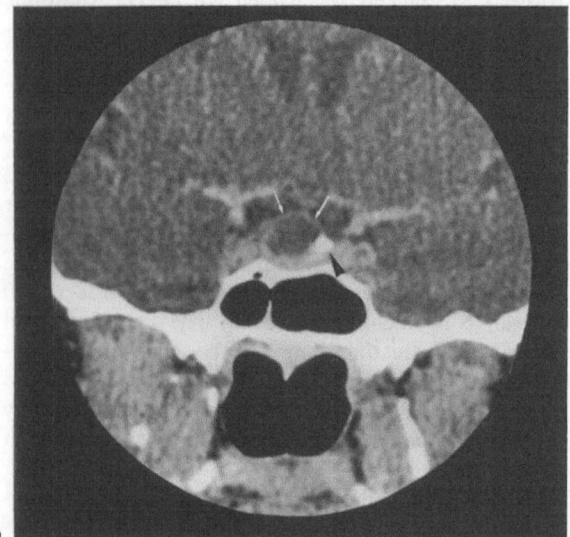

Fig. 1.1a, b. Advantage of high-quality CT scan; history of hyperprolactinemia in a young man. **a** Contrast-enhanced CT scan with a 8800 GE CT/T. The image is not very sharp. An inhomogeneous and partially calcified intrasellar tumor with small suprasellar extension is demonstrated. A pituitary adenoma is suspected and pergolide mesylate is given; 2 months later, absence of any clinical or biological improvement prompted a new CT examination. **b** Contrast-enhanced CT scan with a 9800 GE CT/T. This time, a cystic formation with a small calcification (*arrowhead*) and ring enhancement, is clearly demonstrated (*arrows*) above a slightly compressed pituitary. A Rathke pouch cyst containing motor-oil fluid was confirmed at surgery

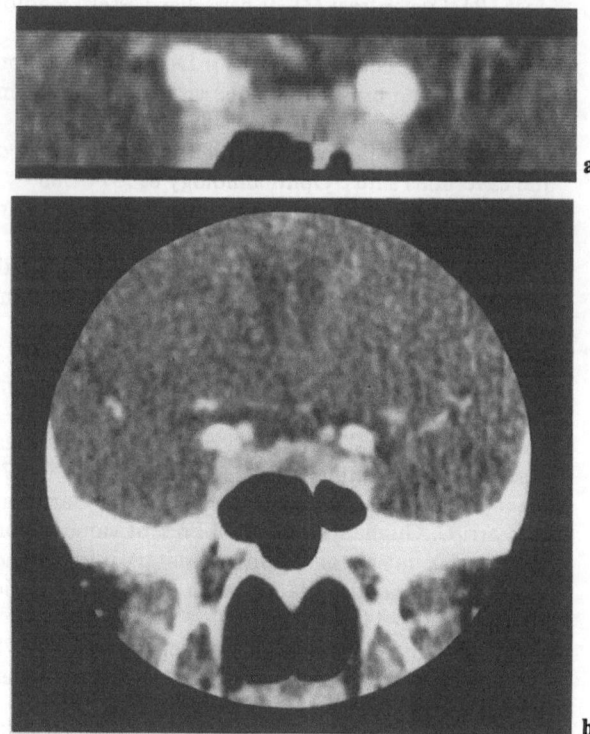

Fig. 1.2a, b. Superiority of direct coronal images (**b**) over coronal reformatted images (**a**). An ACTH-secreting microadenoma is much better visualized on direct contrast-enhanced coronal CT scan; proven at surgery

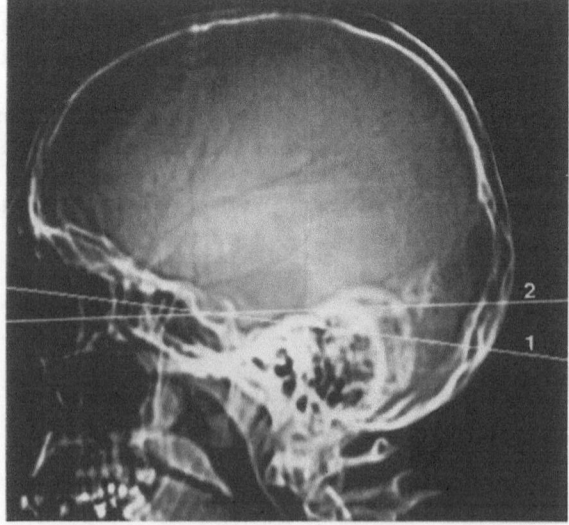

Fig. 1.3. Digital lateral localizer images. Small angulation of the gantry or adequate head positioning (OM-10°) avoids scanning of the petrous bones at the sellar level in axial sections. 1, OM; 2, OM-10°

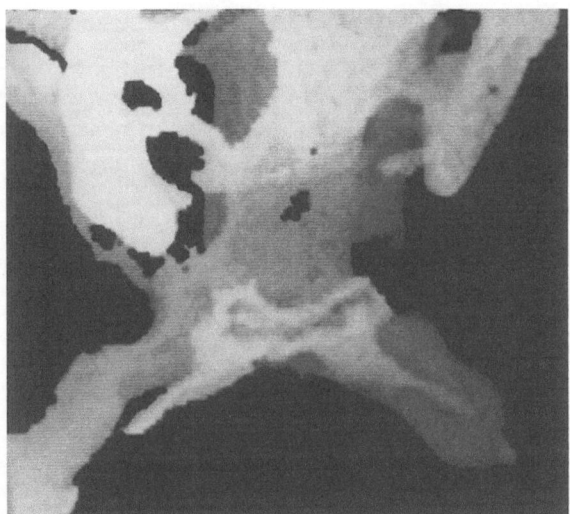

Fig. 1.4. 3-D reconstruction of the sella

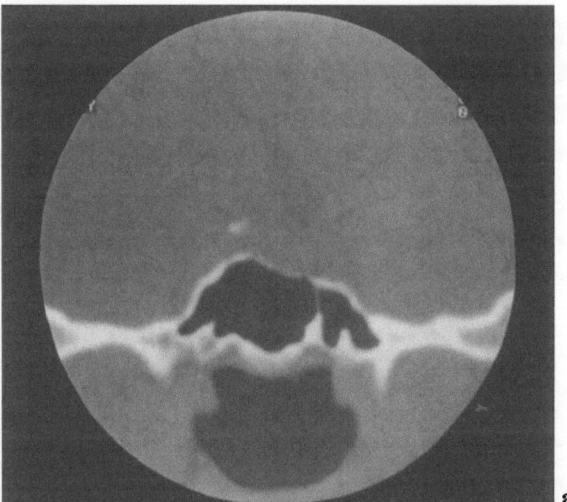

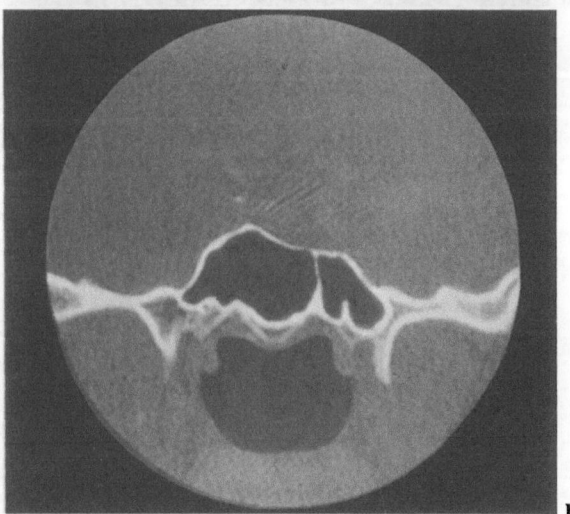

Fig. 1.6a, b. Bone window level (**a**), vs bone review algorithm image (**b**). Sharper demonstration but also underestimation of the sellar floor thinning in **b**

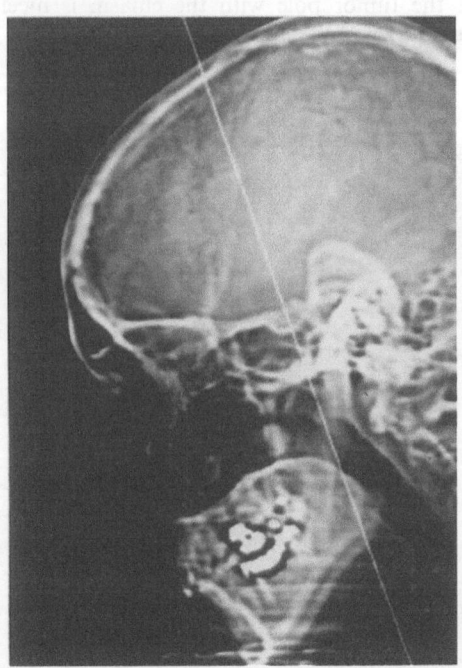

Fig. 1.5. Lateral digital localizer image as a help for precise coronal projection

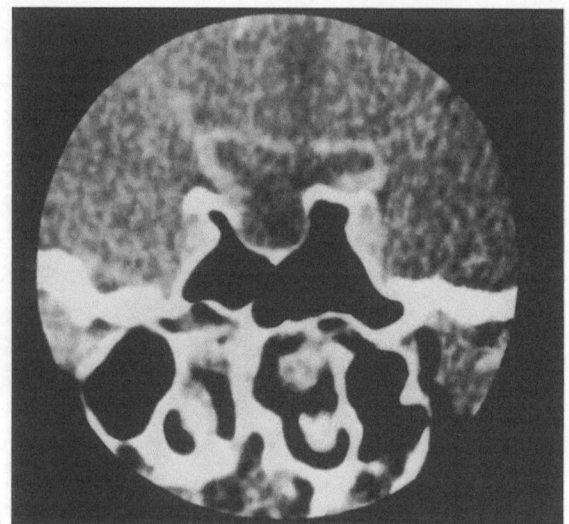

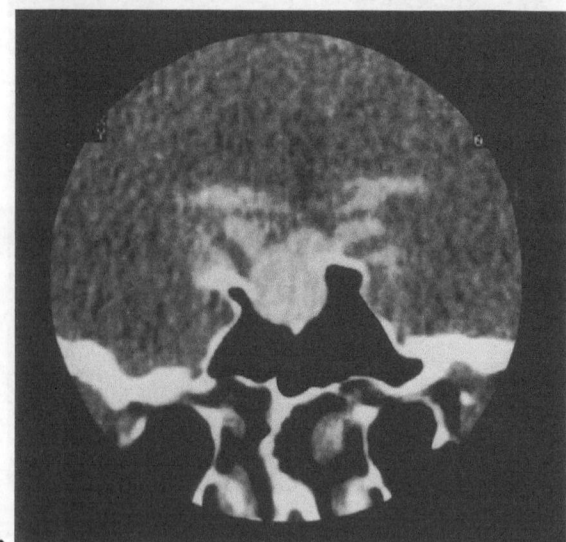

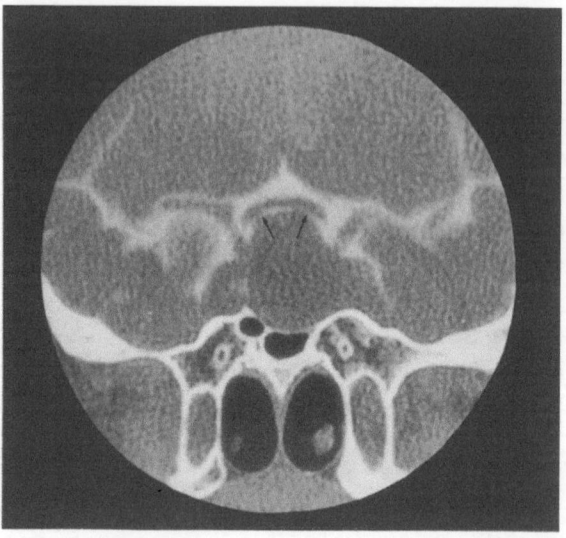

Fig. 1.8. Water-soluble contrast CT cisternography. Pituitary adenoma with suprasellar expansion. The relationship of the tumor pole with the chiasm is nicely demonstrated (*arrows*)

Fig. 1.7a, b. CT water-soluble contrast cisternography. **a** Contrast intravenous enhanced CT scan. Liquid attenuation values within an asymmetrically enlarged pituitary fossa. A cystic formation cannot be excluded. **b** Metrizamide CT cisternography. Free communication between supra- and intrasellar subarachnoid spaces signal an empty sella

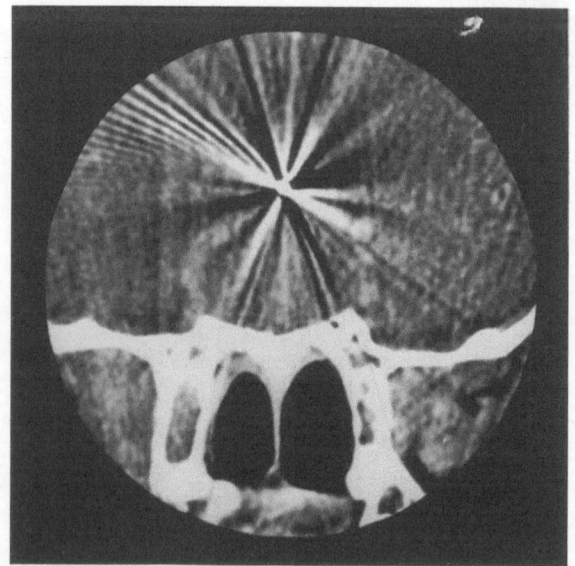

Fig. 1.9. Artifact. Suprasellar surgically placed metallic clip compromises adequate visualization of sellar region

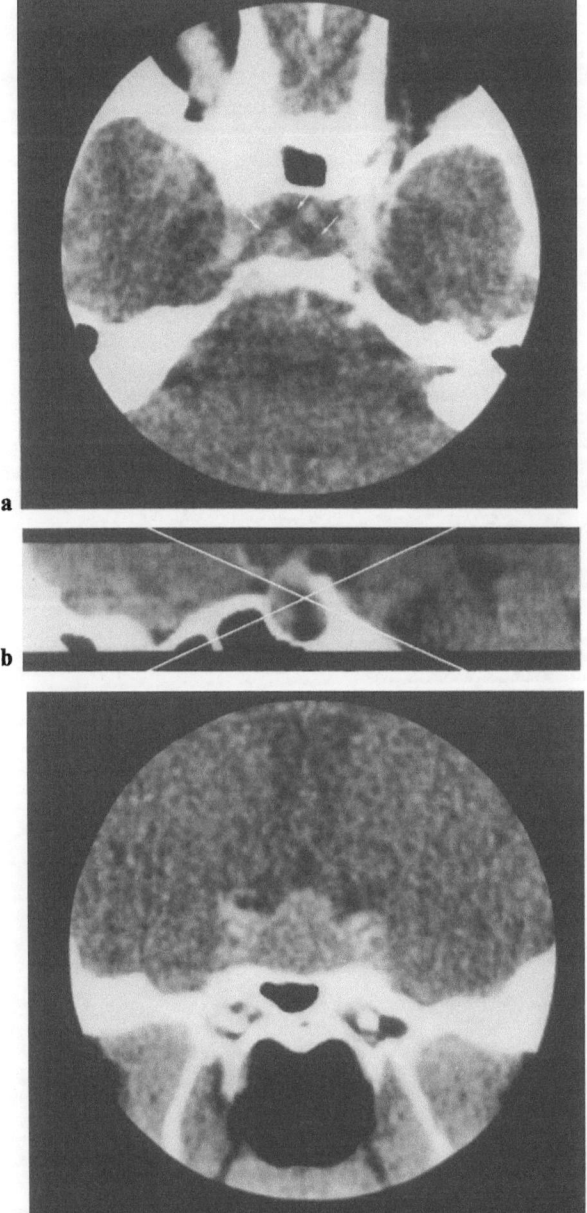

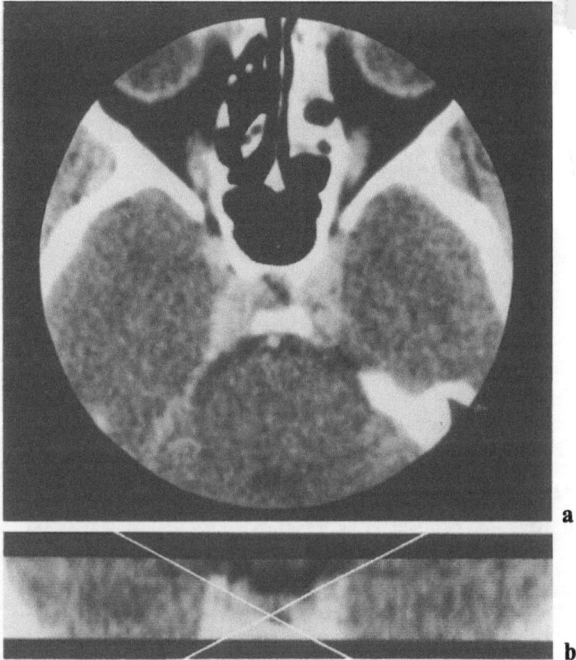

Fig. 1.11a, b. Streak artifact. a Axial CT scan. The left petrous bone is included in this axial cut. Consequently, an oblique streak artifact is present. b Coronal reformatted image. A confusing false-positive image evoking a pituitary microadenoma is visible. Pituitary pattern was normal on direct coronal CT scan (not shown)

Fig. 1.10a–c. Streak artifact. a Axial CT scan. Inadequate head positioning. Petrous bones and sphenoid wings are included in the same cut. A resulting oblique streak artifact is visible at the pituitary level (arrows). b Sagittal reformatted image. Low-dense attenuation values within an enlarged intrasellar content represent the streak artifact and not a necrotic area within the tumor. c Direct coronal enhanced CT scan. An homogeneous pituitary adenoma with small suprasellar extension is demonstrated. No cystic area. At surgery, a solid pituitary adenoma is proven

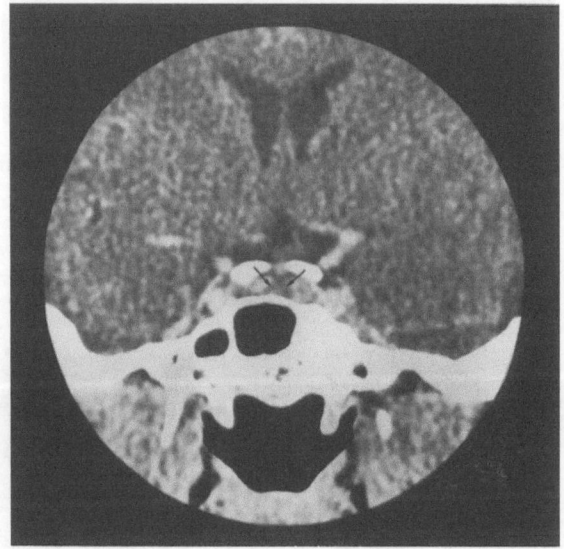

Fig. 1.12. Streak artifacts. X streak artifacts at the dorsum level crossing the pituitary image (arrows)

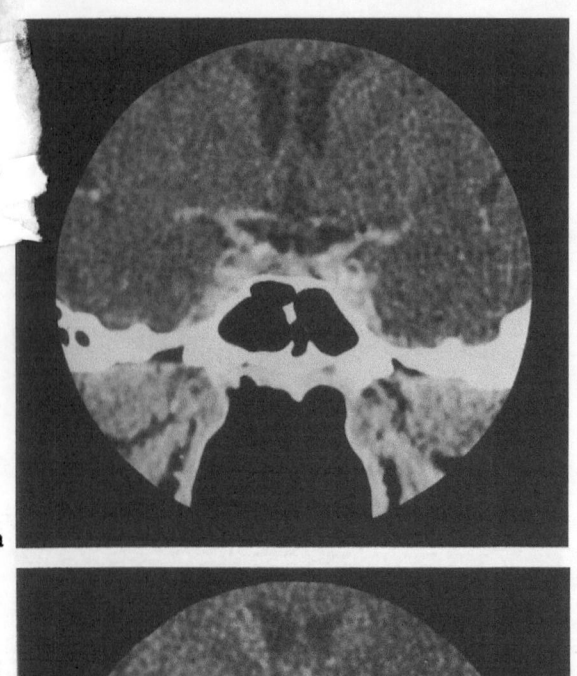

a

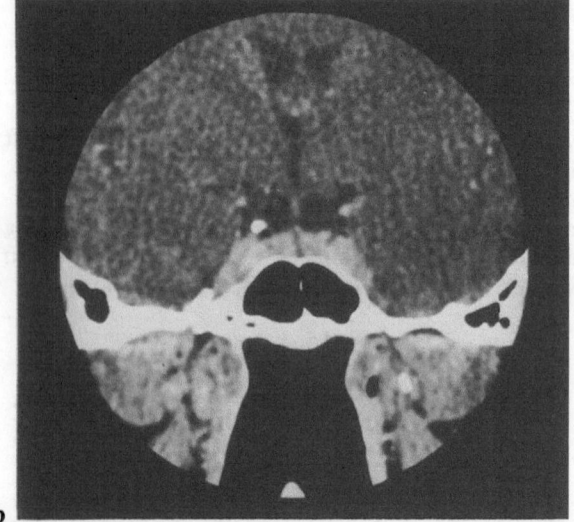

b

Fig. 1.13a, b. Artifacts. Band-like artifacts extending from **a** sphenoid walls and **b** sphenoid septa

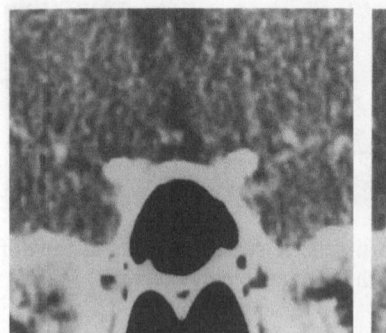 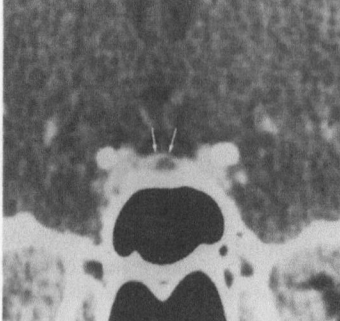 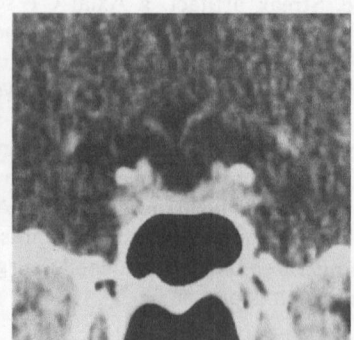

a–c

Fig. 1.14a–c. The "tuberculum sign;" 1.5 mm consecutive contrast-enhanced coronal CT scans. Partial volume averaging at the level of tuberculum sellae and anterior wall of the sella simulates a low-dense intrasellar lesion (**b**) beneath a pseudo-bulging of the diaphragma sellae (*arrows*)

Chapter 2
Radiologic Anatomy of the Sellar Region

We will here briefly describe both the normal CT pattern and the major anatomical variants involving the pituitary gland, the sella turcica, the vascular and nervous structures of the cavernous sinus and the suprasellar cistern. The CT slices can be correlated with anatomical sections presented in the sagittal, coronal, and axial planes (Figs. 2.1, 2.2, and 2.3).

The Pituitary

The pituitary is best studied by 1 or 1.5 mm direct coronal sections. In the absence of contrast medium the gland is easily recognizable between the sellar floor and the chiasmatic cistern (Fig. 2.8 a). Its attenuation value is similar to that of the brain; its superior pole is readily recognized and its height can thus be measured.

After intravenous injection of contrast media, enhancement begins with the pituitary capillary bed, gradually spreading to the periphery of the gland in a centrifugal fashion (see Chap. 3). Approximately 60 s after bolus injection the gland is homogeneous and appears less enhanced than the cavernous sinus (Fig. 2.4). The lateral borders of the pituitary in contact with the cavernous sinus are clearly visualized. The superior pole of the pituitary is virtually always flat or slightly concave superiorly. Nevertheless, a normal convexity of the superior pole is seen in about 5% of the cases, in particular in young women (Swartz) (Fig. 2.14a). A triangular rise in the central part of the superior pole of the gland, at the point of passage of the pituitary stalk, may be present in normal subjects (Fig. 2.14b). Finally, anatomical variants in the sphenoid sinus can cause pituitary asymmetry.

The superior pole of the pituitary, parallel to the sellar floor, can appear tilted or convex upwards (Fig. 2.15); in other types of anatomical variants of the sphenoid sinus, the pituitary can be almost completely asymmetric, displaced to one side of the median sagittal plane (Fig. 2.16). Very occasionally, due to extensive upward pneumatization of the sphenoid sinus, there is no sellar excavation and the pituitary appears to be quite simply set on the roof of the sphenoid sinus. The CT aspect can be misleading, in particular in axial section, if the anatomical variant in the sphenoid sinus is not first identified on plain films (Fig. 18.5). Shortness of the sellar floor from right to left or intrasellar location of the carotid arteries may result in a convexity of the superior limit of the pituitary, frequently giving rise to an increased height of the gland simulating pathological patterns (Fig. 2.17). (Bonneville)

The mean height of the gland is only slightly greater in women (4.4 ± 1.4 mm) than in men (4.4 ± 1.1 mm) (Brown). However, Haughton reported a normal variability of between 1.4 and 6.7 mm: due to this wide anatomical variability, measurement of height of the pituitary gland rarely provides significant diagnostic information. Nevertheless, it is considered highly suspicious when the height of the gland exceeds 8 mm. This is not an absolute rule; we ourselves have seen several cases where the pituitary height was greater than 9 mm in young women in the absence of pregnancy or detectable endocrine disorders. Finally, as shown by Roppolo, there is a slight decrease in height of the gland with increasing age. When the gland is small (2–3 mm), the upper pole frequently is concave superiorly and there is thus free communication between supra- and intrasellar subarachnoid compartments.

The posterior lobe of the pituitary is rarely seen in coronal scans. We have not seen the particular pattern of the pars intermedia described by Roppolo.

In axial section, the pituitary is virtually always less well defined than in coronal section due to partial volume phenomenon seen both in the vicinity of the sellar floor and at the bottom of the chiasmatic cistern (Fig. 2.6).

In more than 50% of cases, the posterior lobe of the pituitary is visible in axial section (Bonneville, Cattin). This visibility is more frequent when the sella turcica is open; however, the posterior lobe can sometimes be visualized even where opening of the sella is narrow. Following intravenous injection of contrast media, the posterior lobe generally appears in only a single section (never in more than two thin contiguous sections) as an oval zone, 4–5 mm in length and 2–3 mm thick; the posterior lobe shows less enhancement than the anterior lobe (Fig. 2.11). The posterior lobe of the pituitary, sometimes pressed against a depression in the dorsum sellae which is not necessarily median (Figs. 2.12, 2.13), always shows a convex anterior border.

The Sella Turcica

The sellar floor is well visualized in coronal sections (Fig. 2.5). We have shown (Fig. 2.17) that a short sellar floor width can change the normal shape of the pituitary gland. Under normal conditions, thickness is not absolutely constant, and there can be some osseous thinning. Localized modifications in thickness of the sellar floor are commonly seen at the points of attachments of sphenoid septa (Fig. 2.18). Passage of the internal carotid artery can sometimes be seen on each side of the sellar floor, with a well-delimited thinning or an imprint of the roof of the sphenoid sinus (Fig. 2.19). Where the anteroposterior diameter of the sella turcica is limited, the same coronal section may show the dorsum sellae and the extremities of one or both anterior clinoid processes which must be differentiated from abnormal calcifications (Fig. 2.21). In 3–4:1000 cases, the sella turcica shows a complete unilateral or bilateral bony wall (Fig. 2.20).

In axial sections, the anterior wall and dorsum sellae are particularly well defined (Fig. 2.7). Extensive or asymmetric pneumatization of the dorsum sellae can be seen. The normal appearance of the middle clinoid processes (Fig. 2.22) or of the less common sellar spine (Fig. 2.23) should be well known. The calcified petroclinoid ligaments are readily identified (Fig. 2.31).

The Cavernous Sinus

The cavernous sinuses are situated outside the sella turcica and the body of the sphenoid (Fig. 2.24). They are generally of symmetrical topography and volume. Under normal conditions the external dural wall is flat or slightly concaved outwards in axial sections. The dural sheath at the external limit of the cavernous sinus frequently shows marked enhancement after contrast injection (Fig. 2.25a, b). On the other hand, the sellar diaphragm itself, also constituted of a dural extension, does not show enhancement, probably due to a lesser degree of vascularization.

The internal carotid artery is well visualized within the cavernous sinus with dynamic scan (Fig. 2.29a), or when the walls are calcified (Fig. 2.29b). Due to its sigmoidal form both in coronal and sagittal planes, it is virtually never completely visualized in a single slice, either in coronal or in axial sections. This artery occupies a medial position within the cavernous sinus, and is only 1 or 2 mm from the lateral pituitary border. In arteriosclerosis, the elderly, or in case of elongation of the carotid, the artery comes into direct contact with the pituitary and can change the shape of the gland. The internal carotid artery is surrounded by a venous plexus, richer laterally than medially, which shows enhancement after intravenous injection of the contrast agent. The vascular structures connecting the right and left cavernous sinuses are generally not visible, with the exception of the basilar sinus situated in an excavation of the posterior surface of the dorsum sellae. Certain vascular structures can sometimes be seen within the sellar diaphragm in dynamic scanning (Fig. 3.9).

Within the venous plexus, nervous structures crossing the cavernous sinus appear as rounded or oval defects in coronal sections, axial sections, or oblique reconstructions. The third nerve (the oculomotor nerve) is always seen in direct coronal sections in the anterior slices, immediately below the anterior clinoid processes (Fig. 2.26). Size and attenuation value of the third nerve are variable; in some cases, a voluminous third nerve presenting marked hypodensity can be mistaken for a fat nodule occupying the most anterior part of the cavernous sinus. A large defect, situated in the anterior part of the cavernous sinus below the anterior clinoid process can also correspond to the presence of the third and fourth (trochlear) nerves in the same dural sheath. In the same anterior section, the sixth nerve (abducens nerve) is frequently visible at the bottom of the cavernous sinus, immediately outside and in contact with the internal carotid artery.

In the middle of the sella turcica, the third cranial nerve is seen at the top of the lateral wall of the sinus. The ophthalmic division of the trigeminal nerve is seen at the inferior part of the lateral wall of the sinus (Fig. 2.26). Finally, in the most posterior coronal sections parallel with the dorsum sellae, the Gasser ganglion appears as an oval structure situated just in front of the tip of the petrous bone (Fig. 2.27).

In axial sections, intracavernous nerves are seen less frequently than in coronal sections; however, the entire course of the nerves can sometimes be visualized up to the entrance to the superior orbital fissure (Fig. 2.25a). The Gasser ganglion is always readily visible within the Meckel cave, at the posteroinferior part of the cavernous sinus (Fig. 2.28).

Fat deposits are normally seen in the anterior part of the sinus, in the immediate vicinity of the superior orbital fissure (Fig. 2.30). The presence of fat deposits in the middle or posterior part of the sinus or between the pituitary and the cavernous sinus (Fig. 9.9) is abnormal, and is generally seen only in Cushing's disease (see Chap. 9) (Bachow).

The Suprasellar Cistern

In coronal sections, the suprasellar cistern is roughly triangular with apex upwards, within which the chiasm is better visualized before enhancement (Fig. 2.8a). The superior pole of the pituitary is shaped by the bordering cistern. The sellar diaphragm per se is not visible. The supraclinoid internal carotid arteries (Fig. 2.4), their branches, and the basilar artery are readily visible within the cistern.

The pituitary stalk is visible between the gland and the infundibular recess of the third ventricle (see Chap. 15).

In axial sections, the suprasellar cistern appears as a five- (Fig. 2.8b) or six-branched (Fig. 2.9a) star, depending on whether the section passes through the pons or through the interpeduncular fossa. The sides of the suprasellar cistern are in contact with the temporal lobe, while its front borders on the frontal lobe. From front to back, the six branches of the star corresponding to the extensions of the suprasellar cistern consist of the interhemispheric fissure, the sylvian fissures, the ambient cistern, and the interpedoncular cistern.

The optic nerves and chiasm are situated at the anterior part of the suprasellar cistern, forming a mass the appearance of which varies with the angle of section. Injection of metrizamide can optimize definition of this structure, as well as the infundibular recess of the third ventricle, the mamillary bodies, the pituitary stalk, and the optic nerves (Fig. 2.10).

The circle of Willis, also situated within the suprasellar cistern, is best studied by dynamic scan (Fig. 2.9b).

References

Aubin ML, Bentson S, Vignaud J (1978) Tomodensito-métrie de la tige pituitaire. J Neuroradiol 51:153–160

Bonneville JF, Poulignot D, Cattin F, Couturier M, Mollet E, Dietemann JL (1982) Apport des méthodes nouvelles dans l'exploration morphologique des tumeurs hypophysaires. Ann Endocrinol (Paris) 43:303–308

Bonneville JF, Cattin F, Moussa-Bacha K, Portha C (1983) Dynamic computed tomography of the pituitary gland: the "tuft sign". Radiology 149:145–148

Bonneville JF, Cattin F, Portha C, Cuenin E, Clere P, Bartholomot B (1985) Computed tomographic demonstration of the posterior pituitary. AJNR 6:889–892

Bonneville JF, Cattin F, Dietemann JL (to be published) The convex pituitary gland. 24th meeting of the American Society of Neuroradiology, San Diego, 22 Jan 1986

Brown SB, Irwin KM, Enzmann DR (1983) CT characteristics of the normal pituitary gland. Neuroradiology 24:259–262

Chambers EF, Turski PA, La Masters D, Newton TH (1982) Regions of low density in the contrast-enhanced pituitary gland: normal and pathologic processes. Radiology 144:109–113

Cohen WA, Pinto RS, Kricheff II (1982) Dynamic CT scanning for visualization of the parasellar carotid arteries. AJNR 3:186–189

Daniels D, Haughton V, Williams A, Gager W, Berns TF (1980) CT of the optic chiasm. Radiology 137:123–127

Di Chiro G, Nelson KB (1962) The volume of the sella turcica. AJR 87:989–1007

Dietemann JL, Lang J, Franck JP, Bonneville JF, Clarisse J, Wackenheim A (1981) Anatomy and radiology of the sellar spine. Neuroradiology 21:5–7

Dietemann JL, Bonneville JF, Cattin F, Poulignot D (1983) Computed tomography of the sellar spine. Neuroradiology 24:173–174

Eresue J, Drouillard J, Philippe JC, Guibert JL, Poux P, Tavernier J (1982) L'exploration des adénomes hypophysaires par scanographie à haute résolution et angioscanographie. Ann Radiol 25:509–517

Flerko B (1980) The hypophysial portal circulation today. Neuroendocrinology 30:56–63

Harris FS, Rhoton AL (1976) Anatomy of the cavernous sinus. A microsurgical study. J Neurosurg 45:169–180

Hayman LA, Evans RA, Hinck VC (1979) Rapid high dose (RHD) contrast CT of perisellar vessels. Radiology 131:121–123

Kline LB, Acker JD, Post JD (1982) CT evaluation of the cavernous sinus. Ophthalmology 89:374–385

Kuuliala I (1980) The normal suprasellar subarachnoid space in CT. Clinical Radiology 31:155–159

La Masters DL, Boggan JE, Wilson CB (1982) Computerized tomography of a sellar spine. Case report. J Neurosurg 57:407–409

Lipman JK, Marshal W (1982) Practical errors in measurement of the pituitary at CT (letter). AJNR 3:87

Mahmoud El Sayed M, Rad PD (1958) The sella in health and disease. The value of the radiographic study of the sella turcica in the morbid anatomical and topographic diagnosis of intracranial tumours. British Journal of Radiology, supp 8

Manelfe C, Giraud B, Espagno J, Kascol A (1978) Cisternographie computérisée au metrizamide. Rev Neurol (Paris) 134:471–484

Muhr C, Bergström K, Grimelius L, Larsson SG (1981) A parallel study of the roentgen anatomy of the sella turcica and the histopathology of the pituitary gland in 205 autopsy specimens. Neuroradiology 21:55–65

Nagagawa Y, Matsumoto K, Fukami T, Takase K (1984) Exploration of the pituitary stalk and gland by high-resolution computed tomography. Comparative study of normal subjects and cases with microadenoma. Neuroradiology 26:473–478

Osborn AG, Anderson RE (1978) Direct sagittal computed tomography scans of the face and paranasal sinuses. Radiology 129:81–87

Parsons C, Hodson N (1979) CT of paranasal sinus tumors. Radiology 132:641–645

Peyster RG, Hoover ED, Adler LP (1984) CT of the normal pituitary stalk. AJNR 5:45–47

Popa G, Fielding U (1930) A portal circulation from the pituitary to the hypothalamic region. J Anat 65:88–91

Rhoton AL Jr, Hardy DG, Chambers SM (1979) Microsurgical anatomy and dissection of the sphenoid bone, cavernous sinus and sellar region. Surg Neurol 12:63–104

Rhoton AL, Harris FS, Renn WH (1977) Microsurgical anatomy of the sellar region and cavernous sinus. Clin Neurosurg 24:54–85

Roberson GH, Tadmor R, Taveras JM, Kleefield J, Ellis G (1977) CT in Metrizamide cisternography. Importance of coronal and axial views. J Comput Assist Tomogr 1:241–245

Roppolo HMN, Latchaw RE, Meyer JD, Curtin HD (1983) Normal pituitary gland: 1. Macroscopic anatomy – CT correlation. AJNR 4:927–935

Roppolo HMN, Latchaw RE (1983) Normal pituitary gland: 2. Microscopic anatomy CT correlation. AJNR 4:937–944

Sage MR, Blumbergs PC, Fowler GW (1982) The diaphragma sellae: its relationship to normal sellar variations in frontal radiographic projections. Radiology 145:699–701

Sage MR, Blumbergs PC, Mulligan BP, Fowler GW (1982) The diaphragma sellae: its relationship to the configuration of the pituitary gland. Radiology 145:703–704

Strother CM, Sackett JF, Appen RE (1977) Anatomic considerations for CT of the optic chiasma. Arch Neurol 34:713–714

Swartz JD, Russel KB, Basile BA, O'Donnell PC, Popky GL (1983) High-resolution CT appearance of the intrasellar contents in women of childbearing age. Radiology 147:115–117

Syvertsen A, Haughton VM, Williams AL, Cusick JF (1979) The computed tomographic appearance of the normal pituitary gland and pituitary microadenomas. Radiology 133:385–391

Taylor S (1982) High resolution computed tomography of the sella. Radiologic clinics of North America 20:207–236

Vincentelli F, Grisoli F, Bartoli JF, Leclerc T, De Schmedt E, Salamon G (1982) Anatomo-radiological basis of sellar surgery and its nasoseptal approach. J Neuroradiol 9:284–303

Wolpert SM, Molitch ME, Goldman JA, Wood JB (1984) Size, shape, and appearance of the normal female pituitary gland. AJNR 5:263–267

Wortzman G, Rewcastle NB (1982) Tomographic abnormalities simulating pituitary microadenomas. AJNR 3:505–512

Xuereb GB, Prichard MML, Daniel PM (1954) The hypophysial portal system of vessels in man. Q J Exp Physiol 39:219–230

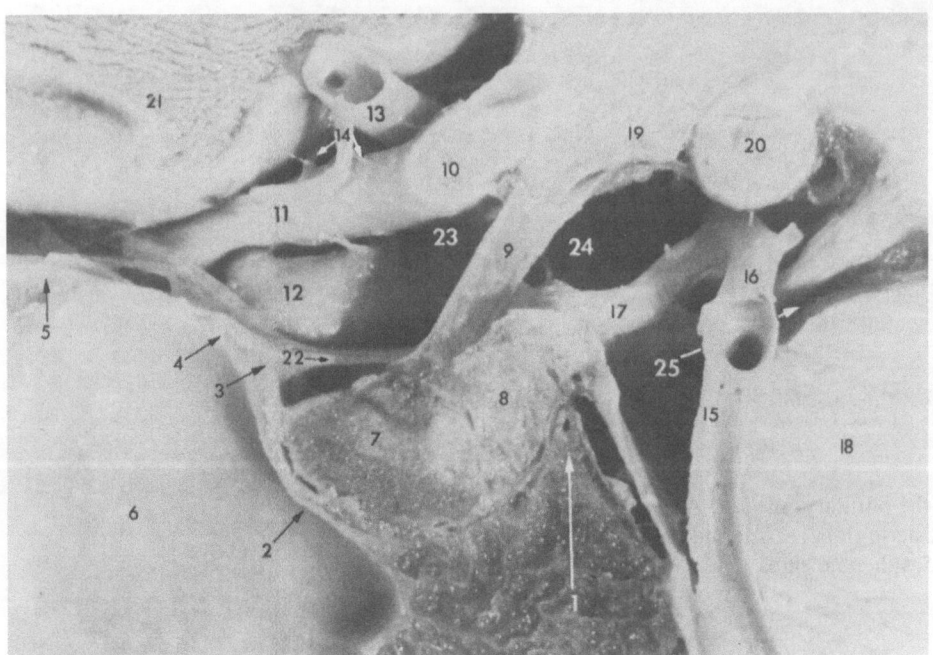

Fig. 2.1. Midsagittal section of the sellar region (courtesy of H. Duvernoy): *1*, Dorsum sellae; *2*, anterior wall of the sella; *3*, tuberculum sellae; *4*, chiasmatic sulcus; *5*, planum sphenoidale; *6*, sphenoid sinus; *7*, anterior pituitary; *8*, posterior pituitary; *9*, pituitary stalk; *10*, optic chiasm; *11*, optic nerve; *12*, internal carotid artery; *13*, anterior cerebral artery; *14*, chiasmal branch of the internal carotid artery; *15*, basilar artery; *16*, posterior cerebral artery; *17*, posterior communicating artery; *18*, pons; *19*, hypothalamus; *20*, mamillary body; *21*, frontal lobe; *22*, diaphragma sellae; *23*, prechiasmatic cistern; *24*, postchiasmatic cistern; *25*, interpeduncular cistern

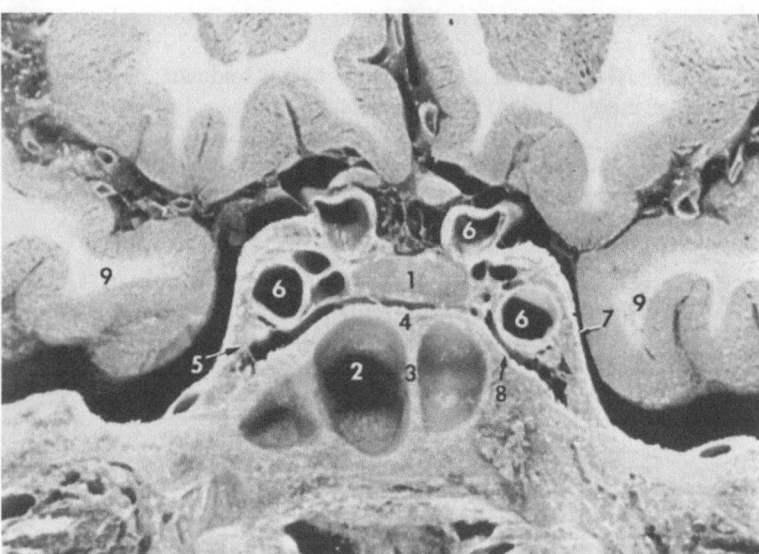

Fig. 2.2. Frontal section of the pituitary fossa (Atlas Anatomique, Sandoz): *1*, Pituitary gland; *2*, sphenoid sinus; *3*, intersinusal septum; *4*, sellar floor; *5*, cavernous sinus; *6*, internal carotid artery; *7*, lateral dural wall of the cavernous sinus; *8*, carotid sulcus; *9*, temporal lobe

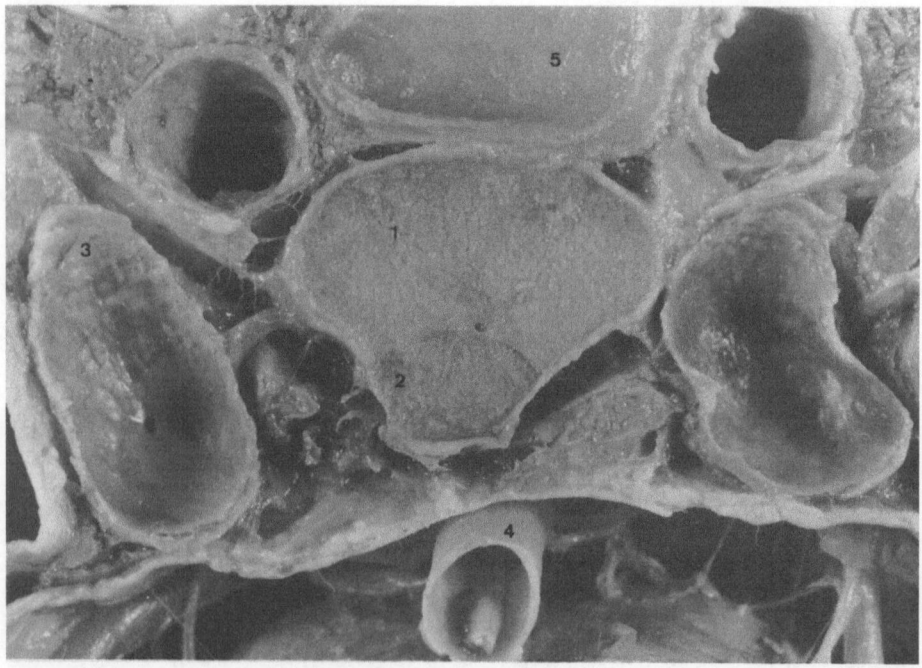

Fig. 2.3. Axial section of the pituitary fossa: *1*, Anterior lobe of the pituitary; *2*, posterior lobe; *3*, internal carotid artery; *4*, basilar artery; *5*, sphenoid sinus

Fig. 2.4. Normal coronal contrast-enhanced CT scan of the sellar region from anterior clinoid processes to dorsum sellae

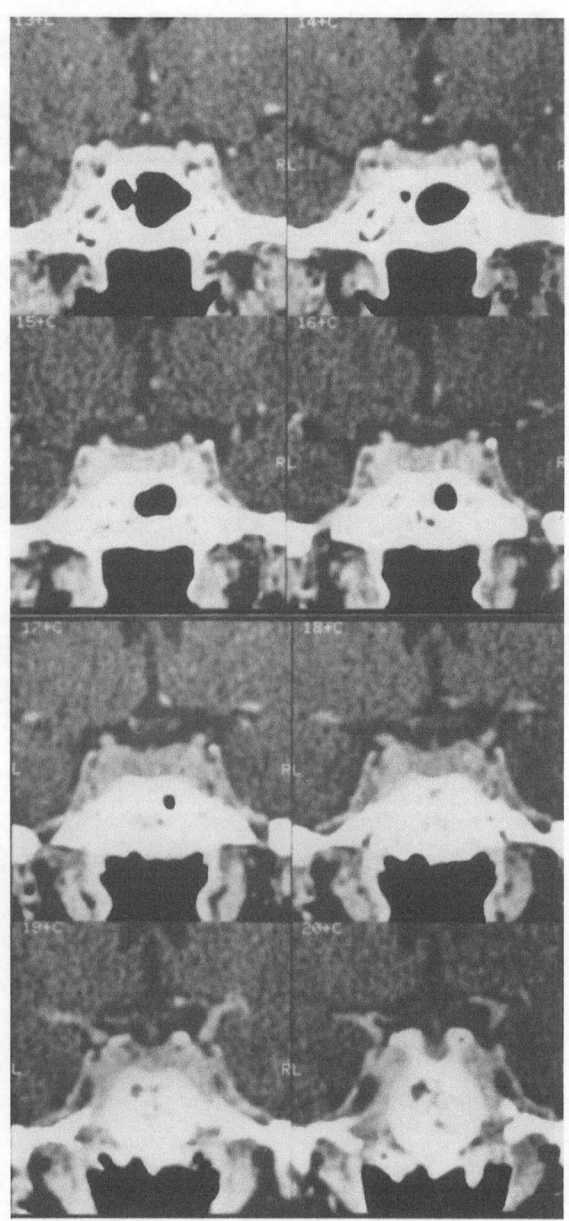

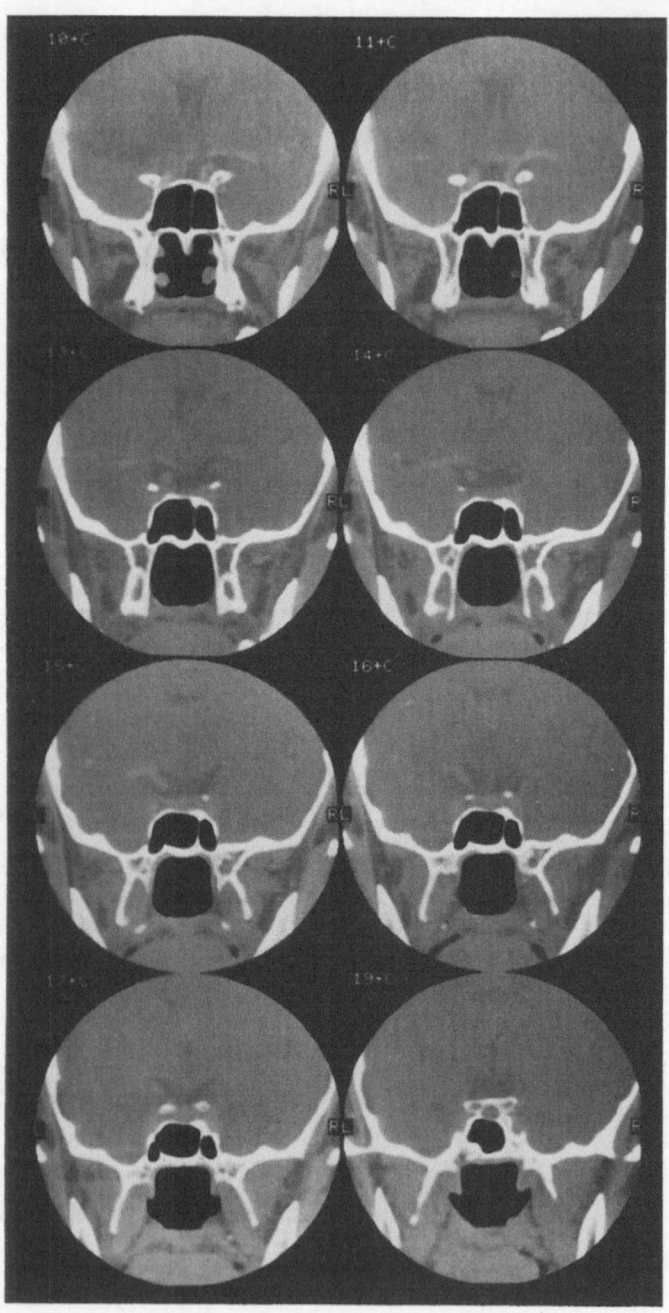

Fig. 2.5. Normal coronal CT scan of the sella turcica (bone window)

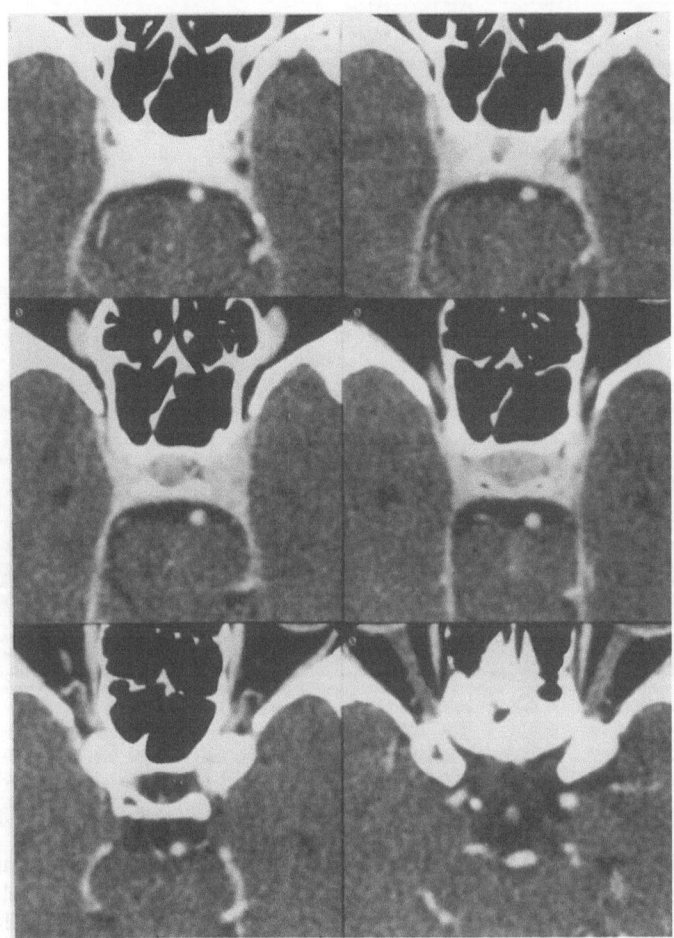

Fig. 2.6. Normal axial contrast-enhanced CT scan of the sellar region from the sphenoid sinus to the suprasellar cistern

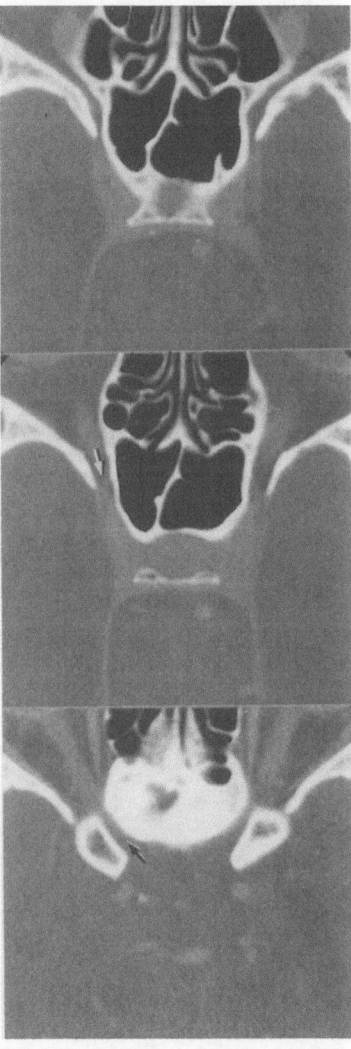

Fig. 2.7. Normal axial CT scan of the sellar region (bone review images); superior orbital fissure (*white arrow*); optic canal (*black arrow*)

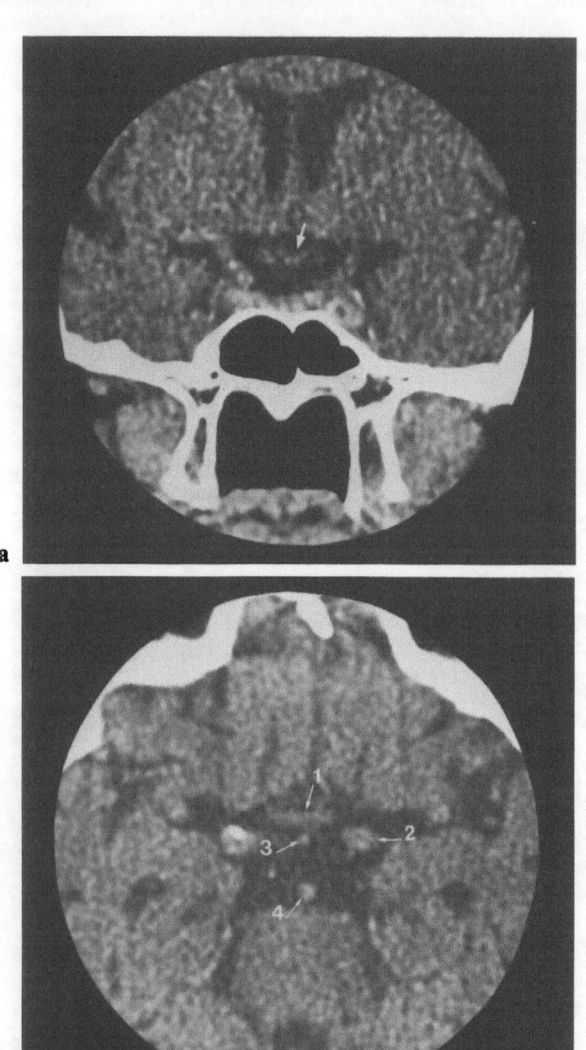

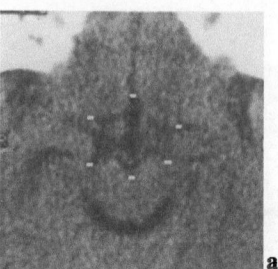

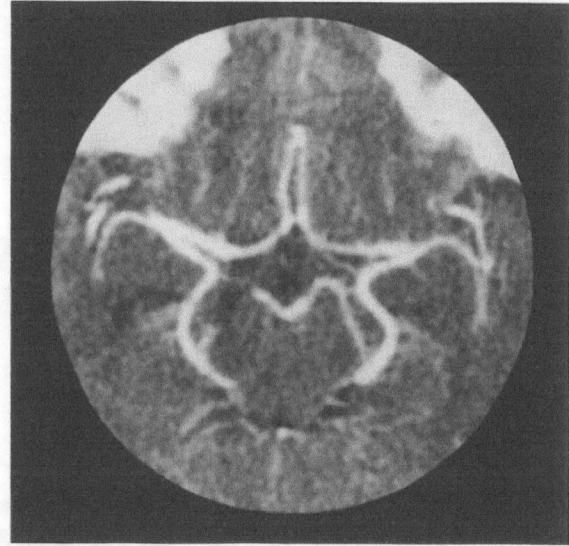

Fig. 2.9. a Native CT scan demonstrating the typical shape of the suprasellar cistern at the peduncular level. **b** Dynamic CT scan at the same level. Excellent visualization of the circle of Willis

Fig. 2.8. a Coronal native CT scan. The optic chiasm (*arrow*) is well shown within the chiasmatic cistern. **b** Axial native CT scan. Normal suprasellar cistern: 1, Optic chiasm; 2, carotid artery; 3, pituitary stalk; 4, basilar artery

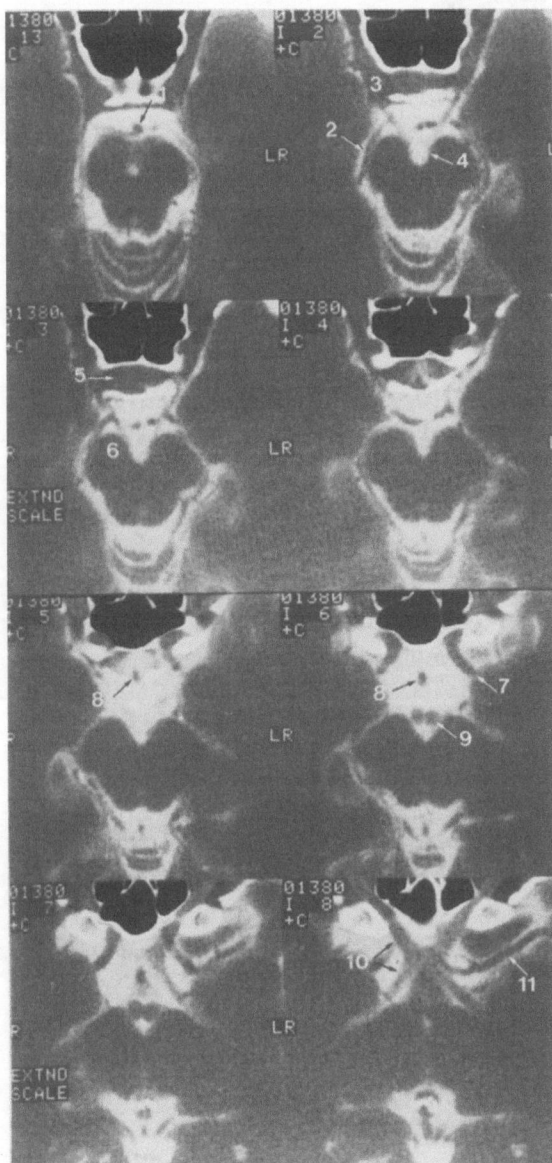

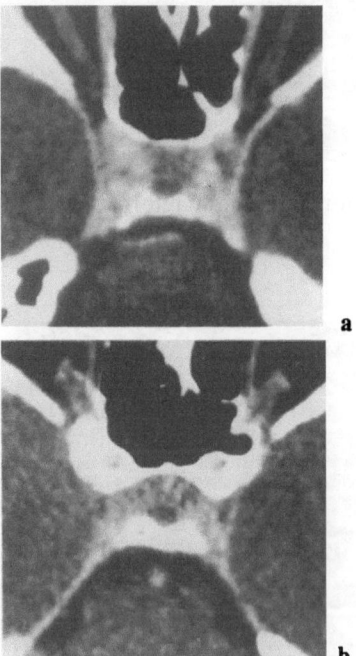

Fig. 2.11a, b. Normal posterior pituitaries in axial sections

Fig. 2.10. Normal axial cisternography of the sellar region: *1*, Basilar artery; *2*, posterior cerebral artery; *3*, cavernous sinus; *4*, interpeduncular cistern; *5*, pituitary gland; *6*, cerebral peduncle; *7*, internal carotid artery; *8*, pituitary stalk; *9*, mamillary bodies; *10*, intracranial optic nerve and optic tract; *11*, middle cerebral artery

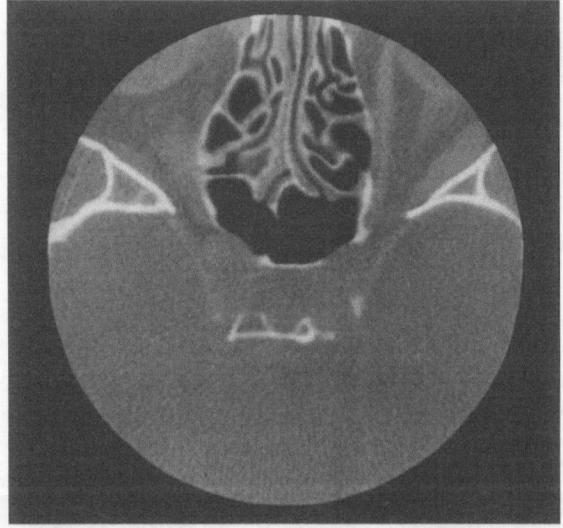

Fig. 2.12. Normal posterior pituitary imprint on the dorsum sellae

Fig. 2.13. Posterior pituitary: sagittal reformatted image

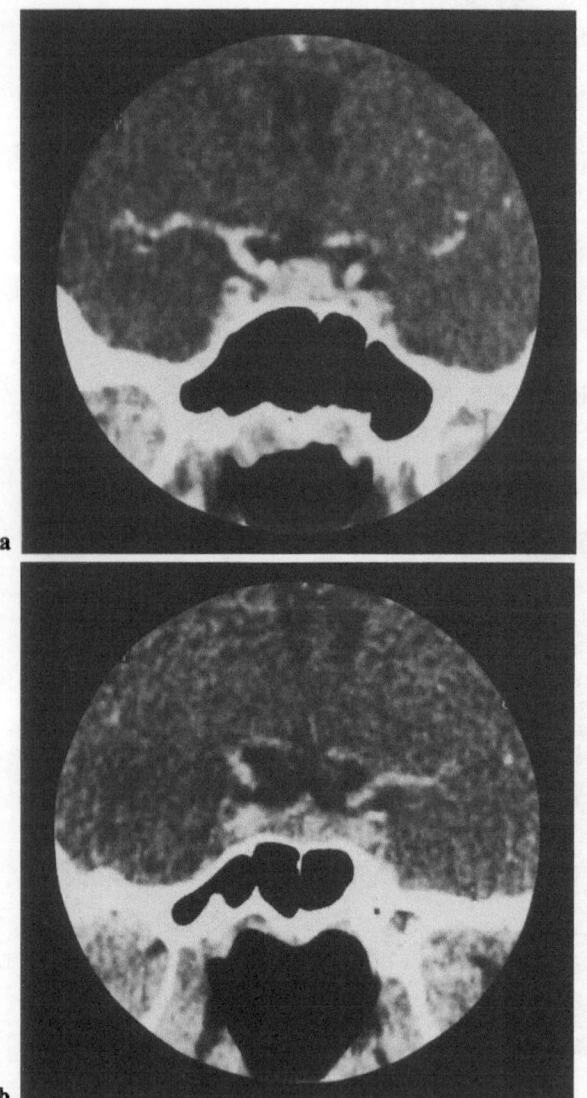

Fig. 2.14. a Increased height of the pituitary with convex superior pole in a healthy young woman. **b** Triangular rise of the sellar diaphragm at the pituitary stalk passage

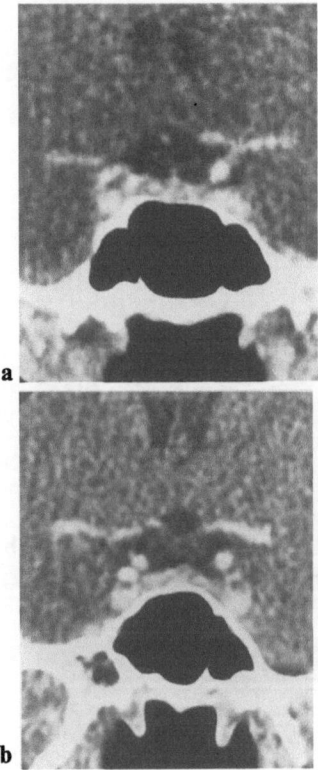

a

b

Fig. 2.15. a Slanted and **b** convex sellar floors: in each case, the superior limit of the pituitary gland parallels the sellar floor

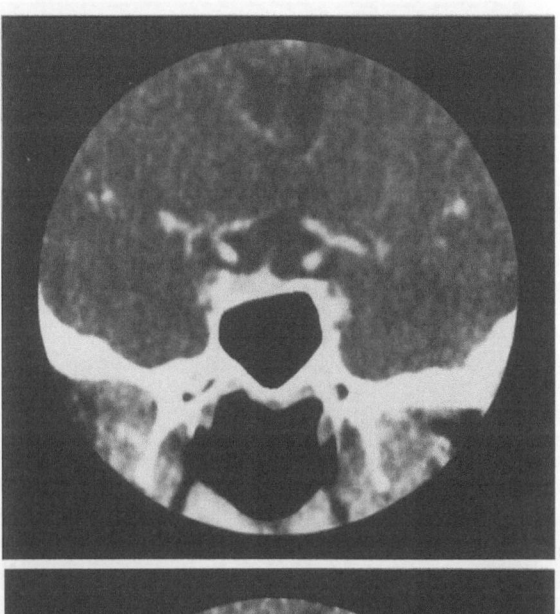

a

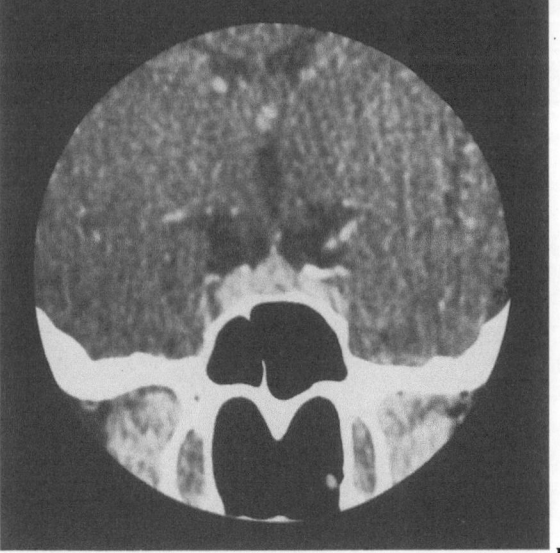

b

Fig. 2.17. a False appearance of increased pituitary volume with convexity of the upper pole of the gland resulting from **b** shortness of the sellar floor from right to left

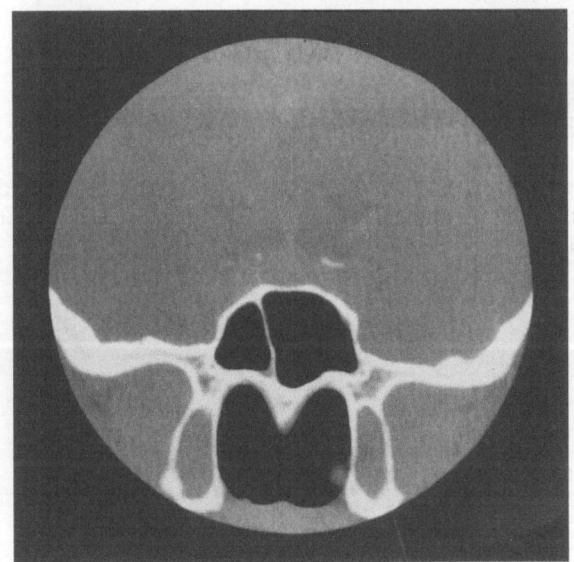

Fig. 2.16. Bulging of the roof of the sphenoid sinus on the left: the entire pituitary gland is located on the right

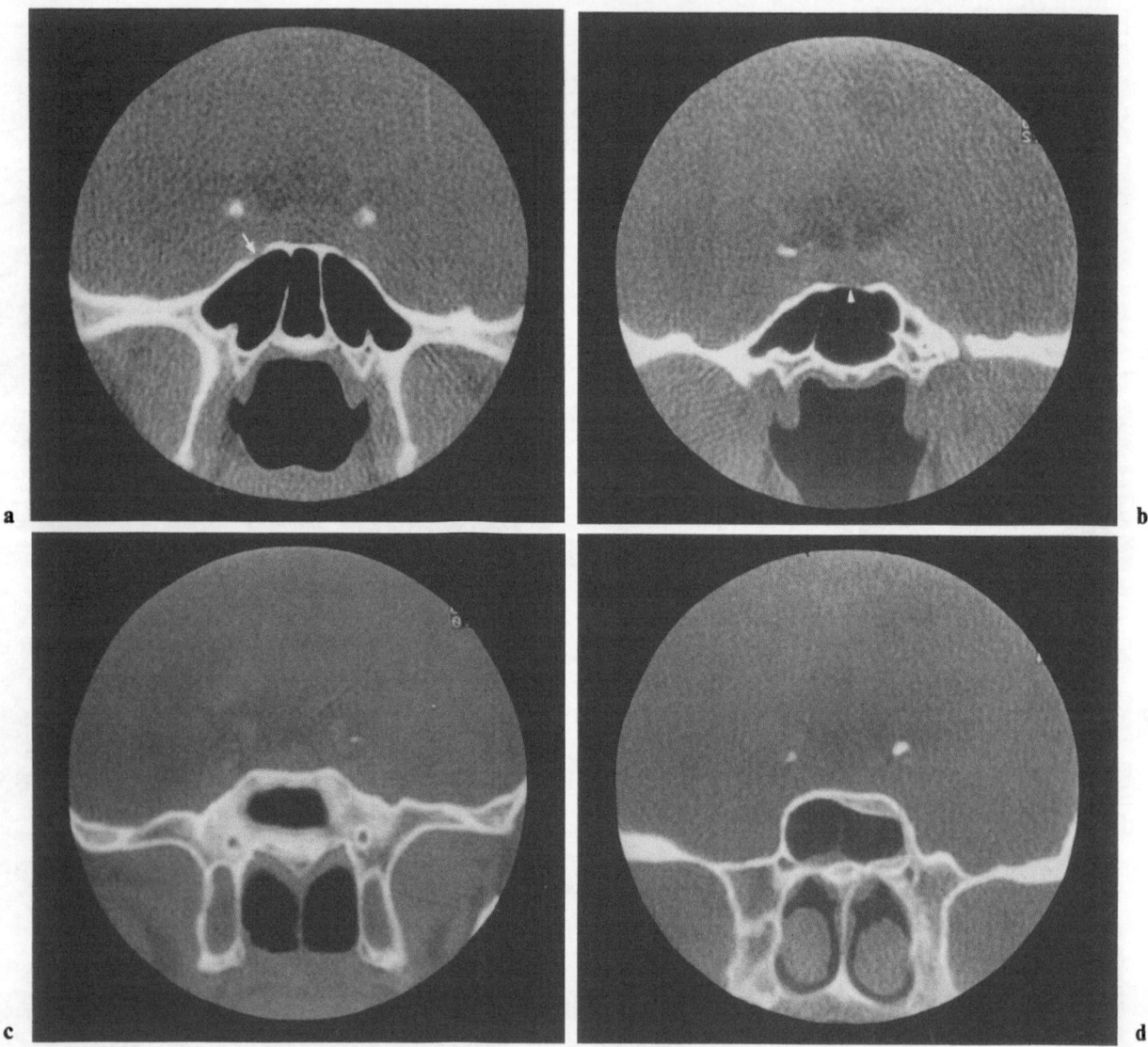

Fig. 2.18a–d. Various normal sellar floors. **a** Imprints of the carotid sulci (*arrow*). **b** Normal cortical thinness of the sellar floor (*arrowhead*): bone changes are more marked on the lower than on the upper aspect of the sellar floor and therefore cannot result from an increased intrasellar content pressure. **c, d** Normal thick sellar floors

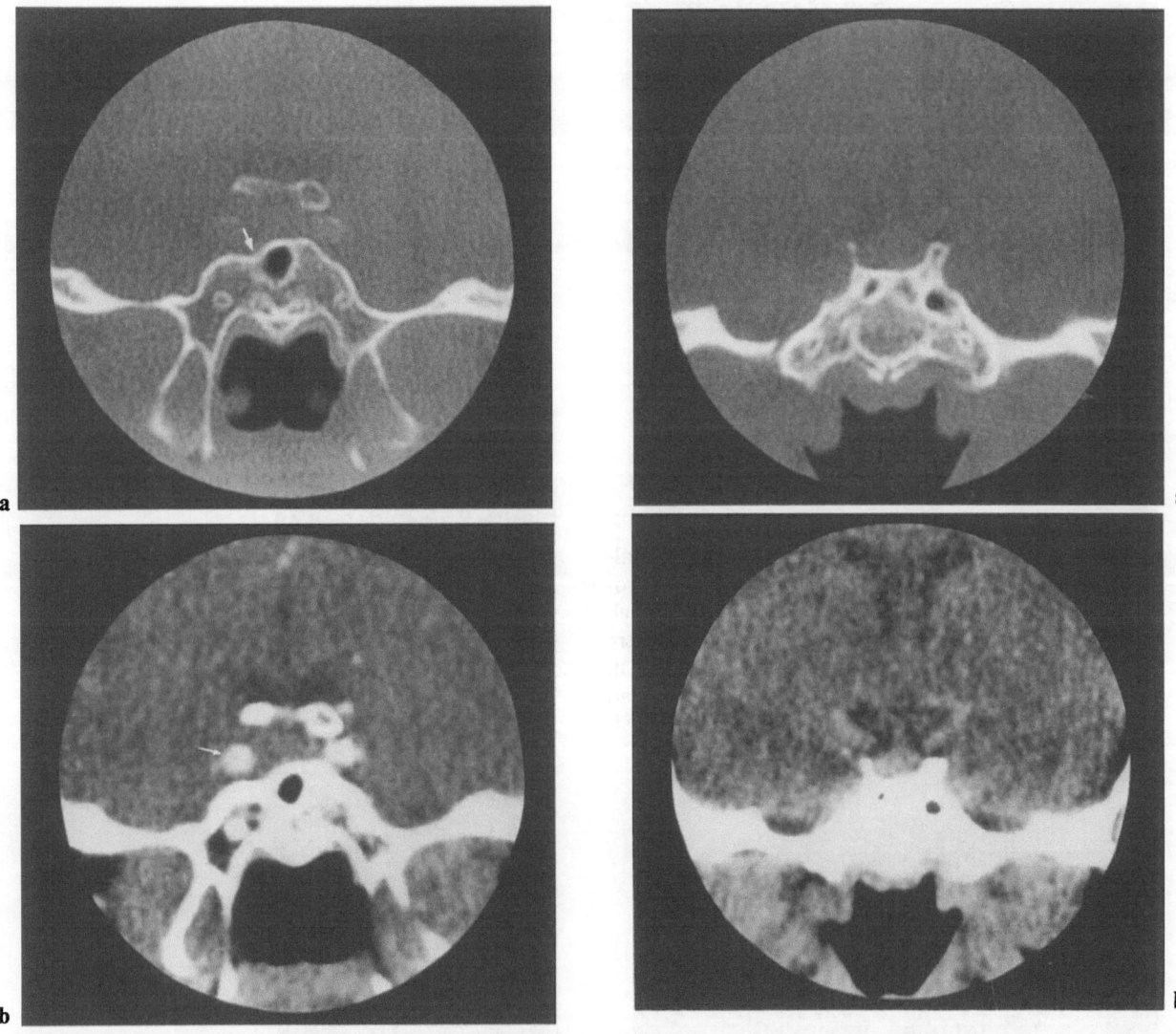

Fig. 2.19a, b. Deep carotid sulcus on the right (*arrow*). The right carotid artery is well shown during dynamic CT scan above the carotid sulcus (*small arrow*)

Fig. 2.20a, b. Bilateral ossification of the lateral wall of the sella. **a** Bone review image. **b** Coronal-enhanced CT scan

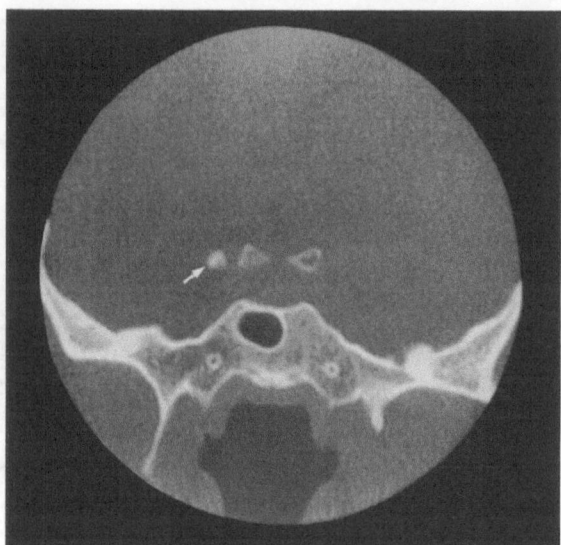

Fig. 2.21. Coronal CT scan. Bone review image. Visualization in the same section of both posterior clinoid processes and right anterior clinoid process (*arrow*). In such cases where the AP diameter of the sella is short or where the anterior clinoid processes are long, visualization in the same section of these bony structures must not lead to an erroneous diagnosis of abnormal suprasellar calcification

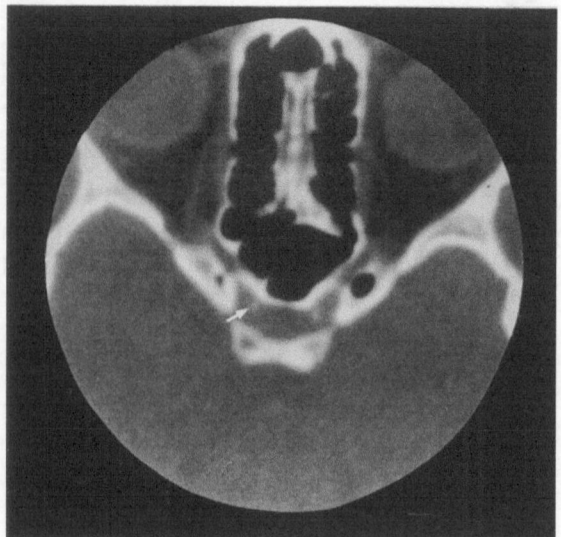

Fig. 2.22. Middle clinoid processes (*arrow*)

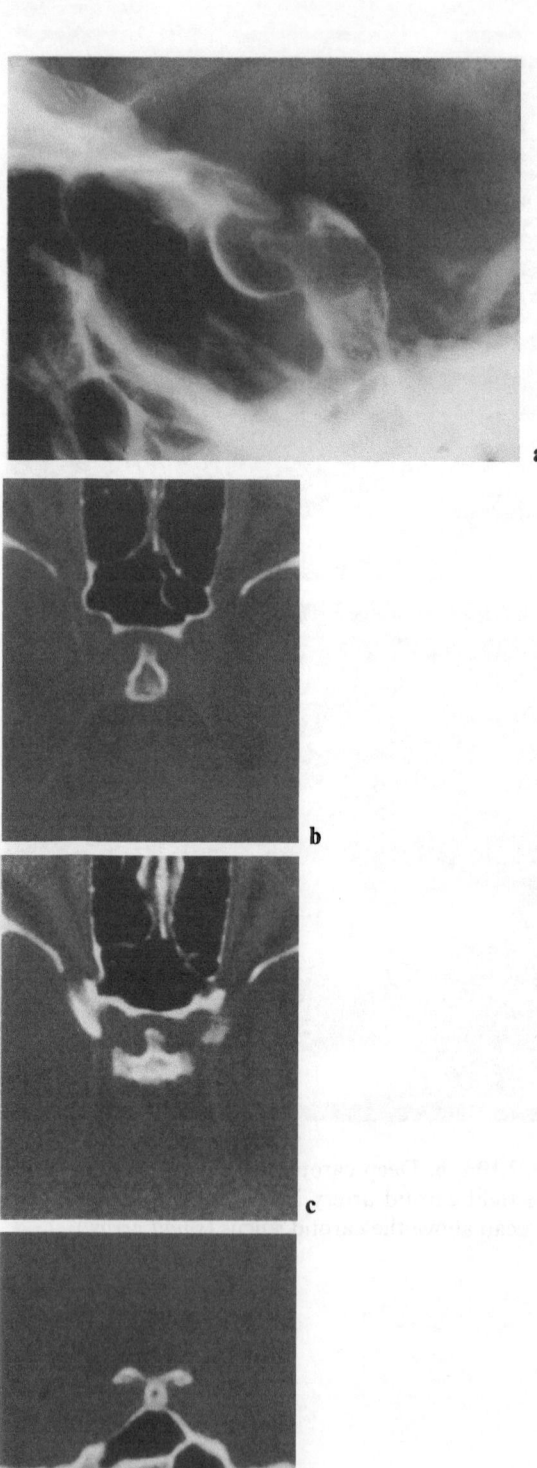

Fig. 2.23a–d. Sellar spine. **a** Magnified lateral view. **b, c** Axial and **d** coronal CT scans. A bony spine arising from the dorsum sellae is pointing forwards into the sellar cavity

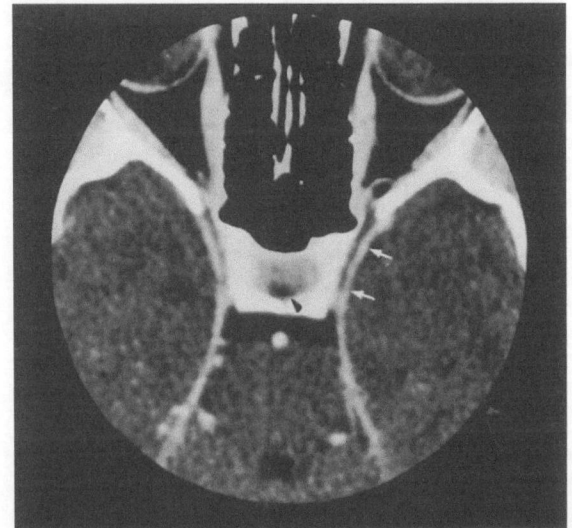

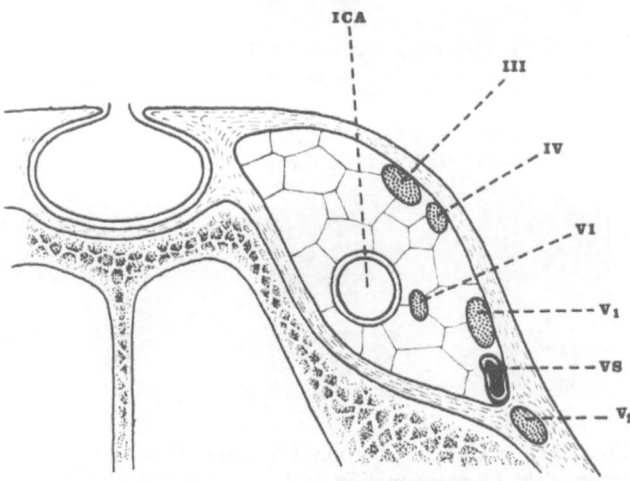

Fig. 2.24. Schematic drawing of the cavernous sinus: *ICA*, internal carotid artery; *III*, oculomotor nerve; *IV*, trochlear nerve; *VI*, abducens nerve; *V1*, ophthalmic nerve; *V2*, maxillary nerve; *VS*, inferior venous space

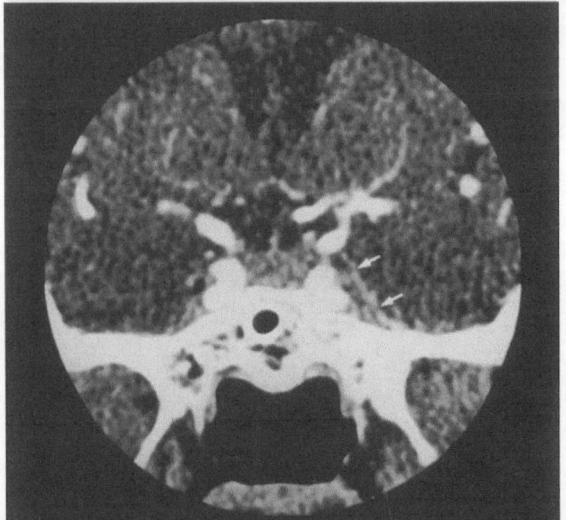

Fig. 2.25. a Cavernous sinus normal dural lateral wall ▶ after contrast enhancement (*arrows*). The posterior pituitary is also well shown (*arrowhead*). **b** Dynamic coronal CT scan. Early enhancement of the dural lateral wall of the cavernous sinus. **c** Normal inferior venous space located between the ophthalmic and the maxillary nerves (*arrow*)

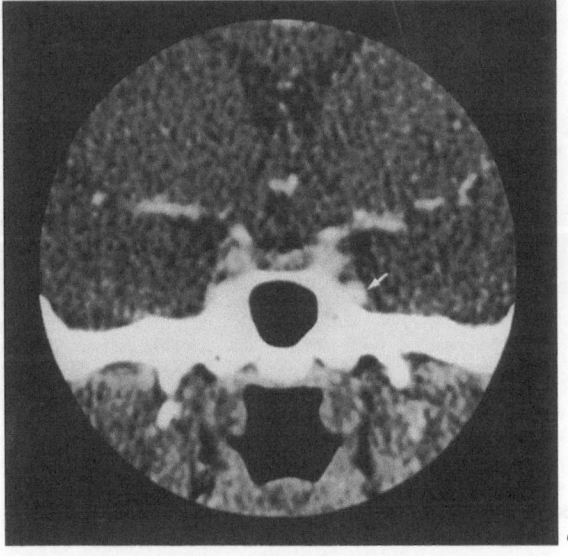

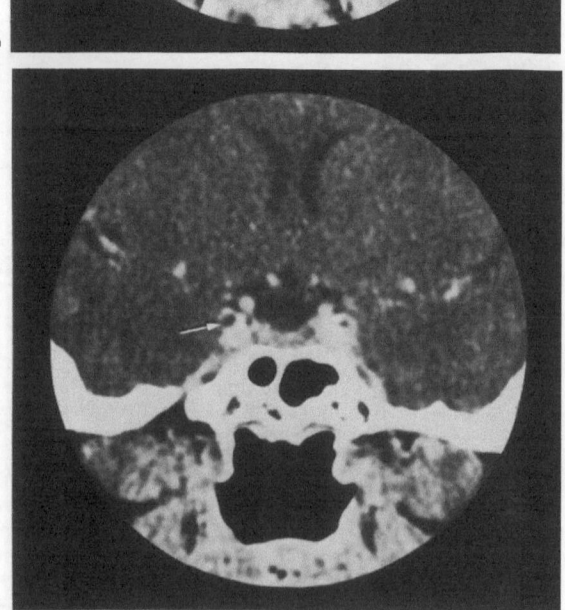

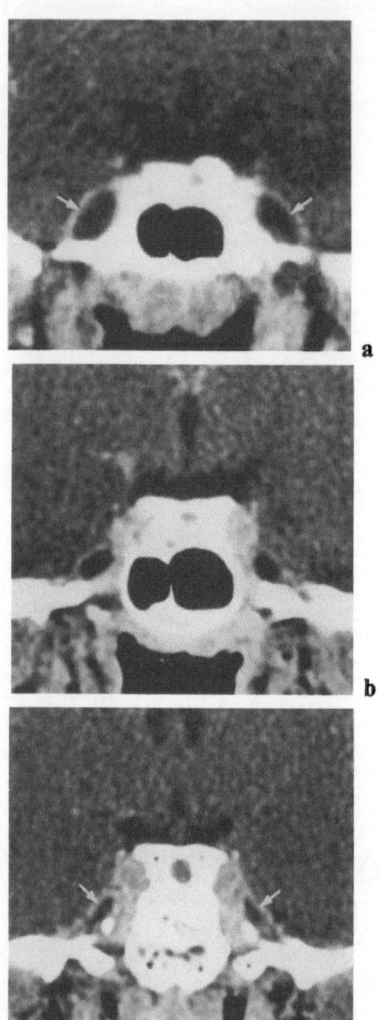

Fig. 2.27a–c. Contrast coronal CT scans. Different patterns of the Gasser ganglion at the dorsum sellae level

Fig. 2.26a–c. Intracavernous cranial nerves. Coronal contrast CT scans. Note the constant visualization of the oculomotor nerve

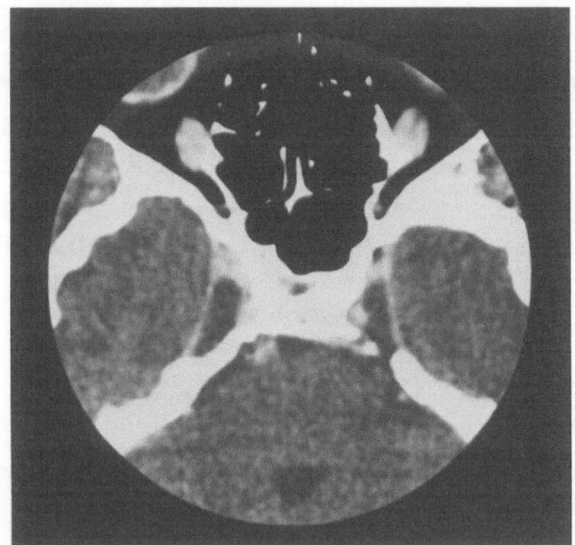

Fig. 2.28. Axial contrast CT scan. Good visualization of Gasser ganglions

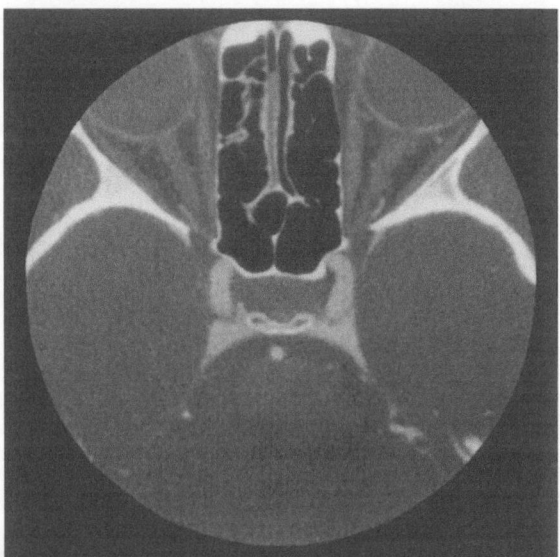

a

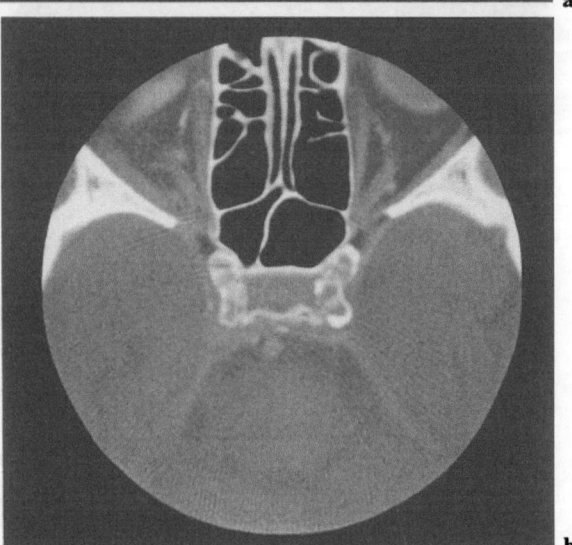

b

Fig. 2.29. a Axial dynamic CT scan. Excellent visualization of both carotid siphons. **b** Bilateral calcified carotid siphons

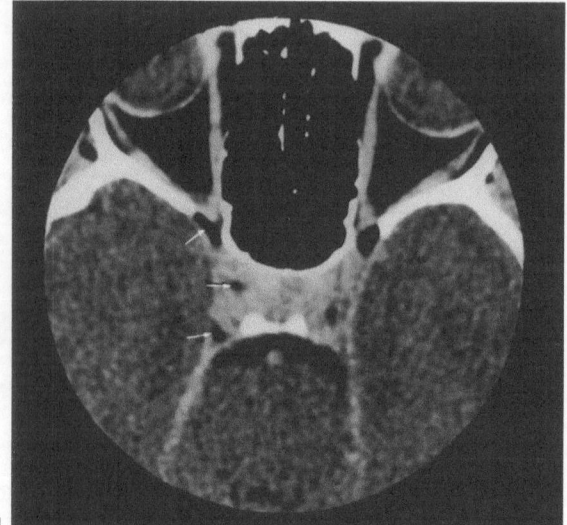

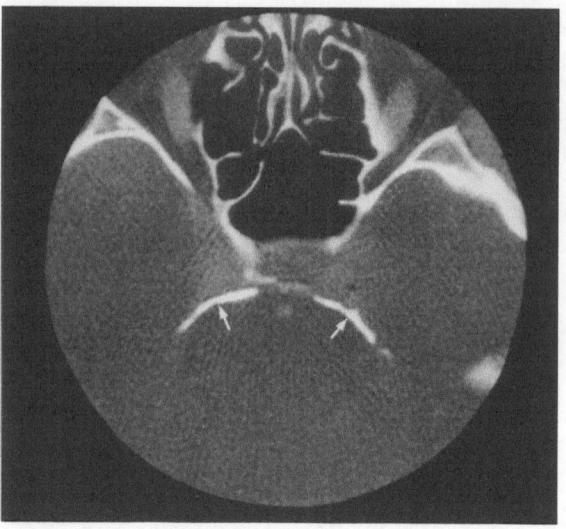

Fig. 2.31. Calcified petroclinoid ligaments (*arrows*)

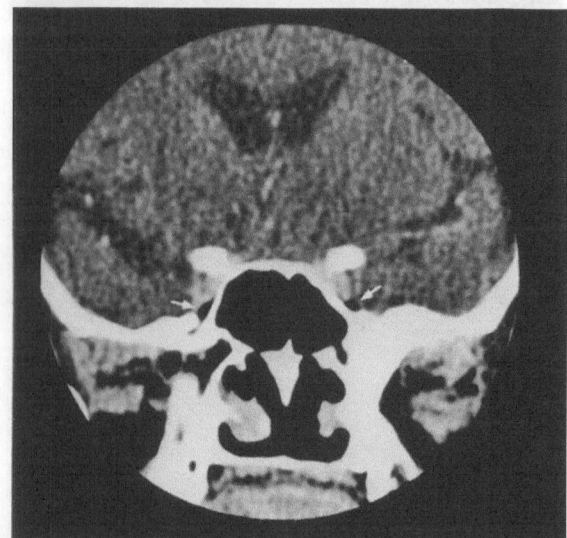

Fig. 2.30 a, b. Normal fat deposits (*arrows*) in the cavernous sinuses and the superior orbital fissures

Chapter 3
Dynamic CT of the Pituitary Gland

General Aspects

Dynamic pituitary CT scan is of major benefit for diagnosis of the smallest pituitary lesions. In 1982, the pituitary dynamic CT technique was suggested in view of the inadequacy of conventional techniques, even using high-resolution systems, for demonstration of the smallest intrasellar adenomas. Visualization of the pituitary capillary bed with this technique is today as fundamental for diagnosis of intrasellar lesions as was demonstration of the calcified pineal gland or internal cerebral vein for identification of the midline of the brain. A brief review of pituitary blood supply is necessary to understand the utility of the dynamic scan (Fig. 3.1).

Review of Pituitary Blood Supply

The various structures comprising the pituitary are directly or indirectly vascularized from two principal sources, the superior and inferior hypophyseal arteries (Xuereb). The inferior hypophyseal artery arises from each side of the intracavernous segment of the internal carotid artery; the superior hypophyseal artery, also paired, arises from the internal carotid immediately after its exit from the cavernous sinus following passage through the dura mater. There are transversal anastomoses between the two superior hypophyseal arteries and between the two inferior hypophyseal arteries. Furthermore, the trabecular artery forms anastomoses with the superior and inferior pituitary networks. The anterior pituitary does not receive any direct arterial blood supply. At the superior part of the pituitary stalk, the hypophyseal arteries give rise to highly convoluted spiral capillary networks (referred to as "gomitoli" by Fumagalli)

which constitute the first capillary bed of the pituitary portal system. The first capillary bed is drained by the long portal vessels which descend the stalk and terminate in the second capillary bed, constituted by the anterior lobe sinusoids (Fig. 3.2).

Popa and Fielding (1930), who first discovered the existence of the pituitary portal system, believed that this system functioned in an ascending fashion, i.e., from the pituitary to the hypothalamus. Proof that blood circulation actually occurs in the opposite direction, i.e., from the hypothalamus to the pituitary, was shown by subsequent morphological studies (Wislocki 1936) and by in vivo observation of function of the portal system in animals (Houssay 1935). Dynamic CT provides confirmation of the Wislocki hypothesis in living subjects, demonstrating that enhancement is progressive and clearly separated in time: initially the carotid siphons are enhanced, followed by the top of the pituitary stalk, the secondary capillary bed, and finally, centrifugal enhancement of the anterior lobe parenchyma (or pars distalis).

Unlike the anterior lobe, the posterior lobe (or infundibular process) has a conventional arterial-capillary-venous circulatory system. The right and left inferior hypophyseal arteries give rise to a capillary network constituted of sinusoids far smaller than those of the anterior lobe. Due to the absence of a portal system, sinusoids of the posterior lobe appear to be enhanced far earlier than those of the anterior lobe (Fig. 3.12). Finally, sinusoids of the anterior and posterior lobe terminate in small venules in the periphery of the gland, draining into the venous sinus which surrounds the pituitary.

The secondary capillary bed of the anterior pituitary is comprised of a very dense network of freely anastomosing vessels, referred to as the sinusoids, dispersed within the parenchymatous cells (Fig. 3.2). These sinusoids, larger than the capillaries, receive no arterial blood and are fed solely by portal vessels carrying blood which has already passed through the first capillary network in the pituitary stalk. There is an obviously functional significance to this vascular disposition: the capillary loops characteristic of the first vascular bed of the pituitary stalk are the site of nerve endings of certain hypothalamo-pituitary tract units; these release substances into the portal system which activate or suppress hormonal secretion by secretory cells of the anterior lobe.

Techniques

After careful determination of the ideal plane, dynamic scanning requires rapid injection (in less than 8 s) of a bolus of 60 cc warm 32% iodinated contrast medium (Hexabrix, Laboratoire Guerbet, Aulnay-Sous-Bois, France). Six to eight 1.5 mm sections are carried out beginning with the start of the injection. With the GE 8800 CT/T, scan time is 9.6 s and the interscan delay 1.4 s (see Chap. 1). This yields seven consecutive sections in approximately 80 s. One further section is taken in the same plane, 30 s later. With the GE 9800 CT/T, scan time is 2 s, and interscan delay time is 2–6 s.

Normal Results

From the morphologic and diagnostic standpoints, dynamic CT allows visualization of:
1. The intracavernous and supraclinoid internal carotid arteries
2. The central part of the secondary pituitary capillary bed, we have called the *pituitary tuft*
3. The centrifugal contrast enhancement of the anterior pituitary
4. The posterior pituitary in axial section
5. The venous spaces of the cavernous sinus (Figs. 2.25c, 3.6a)

The intracavernous internal carotid arteries and the branches of their intracranial division are enhanced first, and are perfectly visible approximately 10 s after the end of injection. The high quality of information thus available concerning topography of the carotid siphons today renders carotid angiography superfluous before transsphenoidal surgery (Figs. 3.14, 4.2, and 4.4).

The secondary pituitary capillary bed is visualized approximately 10 s after optimal enhancement of internal carotid arteries. In all cases, the secondary pituitary capillary bed appears as an unpaired median vascular structure situated below the sellar diaphragm, in a prolongation of the direction of the pituitary stalk. Most frequently its appearance is of a vascular tuft, the surface of which covers approximately one-quarter of the pituitary in the section considered (Figs. 3.3 and 3.4). Much less frequently, the pituitary tuft is far smaller, appearing as a point of variable size (Figs. 3.5 and 3.6). The tuft is virtually always regular and symmetric; however, a moderate degree of asymmetry is possible in the absence of pathological changes (Fig. 3.7). Finally, especially where the pituitary height is small, the pituitary capillary bed may be constituted of a vascular band of variable height running immediately below the superior border of the gland (Fig. 3.8).

Less frequently, dynamic coronal CT scan allows visualization of vascular structures situated within the sellar diaphragm (Fig. 3.9a), immediately above the sellar floor (Fig. 3.9b), or of intraglandular anastomoses (Fig. 3.9c).

Between 20 and 40 s after injection, there is a gradual, regular, and symmetrical centrifugal enhancement of the anterior pituitary from the capillary bed. Beginning 60 s after bolus injection, pituitary enhancement appears normally homogeneous, and the pituitary capillary bed is no longer visible (Fig. 3.10).

In axial sections the capillary bed, probably due to the partial volume phenomenon, is rarely as well visualized as in coronal sections (Fig. 3.11). On the other hand, in dynamic axial sections posterior pituitary capillary enhancement occurs very soon after enhancement of the intracavernous carotid arteries due to the direct arterial blood supply of the posterior pituitary by the inferior hypophyseal arteries (Figs. 3.12 and 3.13). At some distance of the bolus injection the posterior pituitary appears less enhanced than the anterior pituitary (Fig. 3.12). Indeed, when contrast media equilibrates between the intravascular and extravascular interstitial spaces, a lesser enhancement of the posterior lobe is seen, correlating with the lesser volume of the interstitial spaces of the posterior lobe.

Pathological Results

Dynamic CT allows demonstration of raising of the anterior cerebral arteries in adenomas with suprasellar extension (Fig. 6.3a), or displacement of the intracavernous carotid arteries in cases of intrasellar lesions (Fig. 5.31).

Due to compression when the pituitary lesion is larger than 8–10 mm in diameter the pituitary capillary bed is virtually never visible. The absence of visualization of the capillary bed thus constitutes a CT sign of pituitary lesion. In exceptional cases, in the presence of an adenoma greater than 10 mm, the pituitary bed can still be visible in the form of a vascular opacity at the superior pole of the gland (Fig. 3.15).

The capillary bed is regularly displaced or deformed (tuft sign) in cases of lateralized intrasellar lesion less than 6–8 mm in diameter (Fig. 3.16). This sign is fundamental for diagnosis of pituitary adenomas too small to deform the sellar floor or the sellar diaphragm. It is of course also essential for the infrequent isodense adenomas (Fig. 3.17), or for adenomas with poorly defined borders.

In adenomas with a median topography, the pituitary capillary bed cannot be displaced; nevertheless, localized compression of the inferior part of the pituitary tuft may occur (Fig. 3.18).

When the pituitary capillary bed is band-like, a localized absence of opacification has the same diagnostic significance as the usual displacement of the pituitary tuft (Fig. 6.12g).

Finally, after opacification of the pituitary bed, during slow and centrifugal enhancement of the glandular parenchyma, asymmetrical enhancement of the pituitary constitutes a further essential sign for presence of intrasellar lesion (Figs. 3.19 and 3.20).

In general, intrasellar microadenomas are enhanced more slowly and to a lesser extent than is the healthy pituitary tissue.

Other examples of the utility of dynamic CT will be presented in the chapters concerning CT of intrasellar pathology.

References

Aubin ML, Bentson S, Vignaud J (1978) Tomodensitométrie de la tige pituitaire. J Neuroradiol 51:153–160

Bonneville JF, Poulignot D, Cattin F, Couturier M, Mollet E, Dietemann JL (1982) Apport des méthodes nouvelles dans l'exploration morphologique des tumeurs hypophysaires. Ann Endocrinol (Paris) 43:303–308

Bonneville JF, Cattin F, Moussa-Bacha K, Portha C (1983) Dynamic computed tomography of the pituitary gland: the "tuft sign". Radiology 149:145–148

Bonneville JF, Cattin F, Moussa Bacha K, Portha C (1983) Un plus dans l'exploration de l'hypophyse: l'angioscan. Presse Méd 12:1669

Bonneville JF, Cattin F, Portha C, Cuenin E, Clere P, Bartholomot B (1985) Computed tomographic demonstration of the posterior pituitary. AJNR 6:889–892

Cohen WA, Pinto RS, Kricheff II (1982) Dynamic CT scanning for visualization of the parasellar carotid arteries. AJNR 3:186–189

Daniels D, Haughton V, Williams A, Gager W, Berns TF (1980) CT of the optic chiasm. Radiology 137:123–127

Flerko B (1980) The hypophysial portal circulation today. Neuroendocrinology 30:56–63

Harris FS, Rhoton AL (1976) Anatomy of the cavernous sinus. A microsurgical study. J Neurosurg 45:169–180

Hayman LA, Evans RA, Hinck VC (1979) Rapid high dose (RHD) contrast CT of perisellar vessels. Radiology 131:121–123

Kline LB, Acker JD, Post JD (1982) CT evaluation of the cavernous sinus. Ophthalmology 89:374–385

Peyster RG, Hoover ED, Adler LP (1984) CT of the normal pituitary stalk. AJNR 5:45–47

Popa G, Fielding U (1930) A portal circulation from the pituitary to the hypothalamic region. J Anat 65:88–91

Rhoton AL, Harris FS, Renn WH (1977) Microsurgical anatomy of the sellar region and cavernous sinus. Clin Neurosurg 24:54–85

Rhoton AL Jr, Hardy DG, Chambers SM (1979) Microsurgical anatomy and dissection of the sphenoid bone, cavernous sinus and sellar region. Surg Neurol 12:63–104

Sage MR, Blumbergs PC, Mulligan BP, Fowler GW (1982) The diaphragma sellae: its relationship to the configuration of the pituitary gland. Radiology 145:703–704

Wing SD, Anderson RE, Osborn AG (1980) Dynamic cranial CT; preliminary results. AJNR 1:135–139

Wislocki GB (1938) The vascular supply of the hypophysis cerebri of the rhesus monkey and man. Res Publ Assoc Nerv Ment Dis 17:48–68

Xuereb GB, Prichard MML, Daniel PM (1954) The hypophysial portal system of vessels in man. Q J Exp Physiol 39:219–230

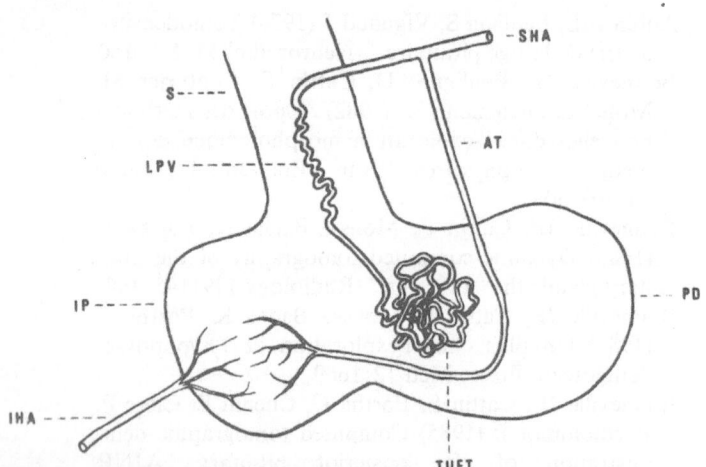

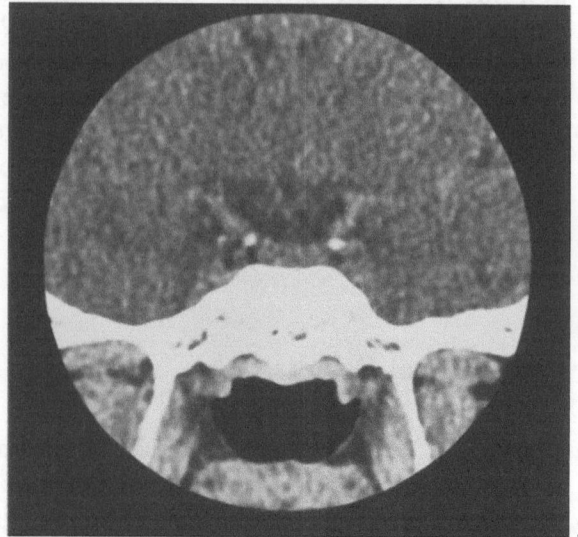

Fig. 3.1. Schematic drawing of the pituitary blood supply. *SHA*, superior hypophyseal arteries; *IHA*, inferior hypophyseal arteries; *AT*, trabecular artery; *S*, stalk; *PD*, pars distalis; *LPV*, long portal vessels; *IP*, infundibular process

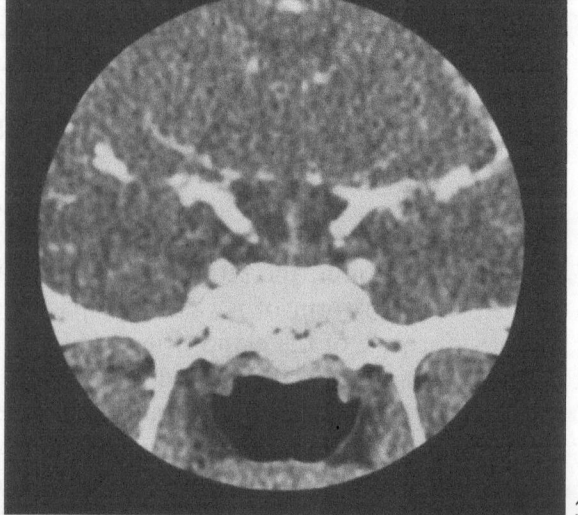

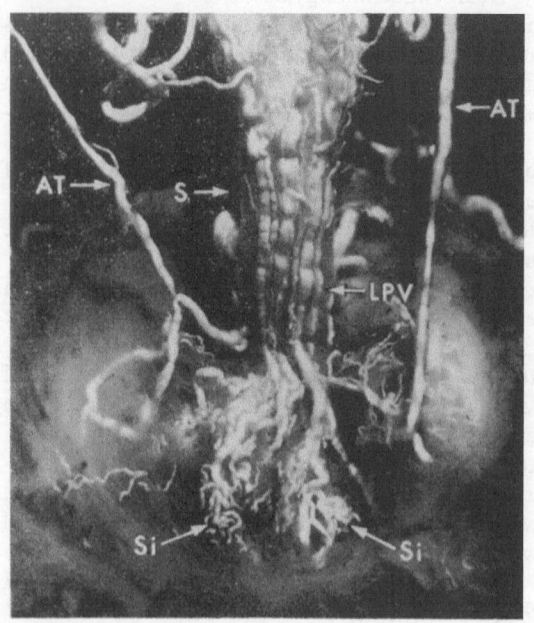

3.2

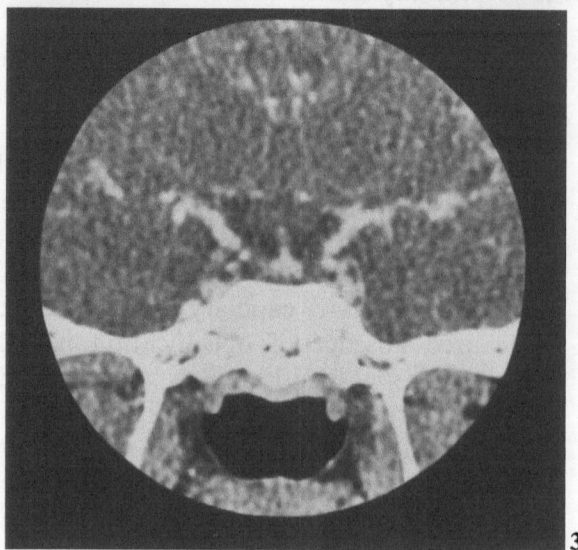

3.3a

3.3b

3.3c

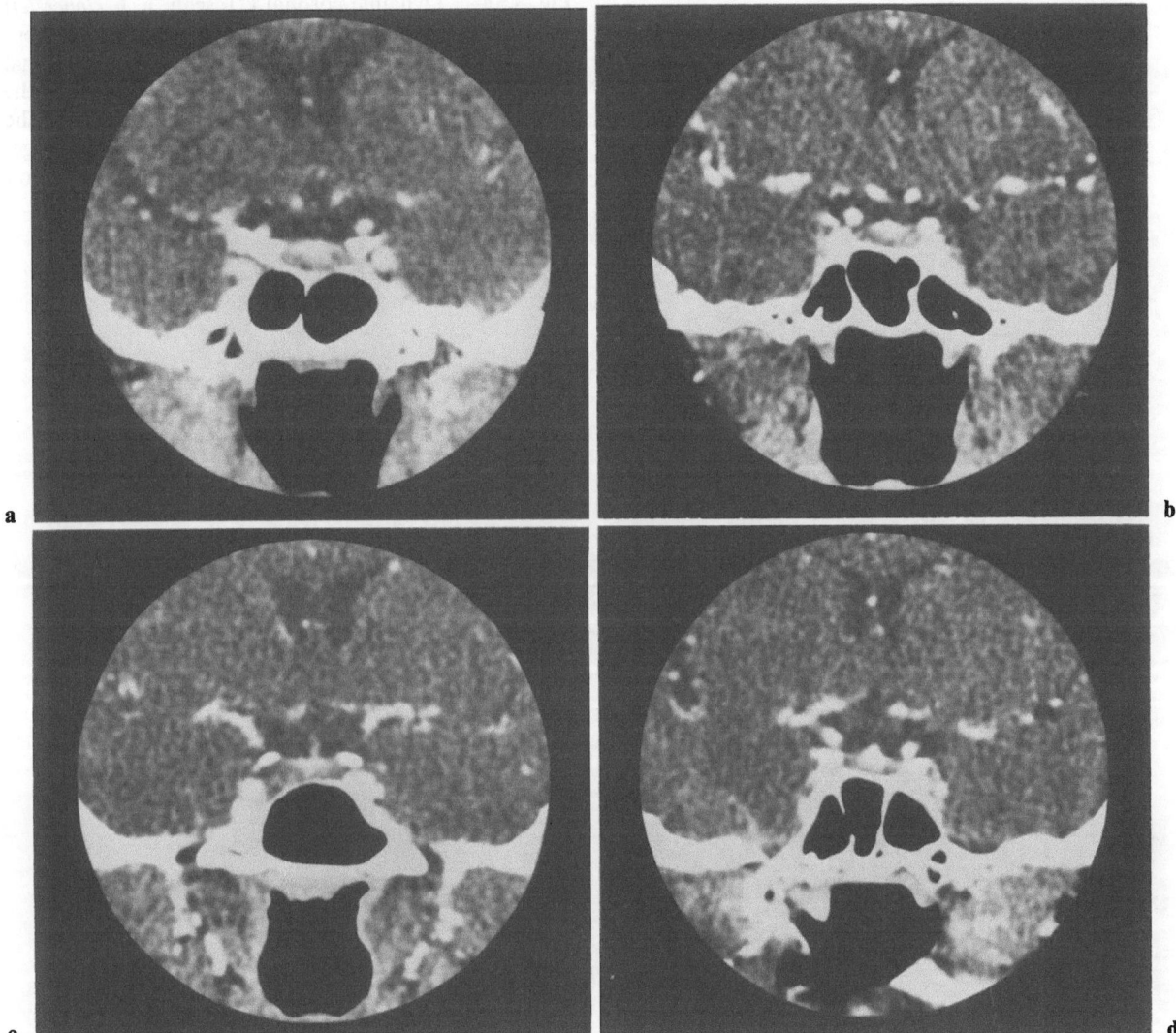

Fig. 3.4a–d. Different patterns of the pituitary capillary bed (pituitary tuft)

◀◀ **Fig. 3.2.** Anterosuperior view of the pituitary gland after partial removal of the central and upper portions of the pars distalis. On each side, a trabecular artery (*AT*) is seen entering the pars distalis. Note the prominent hypophyseal portal vessels (*LPV*) coursing down the stalk (*S*) and branching to form sinusoids in the pars distalis (*PD*), producing the pituitary "tuft." (Reprinted from Xuereb et al., with permission of the author and publisher)

◀ **Fig. 3.3a–c.** Dynamic coronal CT scan. **a** Before contrast, **b** 10 s after bolus injection, optimal opacification of both carotid arteries, and **c** 10 s later, visualization of the pituitary tuft on the midline, in the axis of the pituitary stalk. The pituitary tuft is located within the glandular tissue, below the sellar diaphragm level

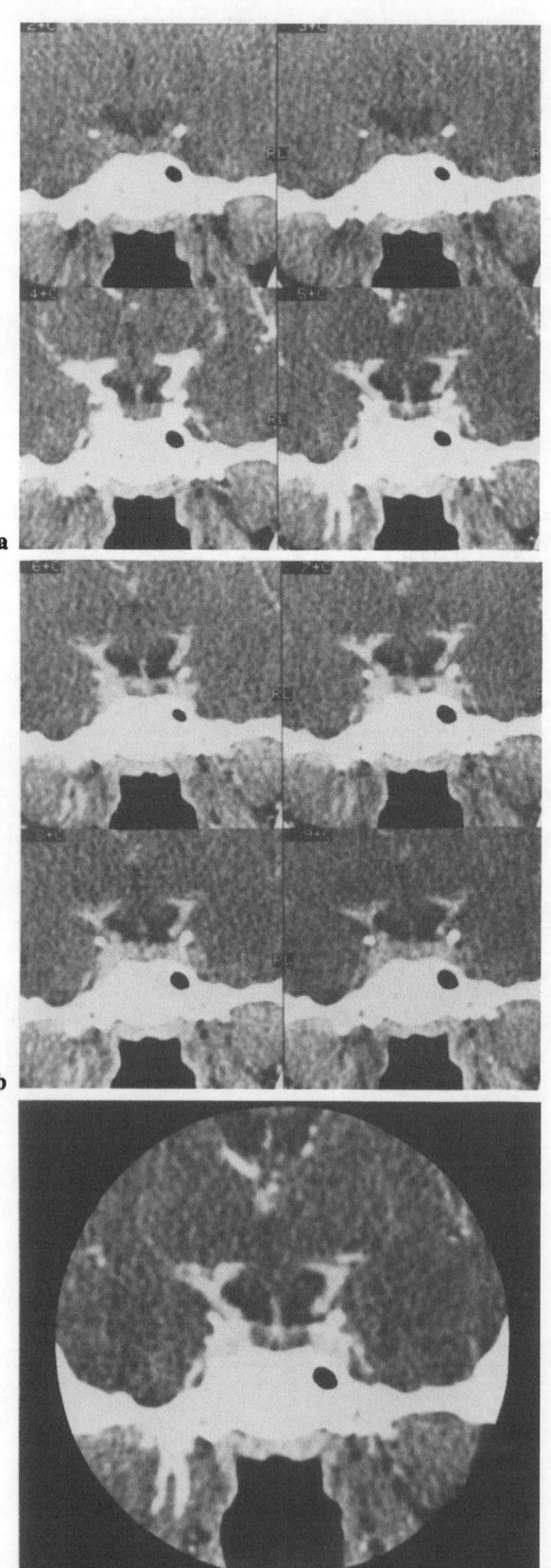

a

b

c

Fig. 3.5a–c. Dynamic coronal CT scan. **a, b** *Upper left* (2, 3): before bolus injection. A scan is performed every 11 s. *Lower right* (9) is 66 s after (3). Observe optimal visualization of carotid arteries (4), opacification of the pituitary tuft (5), and centrifugal enhancement of the gland (6, 7, 8). Pituitary enhancement is almost homogeneous in (9). **c** Magnified view of (5); point-like pituitary tuft

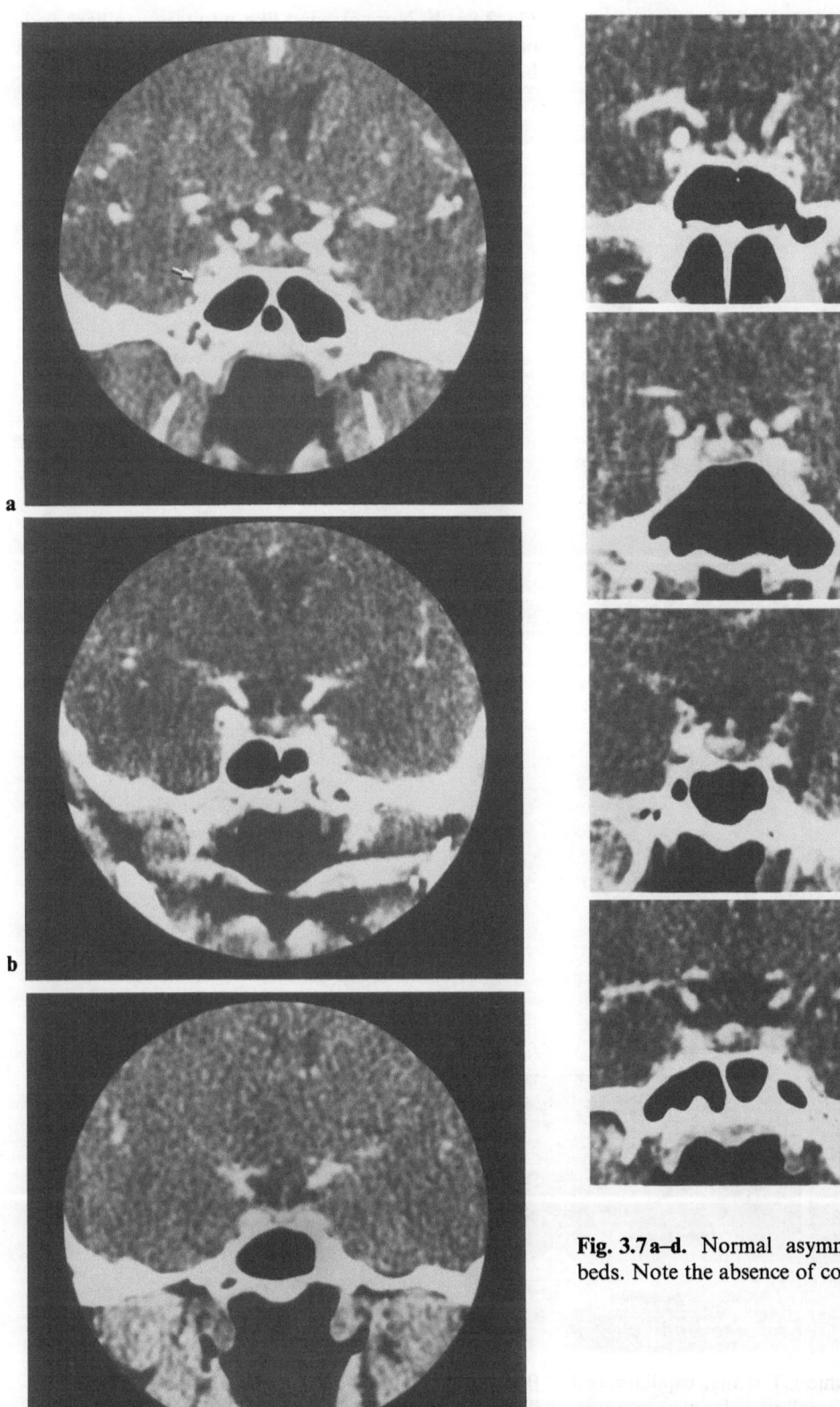

Fig. 3.7 a–d. Normal asymmetrical pituitary capillary beds. Note the absence of compression of the tuft limits

◀ **Fig. 3.6 a–c.** Normal point-like pituitary tufts demonstrated by dynamic coronal CT. Note in **a** a prominent inferior venous space (*arrow*)

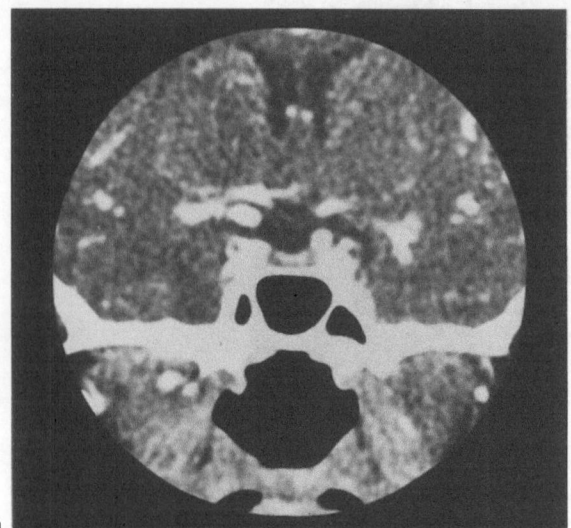

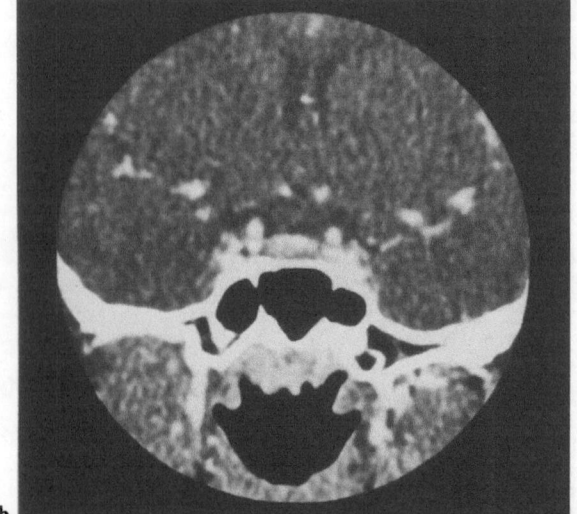

Fig. 3.8a, b. Normal band-like pituitary capillary beds demonstrated by dynamic coronal CT. **a** Thin band and **b** thick band

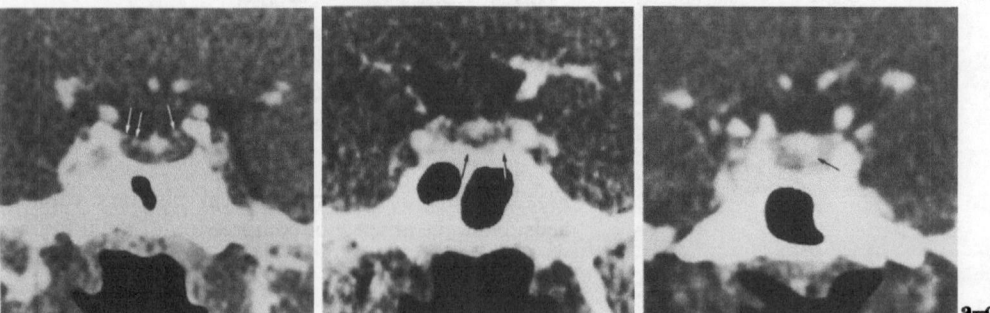

Fig. 3.9a–c. Coronal dynamic CT scans; capillary bed phase. **a** Demonstration of small vascular elements running at the upper surface of the pituitary (*arrows*) (sellar diaphragm vessels?). **b** Pituitary capillary bed last phase. Band-like enhancement immediately above the sellar floor (*arrows*) (anatomical variant of the secondary capillary bed? venous sinuses?). **c** Pituitary capillary bed phase. Vascular anastomosis running within the pituitary (*arrow*)

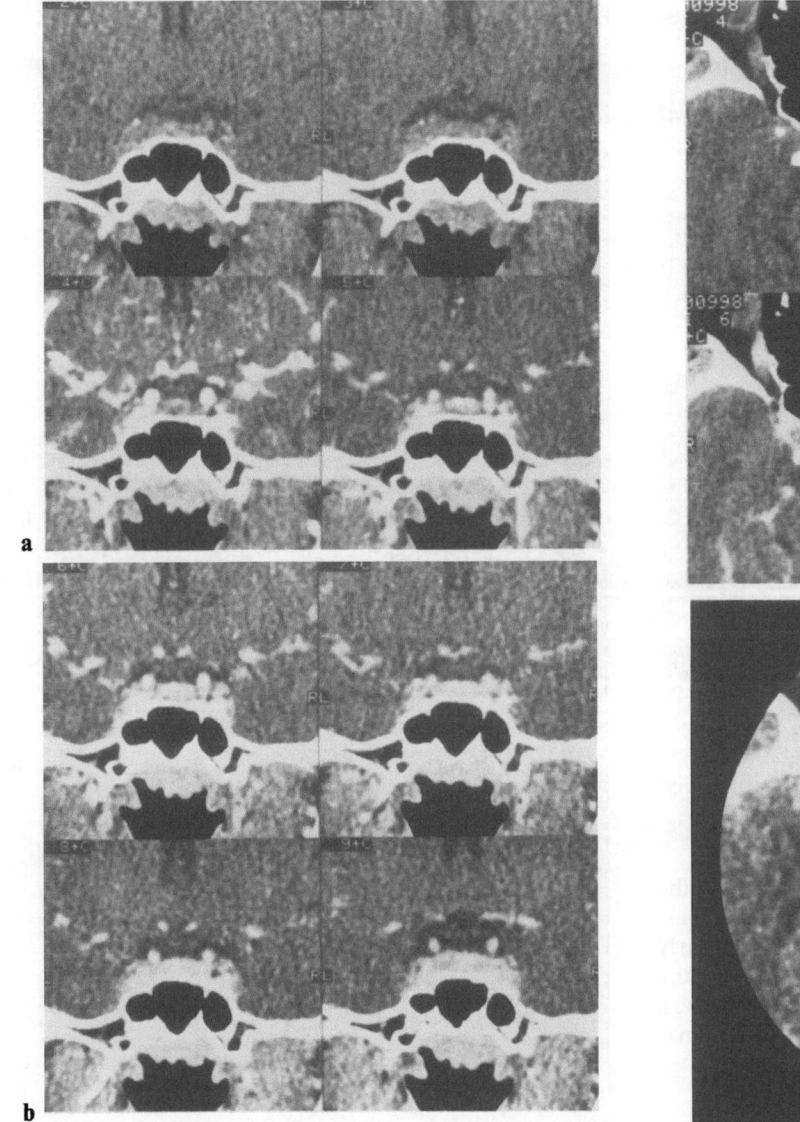

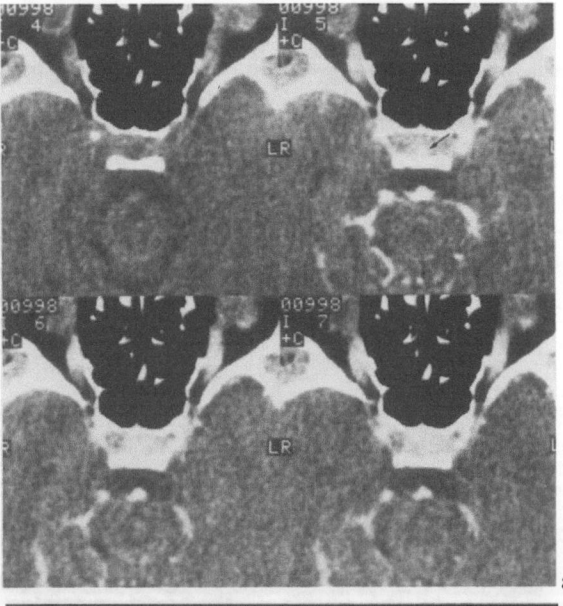

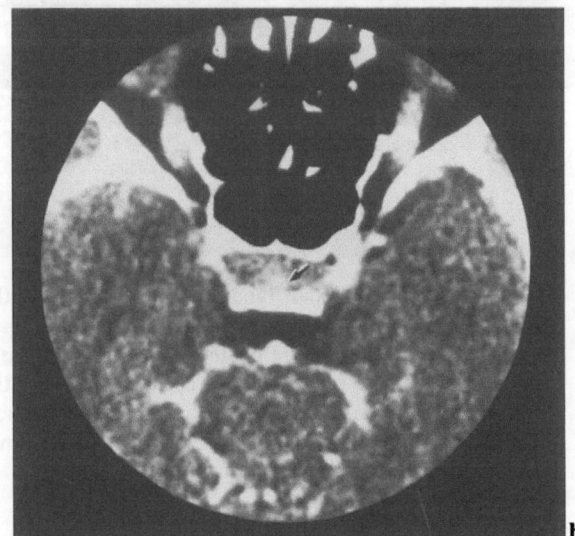

Fig. 3.10a, b. Normal dynamic coronal CT scan. A scan is performed every 11 s. Note the progressive enhancement of the pituitary gland

Fig. 3.11. a Normal axial dynamic CT scan. Demonstration of the pituitary tuft in axial section (*upper right*) (*arrow*). **b** Magnified view of part of **a**

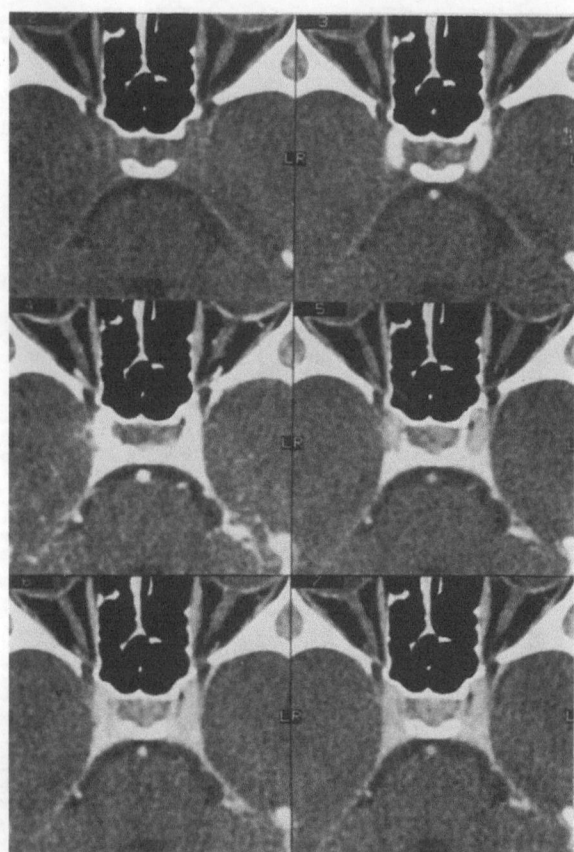

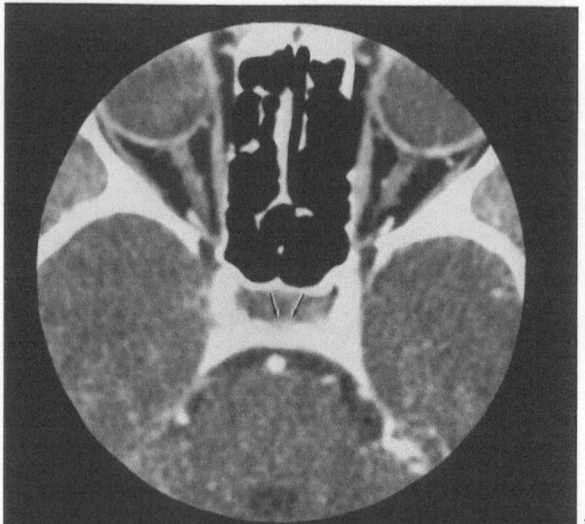

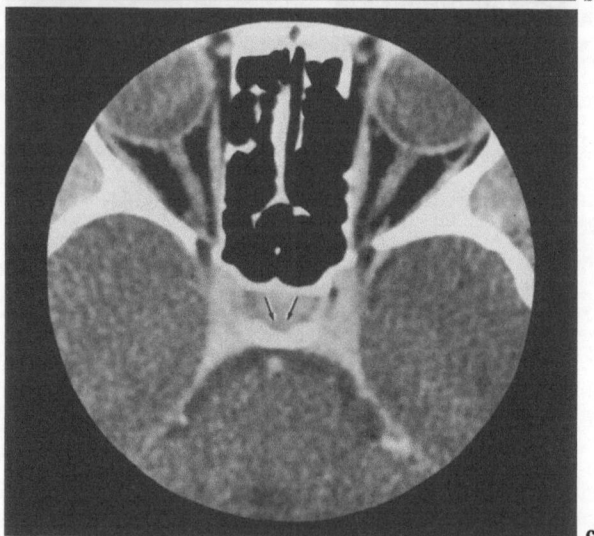

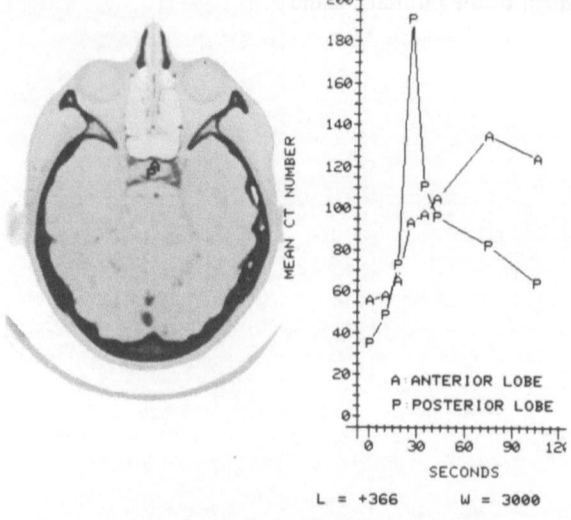

Fig. 3.12. a Normal axial dynamic CT scan below the level of the pituitary tuft. **b, c** Magnified views of part of **a**: Demonstration of the posterior pituitary. **b** Early enhancement of the posterior lobe contemporary of the optimal opacification of the carotid arteries (*arrows*). **c** 30 s later, the posterior pituitary appears less dense than the anterior pituitary (*arrows*)

Fig. 3.13. Representative time/density curves for the anterior and posterior pituitaries

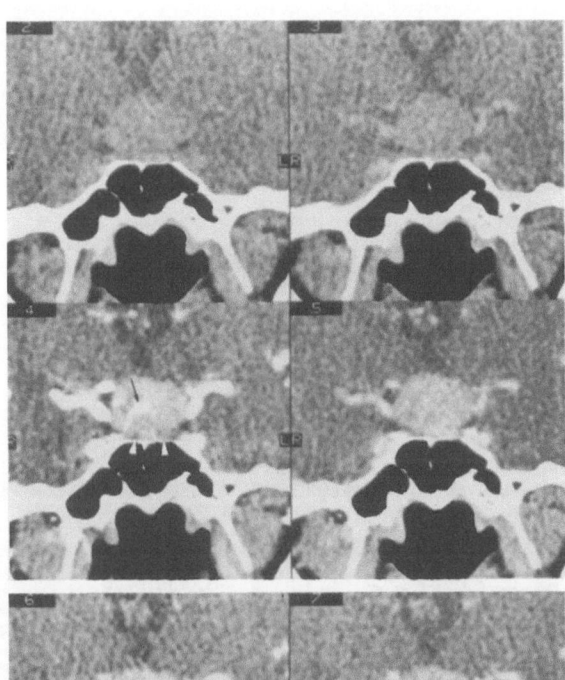

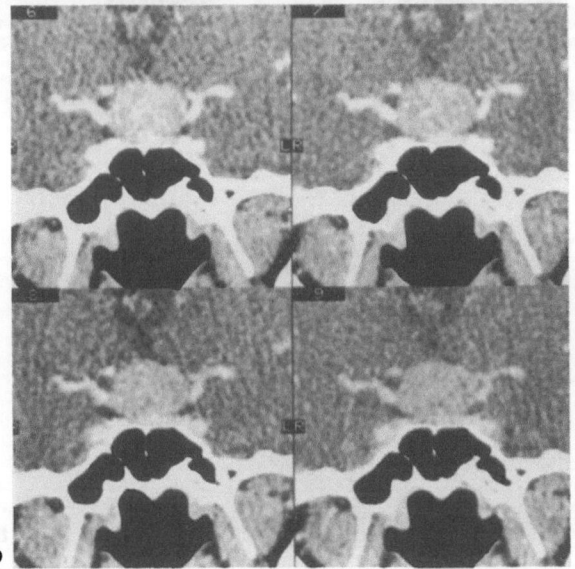

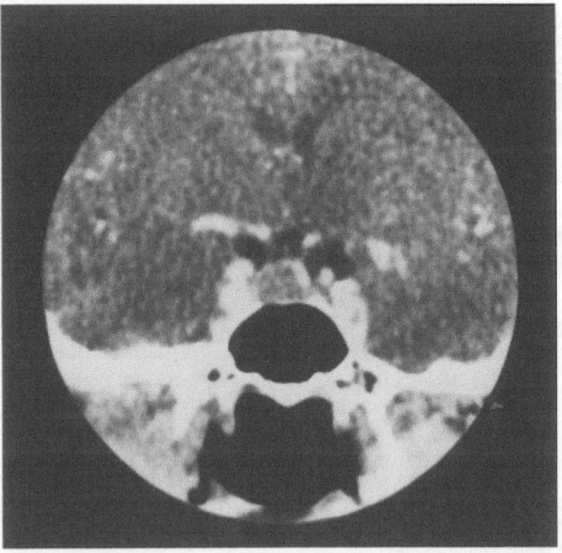

Fig. 3.15. Intrasellar prolactinoma. Dynamic coronal CT scan. Displacement of the pituitary tuft

Fig. 3.14a, b. Dynamic CT scan of a suprasellar tumor. Note the perfect visualization of both carotid arteries. **a** Immediately after bolus injection (*lower left*), intense enhancement of the tumor with visualization of intratumoral vessels (*black arrow*). At this time, the pituitary gland itself is still not enhanced (*arrowheads*). **b** Differentiation of the tumor from the pituitary is more difficult after dynamic CT scan (*lower right*). At surgery a hypothalamic choristoma was found

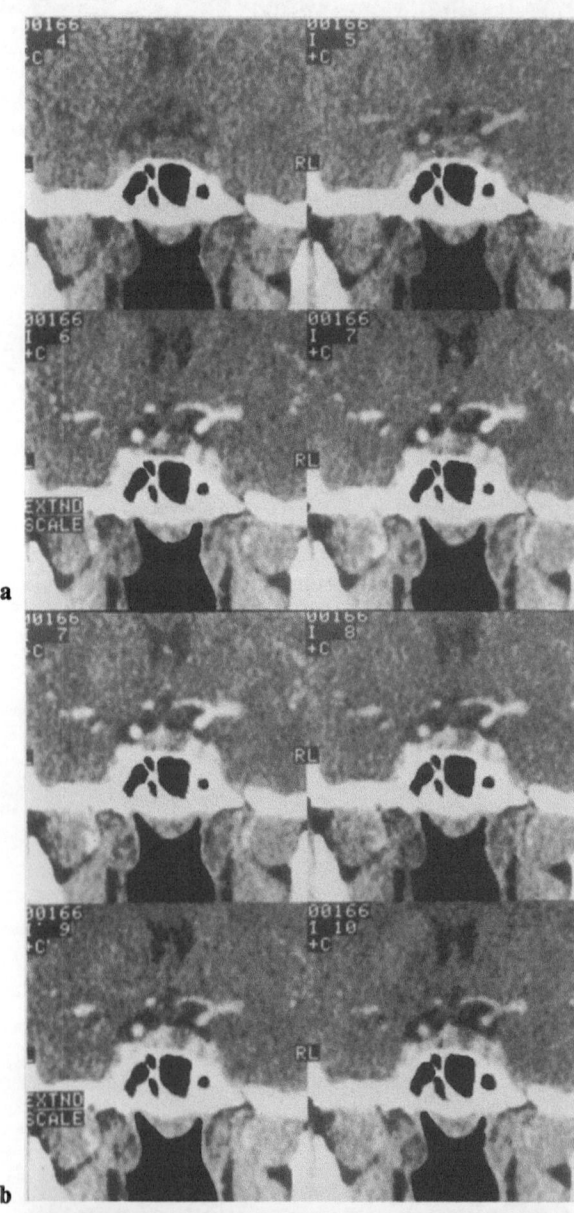

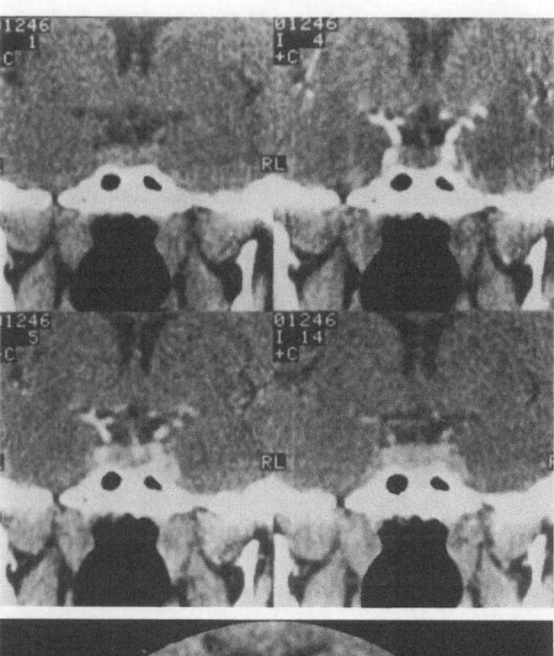

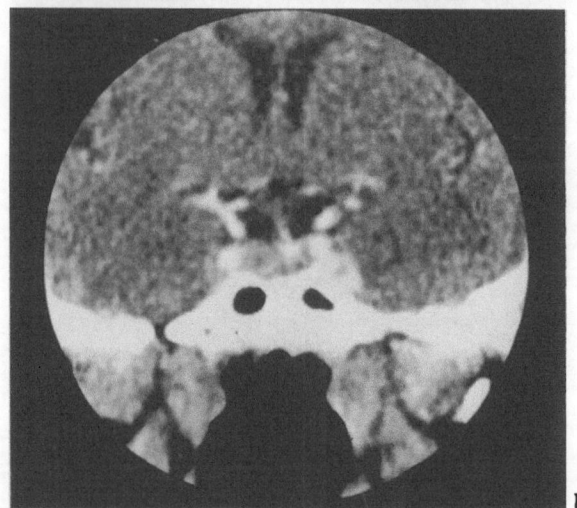

Fig. 3.17. a Dynamic coronal CT scan. Usefulness in a case of isodense ACTH-secreting pituitary microadenoma proven at surgery. *Lower left*, displacement and compression of the secondary capillary bed (tuft sign). *Lower right*, 60 s later, conventional coronal CT scan appears normal. **b** Magnified view of tuft's phase of **a**

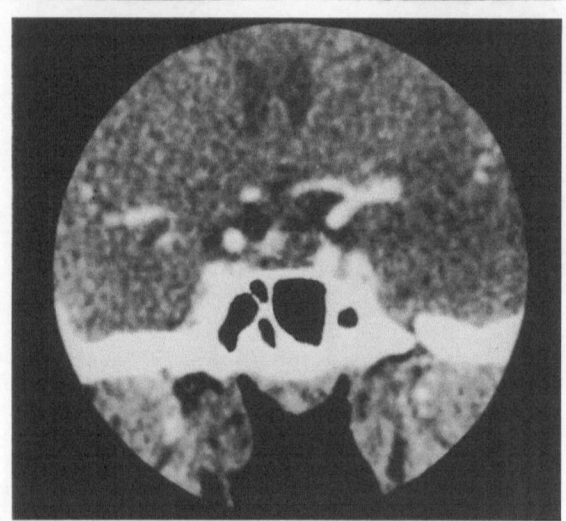

◄ **Fig. 3.16a–c.** Pituitary microprolactinoma. **a, b** Dynamic coronal CT scan. **c** Magnified view of part of **a**. The pituitary tuft is displaced from the midline by a rounded hypodense microadenoma better seen during dynamic CT scan

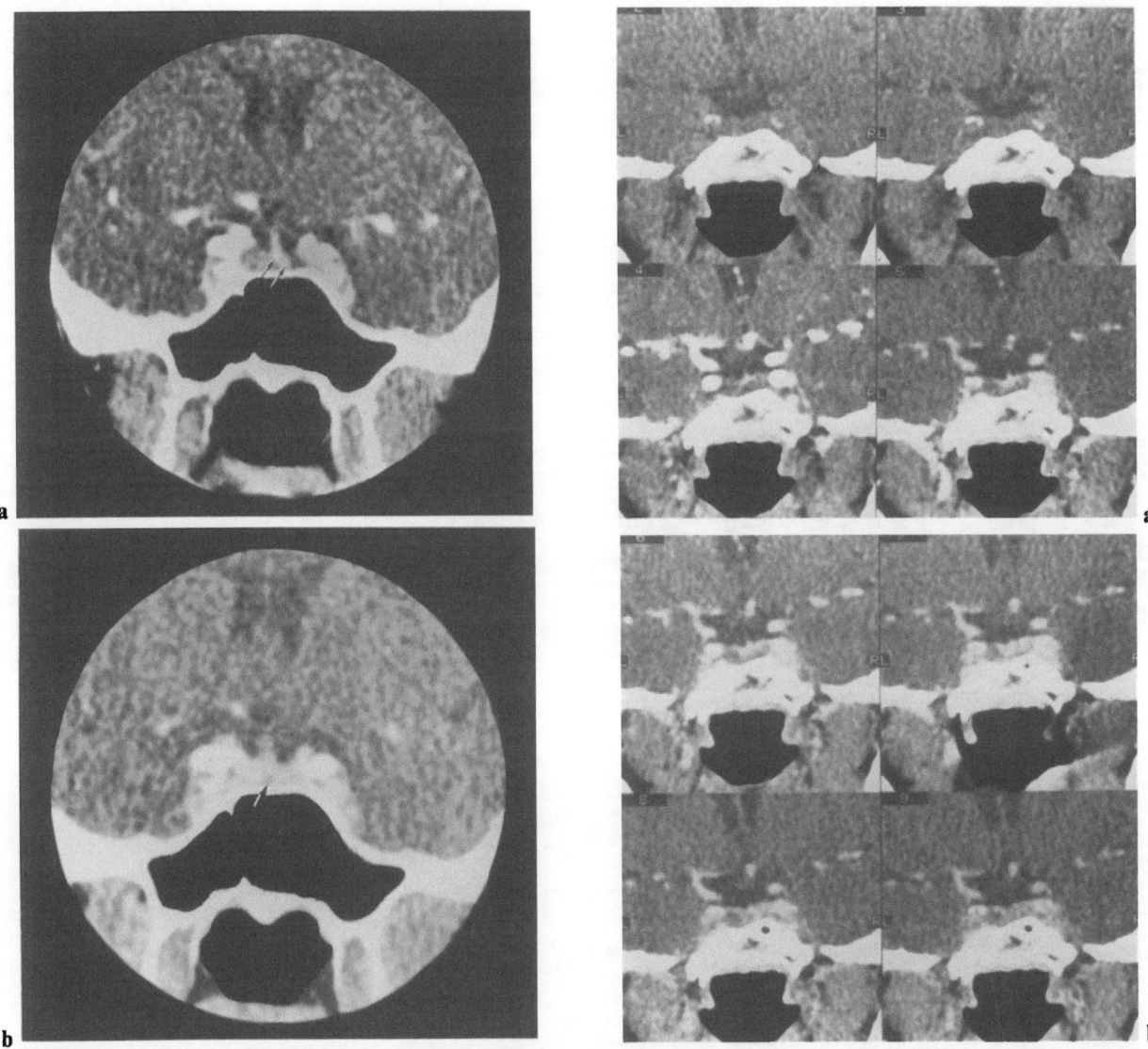

Fig. 3.18. a Dynamic coronal CT; Cushing's disease. Slight localized compression of the pituitary tuft (*arrows*). **b** Conventional coronal CT, 60 s later. Tiny midline low dense area (*arrow*) just opposite the flattened tuft

Fig. 3.19a, b. Microprolactinoma demonstrated by dynamic coronal CT scan. The enhancement of the pituitary is delayed on one side. The microadenoma is then much better seen during dynamic CT scan. At the end of dynamic CT scan, the adenoma's limits are not as well defined

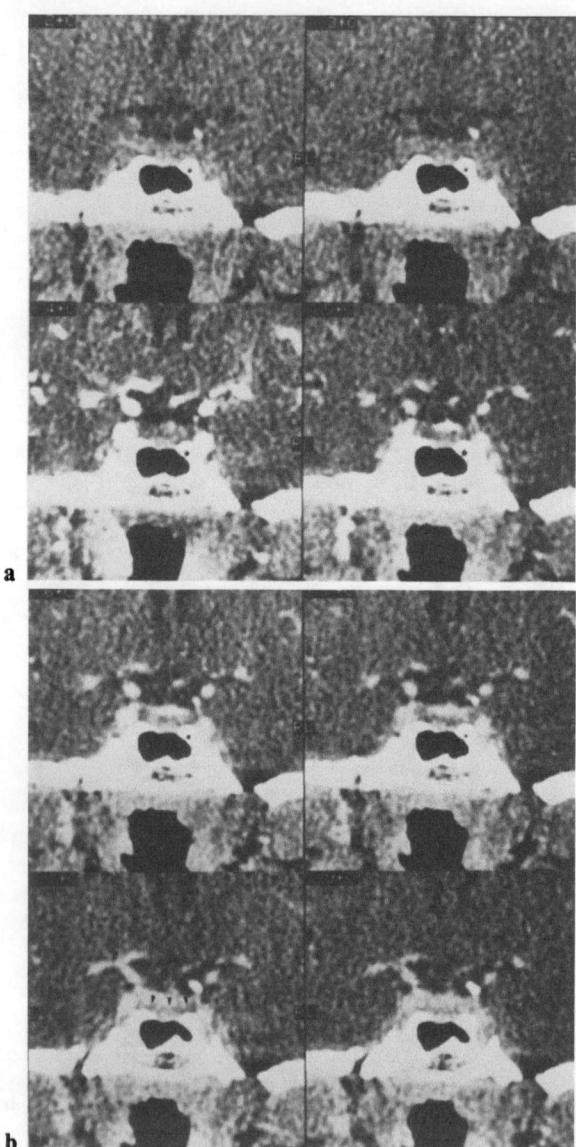

Fig. 3.20a, b. Microprolactinoma. Dynamic CT scan. Marked delayed enhancement of the inferior part of the pituitary, just above the sellar floor. The compressed pituitary capillary bed is visible longer than usual

Chapter 4
Pituitary Adenomas with Suprasellar Extension

Since the old category of "chromophobe" adenomas has been abandoned, prolactinomas appear to be the most frequent adenomas with extrasellar extension. Nonsecretory or nonfunctional adenomas are seen less commonly.

Radiologic signs of adenomas with suprasellar extension are generally identical, regardless of the class, and an overall description will be given in this chapter.

General Aspects

Nonsecreting Pituitary Adenomas

Nonsecreting pituitary adenomas are diagnosed late, the presenting signs being changes in the visual field such as bitemporal hemianopia due to suprasellar development, or anterior pituitary insufficiency due to compression of anterior pituitary secretory cells. In rare cases, the accidental discovery of sellar modifications during X-ray studies for trauma or sinusitis is indicative of nonfunctional adenomas too small to cause visual signs or anterior pituitary insufficiency. Nevertheless, experience shows that sellar modifications discovered "by accident" are virtually always secondary to intrasellar expansion of the subarachnoid spaces, sometimes to triventricular hydrocephaly by stenosis of the sylvian aqueduct, and very rarely to a silent, nonfunctional adenoma.

Prolactinomas with Suprasellar Development

Prolactinomas with suprasellar development are as frequent in males as in females; in males, the tumoral syndrome is virtually always the presenting sign and hypogonadism is discovered only upon careful interview. Long-standing, neglected, or incorrectly treated disturbances in menstruation or sterility (see Chap. 5) are virtually always seen when suprasellar prolactinomas are discovered in women; nevertheless, diencephalic symptoms such as diabetes insipidus occurring during pregnancy can also lead to diagnosis of the adenoma (see Chap. 7).

Mixed Adenomas

Mixed adenomas which secrete both prolactin and GH are generally seen in women; the dysmorphic syndrome is usually secondary, with amenorrhea-galactorrhea being the presenting syndrome; nevertheless, the opposite can also hold.

Mixed adenomas have a greater growth potential than prolactin adenomas; suprasellar expansion is frequent, and when they are refractory to medical treatment (see Chap. 6) surgery is frequently necessary; surgery can provide no guarantee against recurrences.

Gonadotropic Adenomas

Gonadotropic adenomas are infrequent; they are generally large upon discovery, evolve rapidly, and can be invasive. They can present either as voluminous suprasellar expansions, inferior expansions, or they can be situated in the posterior fossa (Fig. 10.3). Their radiologic profile does not differ from that of other adenomas (Fig. 4.3) (see Chap. 10).

GH-Secreting Pituitary Adenomas

GH-secreting pituitary adenomas generally expand downwards. There is frequently also a moderate superior extension, occupying the inferior part of the chiasmatic cistern. These lesions will be discussed in Chap. 8. Note that mixed GH-prolactin secreting adenomas are more prone to frank suprasellar expansion.

TSH-Secreting Pituitary Adenomas

TSH-secreting pituitary adenomas can present a suprasellar extension (Fig. 10.2) and do not have a specific radiologic profile. While the radiologic distinction between TSH adenomas and reactive hyperplasia to peripheral hypothyroidism is difficult, hyperplasia generally responds favorably to treatment with thyroid extracts.

ACTH-Secreting Pituitary Adenomas

ACTH-secreting pituitary adenomas do not present suprasellar extension.

CT of Pituitary Adenomas with Suprasellar Extension

Plain films are usually taken before CT. In general, they allow rough estimation of the volume and direction of extension of the adenoma, thus permitting adaptation of technique, in particular thickness of sections. The plain films generally show an enlargement of the sella turcica with a thinning of sellar walls (Fig. 4.6). Nevertheless, even a normal sella turcica does not formally eliminate the possibility of an adenoma with suprasellar extension (Fig. 4.1). In general, osseous changes in the sellar region as well as calcifications are better evaluated by CT: tomography of the sellar region is no longer required.

Plain films can still facilitate differential diagnosis: craniopharyngiomas are frequently calcified, and osseous changes predominate in the

chiasmatic sulcus and the dorsum sellae; the sellar floor is generally better conserved than in pituitary adenomas (Fig. 16.1). In presellar meningiomas, insertion osteomas or blistering of the planum sphenoidale are indicative (Fig. 16.4).

Native CT Without Contrast

In usual practice, a native scan without injection of contrast medium is not essential for study of adenomas with suprasellar extension. When a dynamic coronal CT is carried out, the preliminary section before bolus injection is sufficient to evaluate volume of the suprasellar mass in the chiasmatic cistern. The adenoma is generally isodense (Figs. 4.2 and 4.15), or only slightly denser (Fig. 4.3) than the brain; in some cases it may be hypodense (Fig. 4.4). A spontaneous hyperdensity can be indicative of recent intra-adenomatous hemorrhage. Calcifications are uncommon (Fig. 4.25), and rarely of large size (Fig. 5.14). Thinning or destruction of the sellar floor is mostly present (Figs. 4.5 and 4.17).

Contrast CT

Evaluation of pituitary adenomas with suprasellar extension generally requires coronal sections with dynamic CT to evaluate the relationship of the tumor with the internal carotid arteries and their branches, as well as other elements of the cavernous sinus (Figs. 4.2 and 4.4). The secondary pituitary capillary bed is generally not identifiable, and where it is, suprasellar extension of the adenoma is very moderate (Fig. 3.15).

Bolus injection is followed by infusion so as to maintain plasma iodine levels constant throughout the examination; along with direct coronal sections, axial sections are taken for reformatting.

Topographical Aspects
Suprasellar extension is variable.

At the First Stage. The first stage is intermediate

between strictly intrasellar adenomas and adenomas with true suprasellar extension.

In direct coronal sections, these adenomas occupy the lower part of the chiasmatic cistern; their upper pole is separated from the inferior surface of the optic chiasm by a variably thick layer of cerebrospinal fluid. Contrast between the opacity of the tumor and the hypodense cisterns allows precise anatomical definition of the adenoma, superior extension of which is most of the time vertical along the axis of the sella turcica (Fig. 4.6 and 4.7)

In axial sections, the perfectly round tumor occupies the central part of the chiasmatic cistern (Fig. 4.8 and 4.1). At this stage, moderate dilatation of the crural and lateropeduncular cisterns is not uncommon.

At the Intermediate Stage. Large pituitary adenomas with major superior extension completely fill the chiasmatic cistern in coronal sections.

The suprasellar extension can be median and symmetrical, or it can be lateralized (Fig. 4.10). A dilatation of the cistern can still be seen (Fig. 4.10). Stricture of the tumor at the sellar diaphragm level can yield an hourglass-like pattern (Fig. 4.11 and 4.12).

The inferior part of the third ventricle is amputated when the vertical diameter of the tumor reaches 3 cm (Figs. 4.9 and 4.19).

In axial sections, pituitary adenomas fill the anterior part of the suprasellar cistern. They are round or oval in shape, with a major anteroposterior axis (Fig. 4.13).

At a Later Stage. The entire cistern is filled. Compression of the medial wall of the anterior horns of the lateral ventricles is possible. Posterior extension with compression of the cerebral peduncles is less common (Fig. 4.14). When the adenoma is very large, it frequently presents polycyclic contours (Fig. 4.15).

Invasive pituitary adenomas can grow into the frontal lobe (Fig. 4.16). Cushing stated that the largest frontal tumors were generally pituitary adenomas. Invasion of the temporal lobe, cavernous sinus, the ethmoid and sphenoid sinuses, and the cavum, with filling of the interpeduncular and pontic cisterns can also be seen

Figs. 4.12, 4.17, 4.18 and 10.3). Unlike the situation with suprasellar meningiomas, even when there is major suprasellar extension, pituitary adenomas are rarely surrounded by a ring of edema.

Types of Enhancement
After intravenous injection of contrast medium, three principal types of enhancement can be seen:

1. Most frequently, enhancement is homogeneous and intense, however less than that seen with meningiomas (Figs. 4.2, 4.3, and 4.19)

2. Enhancement can be annular, with a central hypodensity not necessarily corresponding to a liquid content (Fig. 4.20). Peripheral enhancement can be thin or thick (Fig. 4.21), and if the adenoma is truly cystic, one can see enhancement of the central part of the adenoma at some distance from the intravenous injection (Fig. 4.22)

3. Finally, rounded areas less dense than the adenoma are sometimes seen, corresponding to cysts when the hypodensity is close to that of CSF, or more frequently, to areas of necrosis or old hemorrhage (Figs. 4.23 and 4.24). These hypodense areas are more frequently situated in the intrasellar than the suprasellar part of the adenoma. Calcifications are possible, though more frequently peripheral than central (Fig. 4.25); these are sometimes indistinguishable from those of craniopharyngiomas (Fig. 6.4).

CT Cisternography

Indications for CT cisternography are currently very limited. It is used only in the rare cases where the resolution of the system does not allow differentiation between the empty sella and a hypodense adenoma (uncommon with currently available systems), or so as to better evaluate the relationship between the adenoma and the chiasm (Fig. 4.26).

Although coronal or axial sections, or coronal or especially sagittal reformatted images clearly provide an exceptional quality of visualization, in view of the aggressive nature of this procedure, even with modern water-soluble media, CT cisternography should be reserved for a few exceptional cases.

References

Azar Kia B, Palacios E, Churchill RJ (1977) Diagnosis of sellar and parasellar lesions by CT and other diagnostic modalities. CT 1:249–256

Banna M, Baker HL, Jr, Houser OW (1980) Pituitary and parapituitary tumours on CT. Brit J Radiol 53:1123–1143

Bonneville JF, Dietemann JL (1981) Radiology of the sella turcica. Springer Verlag, Berlin Heidelberg New York

Bonneville JF, Poulignot D, Cattin F, Couturier M, Mollet E, Dietemann JL (1982) Apport des méthodes nouvelles dans l'exploration morphologique des tumeurs hypophysaires. Ann Endocrinol (Paris) 43:303–308

Brennan TG, Krishna-Rao CNG, Robinson W (1977) Tandem lesions: chromophobe adenoma and meningioma. J Comp Ass Tomog 1:517–520

Cabanis EA, Van Effenterre R, Iba-Zizen MT (1979) CT in parasellar space-occupying lesions and therapeutic decision. Acta Neurochirurgica (suppl. 28) 28:329–333

Citrin CM, Davis DO (1977) CT in the evaluation of pituitary adenomas. Invest. Radiol. 12:27–35

Cusick JF, Haughton VH, Hagen TC (1980) Radiological assessment of intrasellar prolactin-secreting tumors. Neurosurgery 6:376–379

Daniels DL, Williams AL, Thornton RS, Meyer, GA, Cusick JF, Haughton VM (1981) Differential diagnosis of intrasellar tumors by CT. Radiology 141:697–701

Danoff BF, Pripstein S, Croce N (1978) Value of CT in delineating suprasellar extension of pituitary adenoma for radiotherapeutic management. Cancer 42:1066

Derome PJ (1982) Les adénomes hypophysaires. Encycl Med Chir 17340 A 10

Derome PJ, Jedynak CP, Peillon F (1980) Pituitary adenomas. Biology, physiopathology and treatment. Asclepios Publishers, Paris

Di Chiro G, Nelson KB (1962) The volume of the sella turcica. AJR 87:989–1007

Dietemann JL, Bonneville JF (1985) Radiological Diagnosis of Pituitary Diseases. In: The pituitary gland. Ed H Imura, pp 341–361 Raven Press, New-York

Drayer BP, Rosenbaum AE, Kennerdell JS, Robinson AG, Bank WO, Deeb Z.L. (1977) CT diagnosis of suprasellar masses by intrathecal enhancement. Radiology 123:339–344

Drayer BP, Rosenbaum AE, Riegel DB, Bank WO, Deeb ZL (1977) Metrizamide CT cisternography: pediatric applications. Radiology 124:349–357

Drayer BP, Kattah J, Rosenbaum A, Kennerdell J, Maroon J (1979) Diagnostic approaches to pituitary adenomas. Neurology 29:161–169

Eresue J, Drouillard J, Philippe JC, Guibert JL, Poux P, Tavernier J (1982) L'exploration des adénomes hypophysaires par scanographie à haute résolution et angioscanographie. Ann Radiol 25:509–517

Gardeur D, Metzger J (1982) Adénomes hypophysaires intra-sellaires. In: Tomodensitométrie intra-crânienne. Livre V: Pathologie sellaire. Ellipses, Paris, p 41

Gardeur D, Nachanakian A, Kulesza E, Metzger J (1979) La tomodensitométrie dans les adénomes hypophysaires. Ann Radiol 22:489–499

Gemzell C, Wang CF (1979) Outcome of pregnancy in women with pituitary adenoma. Fertil Steril 31:363–372

Ghoshhajra K (1981) High-resolution metrizamide CT cisternography in sellar and suprasellar abnormalities. J. Neurosurg 54:232–239

Gyldensted C, Karle A (1977) CT of intra- and juxtasellar lesions: a radiological study of 108 cases. Neuroradiology 14:5–13

Hall K, Mc Allister VL (1980) Metrizamide cisternography in pituitary and juxta-pituitary lesions. Radiology 134:101–108

Hardy J, Mohr G (1981) Prolactinoma: surgical aspects. In: Hardy J (ed) Prolactinoma. Masson, Paris New York

Hashimoto N, Handa H, Takeuchi J, Ishikawa M, Nakano Y (1982) Collection of gas within a huge chromophobe adenoma. Neuroradiology 23:289–290

Hatam A., Bergström M., Greitz T. (1979) Diagnosis of sellar and parasellar lesions by CT. Neuroradiology 18:249–258

Khangure MJ, Apsimon HT (1981) Some pitfalls in the diagnosis of pituitary tumors: the importance of carotid angiography. Surg Neurol 16:300–308

Kovacs K, Ryan N, Howath E, Singer W, Excrin C (1980) Pituitary adenomas in old age. J Gerontol 35:16–22

Kricheff II (1979) The radiologic diagnosis of pituitary adenoma. Radiology 131:263–265

Kuuliala I (1981) CAT of pituitary adenomas Clinical Radiology 32:259–264

Kuuliala I, Katevuo K, Ketonen L (1981) Metrizamide cisternography with hypocycloid and CT in sellar and supraselar lesions. Clin Radiol 32:403–408

Leeds NE, Naidich TP (1977) CT in the diagnosis of sellar and parasellar lesions Semin Roentgenol 12:121–135

Linfoot JA (1979) Recent advances in the diagnosis and treatment of pituitary tumors. Raven Press, New-York

Mac Pherson P, Anderson DE (1981) Radiological differentiation of intrasellar aneurysms from pituitary tumours. Neuroradiology 21:177–183

Moseley IF, Sanders MD (1982). Computerized Tomography in Neuro-ophtalmology. Chapman and Hall, London

Naidich TP, Pinto RS, Kushner MJ, Lin JP, Kricheff II, Leeds NE, Chase NE (1976) Evaluation of sellar and parasellar masses by computed tomography. Radiology 120:91–99

Nakstad PHJ, Skalpe IO (1981) CT in the evaluation of the supraclinoid arteries in suprasellar pituitary gland tumors. Acta Radiol (Diagn) 22:399–402

Numaguchi Y, Kishikawa T, Ikeda J, Fukui M, Kitamura K, Tsukamsto Y, Masuo K, Maatsuura K (1981) Neuroradiological manifestations of suprasellar pituitary adenomas meningomas and craniopharyngomas. Neuroradiology 21:67–74

Peyster RG, Hoover E (1984) Computerized Tomography in orbital disease and Neuro-ophtalmology. Year Book Medical Publishers, Chicago, London

Raymond LA, Tew J (1978) Large suprasellar aneurysms imitating pituitary tumour J Neurol Neurosurg Psychiatry 41:83–87

Richmond IL, Wilson CB (1978) Pituitary adenomas in childhood and adolescence. J Neurosurg 49:163–168

Richmond IL, Newton TH, Wilson CB (1980) Indications for angiography in the preoperative evaluation of patients with prolactin secreting pituitary adenomas. J Neurosurg 52:378–380

Rilliet B, Mohr G, Robert F, Hardy J (1981) Calcifications in pituitary adenomas Surg Neurol 15:249–255

Roberson GH, Tadmor R, Taveras JM, Kleefield J, Ellis G (1977) CT in Metrizamide cisternography. Importance of coronal and axial views. J Comput Assist Tomogr 1:241–245

Robertson HJ, Rose A, Ehmi B, England G, Meriweather R (1981) Trends in the radiological study of pituitary adenoma. Neuroradiology 21:75–78

Ross LS (1981) Routine hormonal screening of infertile men: is it worthwhile? J Urol 126:756–758

Sakoda K, Mukada K, Yonezawa M, Matsumura S, Yoshimoto H, Mori S and Uozimi T. (1981) CT scan of pituitary adenomas. Neuroradiology 20:249–253

Sheldon P, Molyneux A (1978) Metrizamide cisternography and CT for the investigation of pituitary regions. Neuroradiology 17:83–87

Spark RF, O'Reilly G, Wills CA, Ransil BJ, Bergland R (1982) Hyperprolactinemia in males with and without pituitary macroadenomas. Lancet 2:129–131

Taveras JM, Wood EH (1976) Diagnostic neuroradiology. 2nd Ed Williams and Wilkins, Baltimore.

Thieblot Ph, Derome P, Luton JP (1980) Adénomes "chromophobes". Rev Prat (Paris) 30:3045–3050

Valenta LJ, Sostrin RD, Eisenberg H, Tamkin JA, Elias AN (1982) Diagnosis of pituitary tumors by hormone assays and computerized tomography. Am J Med 72:861–873

Vignaud J, Aubin ML, Bories J (1979) Apport de la tomodensitométrie à l'exploration de la région sellaire et supra-sellaire. Rev Neurol 135:41–50

Von Werder K, Fahlbusch R, Rjosk HK (1981) Bromocriptine therapy of macroprolactinomas. Neuroendocrinol Lett 3:328 (abst)

Wass JA, Moult PJ, Moult PJ, Thorner MO (1979) Reduction of pituitary-tumour size in patients with prolactinomas and acromegaly treated with bromocriptine with or without radiotherapy. Lancet 2:66–69

Wass JA, Williams J, Charlesworth M (1982) Bromocriptine in management of large pituitary tumours. Br Med J 284:1908–1911

Wilson CB, Dempsey LC (1978) Transsphenoidal microsurgical removal of 250 pituitary adenomas. J Neurosurg 48:13–22

Wollesen F, Andersen T, Karle A (1982) Size reduction of extrasellar pituitary tumors during bromocriptine treatment. Ann Intern Med 96:281–286

Wolpert SM, Post KD, Biller BJ, Molitch ME (1979) The value of computed tomography in evaluating patients with prolactinomas. Radiology 131:117–119

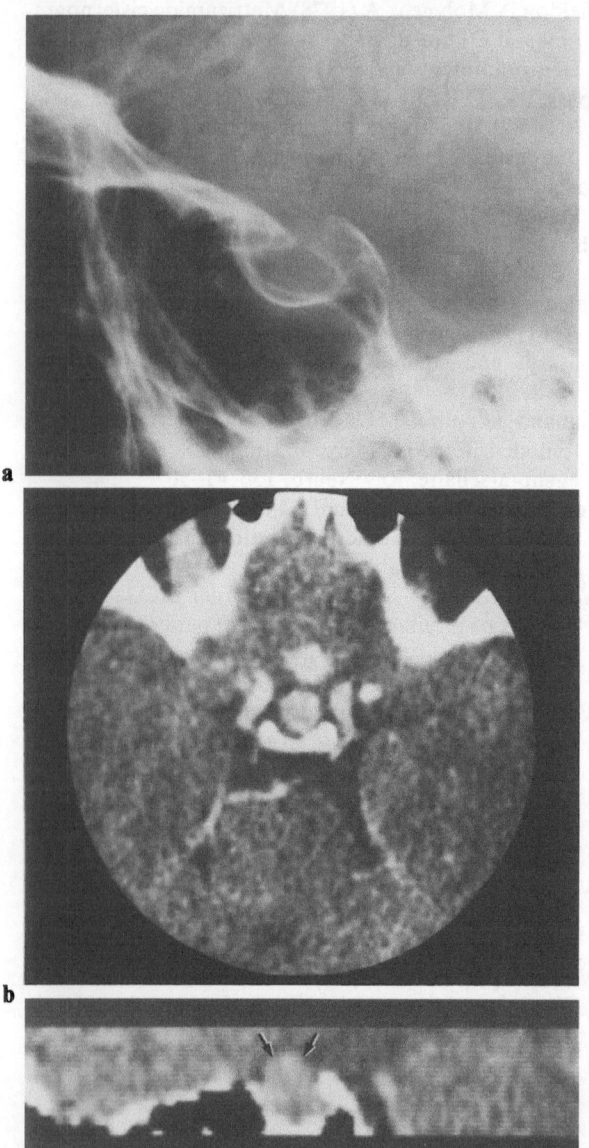

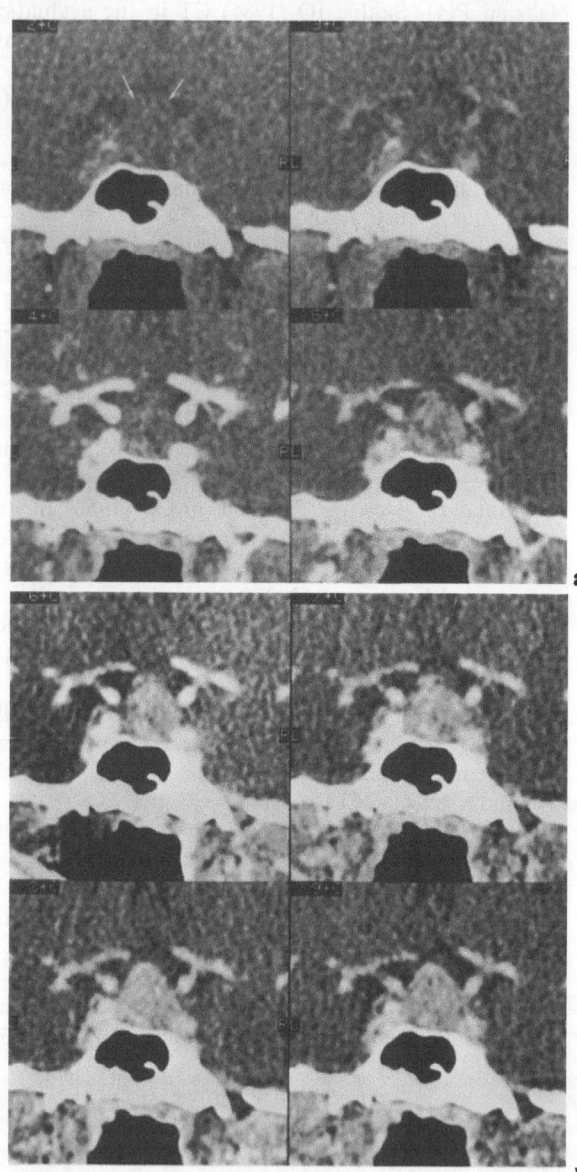

Fig. 4.1a–c. Prolactinoma with small suprasellar extension. Amenorrhea-galactorrhea in a 28-year-old woman. Prolactinemia is 60 ng/ml. **a** Sella turcica magnified lateral view is normal. *No* thinning of the sellar floor. **b** Contrast axial CT scan through the inferior part of the suprasellar cistern: visualization of the upper boundary of the adenoma. **c** Midsagittal reformatted image: the prolactinoma is bulging in the suprasellar subarachnoid space (*arrows*)

Fig. 4.2a, b. Nonsecreting pituitary adenoma with suprasellar extension. Headache and bitemporal hemianopia in a 36-year-old man. Dynamic CT scan. Before contrast, the mass is isodense relative to the brain and occupies the suprasellar cistern (*arrows*). Marked enhancement of the adenoma after contrast

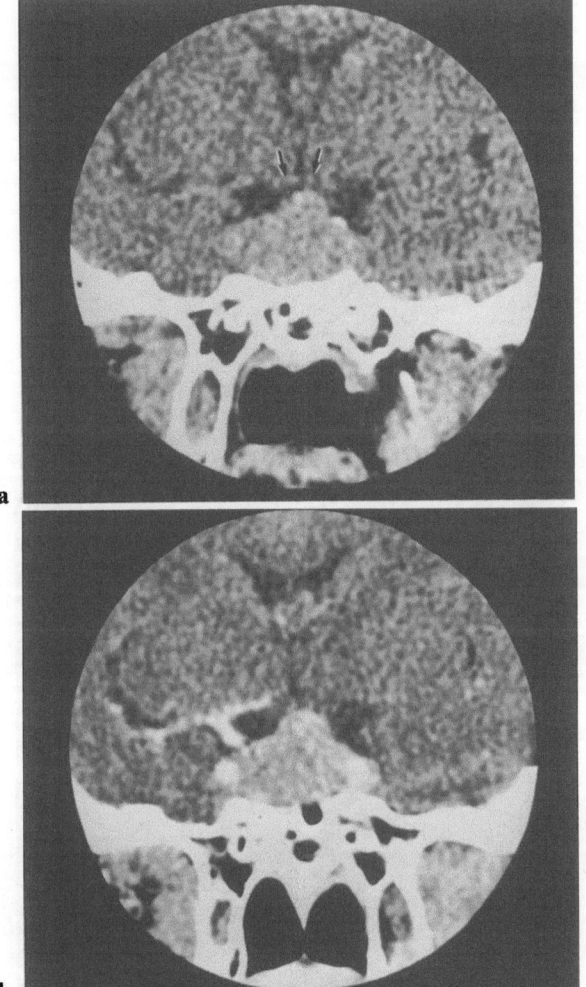

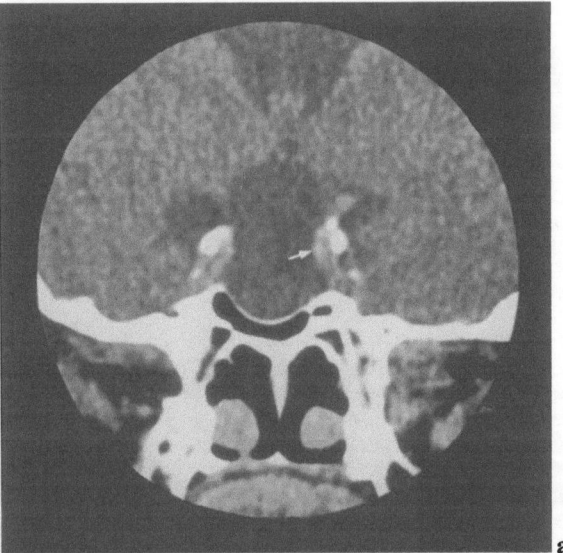

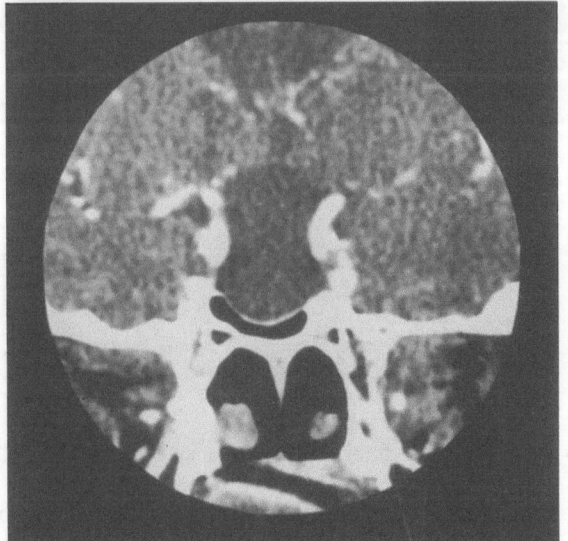

Fig. 4.3a, b. Hyperdense pituitary adenoma with small suprasellar extension. Amenorrhea and hyperprolactinemia in a 43-year-old woman. **a** Native coronal CT scan. Before contrast injection, the adenoma is denser than the brain. The upper limit of the tumor is at the level of the inferior aspect of the optic chiasm. **b** Contrast coronal CT scan. Faint homogeneous enhancement of the adenoma. A gonadotropic adenoma was found at surgery

Fig. 4.4a, b. Hypodense pituitary adenoma with suprasellar extension. Bitemporal hemianopia in a 61-year-old woman. **a** Native coronal CT. Well-defined hypodense intra- and suprasellar lesion. Depressed and thinned sellar floor. Calcified carotid siphons (*arrow*). **b** Dynamic coronal CT. Good visualization of both supraclinoid carotid arteries. No enhancement of the mass

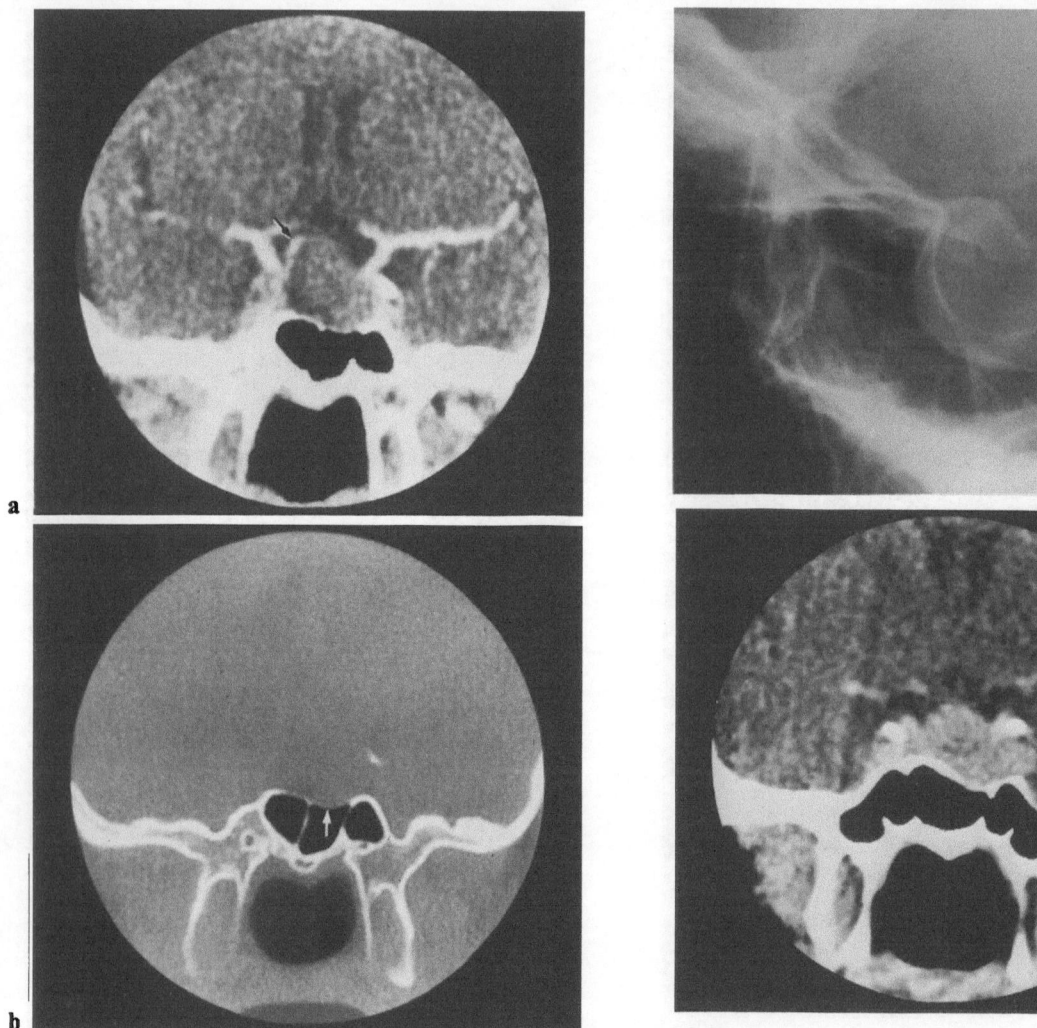

Fig. 4.5a, b. Prolactinoma. Amenorrhea in a 19-year-old woman. Hyperprolactinemia (250 ng/ml). **a** Dynamic coronal CT scan. Good delineation of both supraclinoid carotid arteries and upper limit of the adenoma which is within the chiasmatic cistern. Partial rim enhancement (*arrow*). **b** Bone review image at the same level shows a depressed and eroded sellar floor (*arrow*)

Fig. 4.6a, b. Holosellar prolactinoma. Amenorrhea and hyperprolactine mia (200 ng/ml) in a 24-year-old woman. **a** Sella turcica magnified lateral view demonstrates an asymmetrical thinned sellar floor. **b** Contrast coronal CT scan. Depressed sellar floor on the left. Bulging of the sellar content towards the chiasmatic cistern

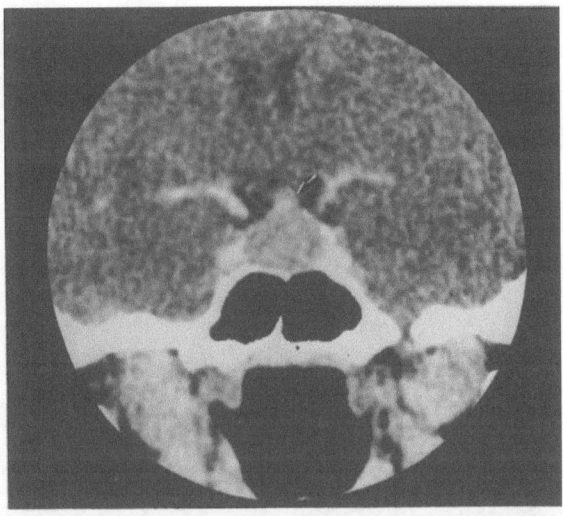

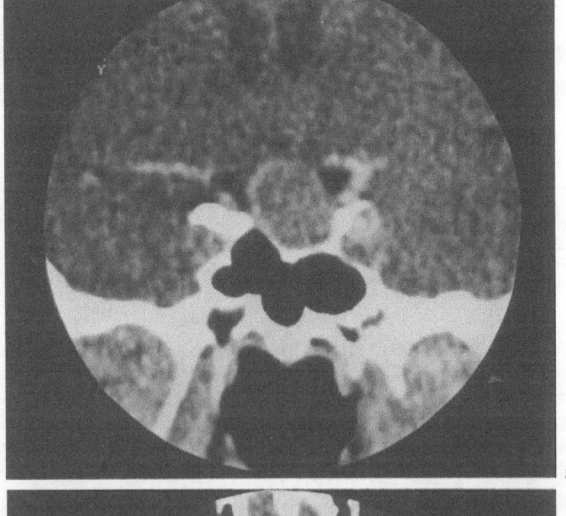

Fig. 4.7. Prolactinoma. Primary amenorrhea and hyperprolactinemia (400 ng/ml) in a 17-year-old woman. Coronal contrast CT scan. Raising and thickening of the inferior portion of the pituitary stalk (*arrow*) by an enlarged sellar content

a

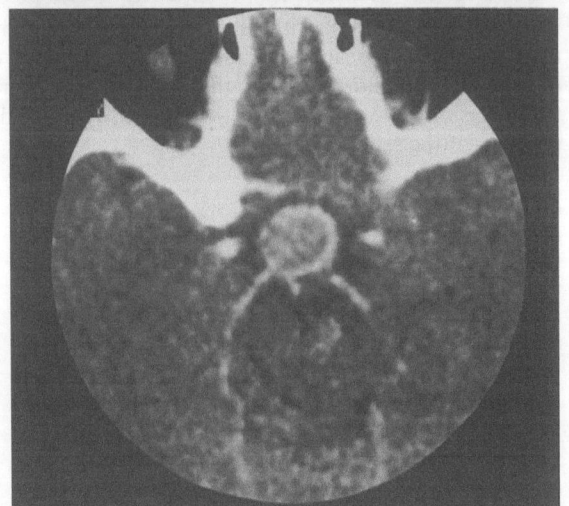

Fig. 4.8a, b. GH- and prolactin-secreting pituitary adenoma in a 45-year-old woman. **a** Contrast coronal CT scan. Asymmetrical development of a pituitary adenoma within the suprasellar cistern. Note a slight rim enhancement. **b** Contrast axial CT scan. The adenoma occupies the central part of the chiasmatic cistern

b

Fig. 4.9. Pituitary adenoma with suprasellar extension. ▶ Delayed puberty in a 15-year-old girl with visual field defect. Dynamic coronal CT scan. The contours of the mass are visible before contrast injection (*arrows*); the adenoma is better delineated after bolus injection. Good visualization of both supraclinoid carotid arteries. The sellar floor is depressed

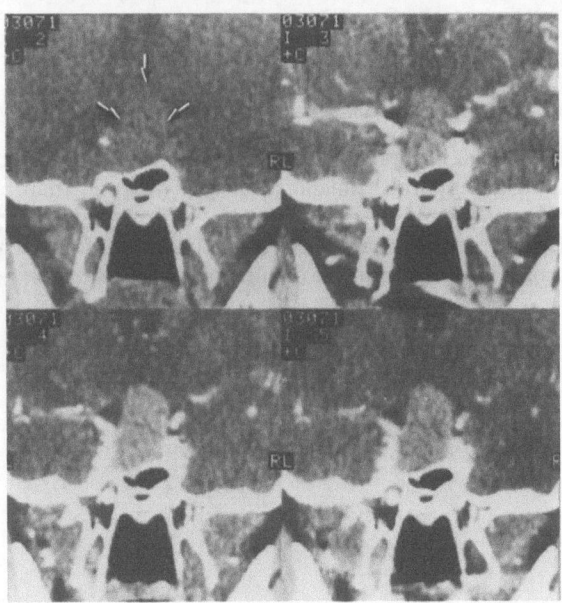

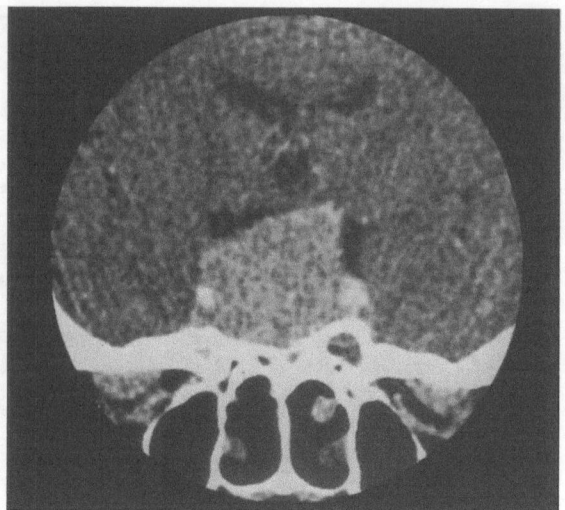

Fig. 4.10. Prolactinoma with suprasellar extension. Contrast coronal CT scan. Asymmetrical development of the pituitary tumor towards the suprasellar cistern. Compression of the left part of the chiasm. Regular depression of the sellar floor

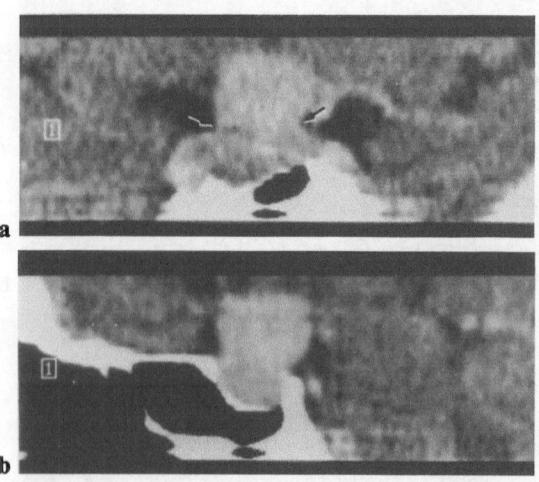

Fig. 4.11a, b. Pituitary adenoma with hourglass expansion. Bitemporal hemianopia in a 63-year-old man. **a, b** Coronal and midsagittal contrast reformatted images. Homogeneous enhancement of the pituitary tumor. The tumor is strangled at the diaphragma level (*arrows*)

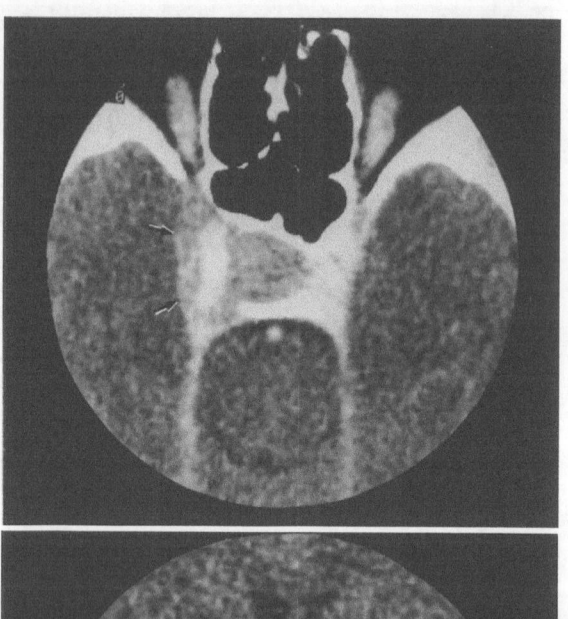

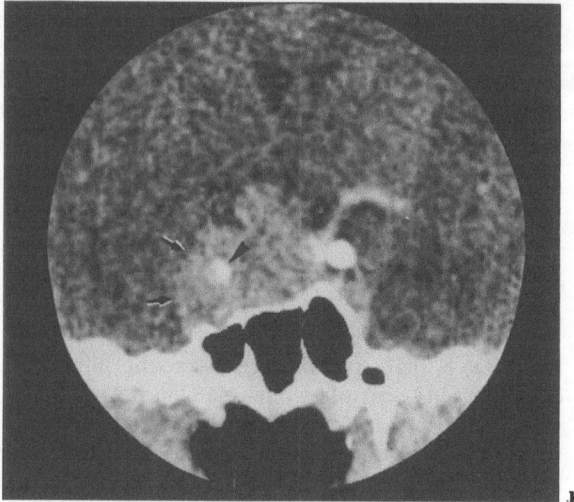

Fig. 4.12a, b. Pituitary adenoma with small hourglass suprasellar extension and invasion of right cavernous sinus. Primary amenorrhea and hyperprolactinemia (76 ng/ml) in a 19-year-old woman. **a** Axial and **b** coronal contrast CT scan. Bulging of the lateral limit of the right cavernous sinus (*arrows*). Intracavernous right carotid siphon is well shown within the enlarged cavernous sinus (*arrowhead*)

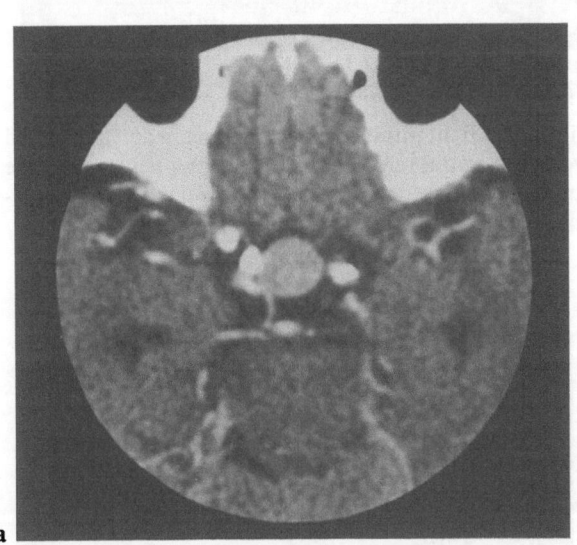

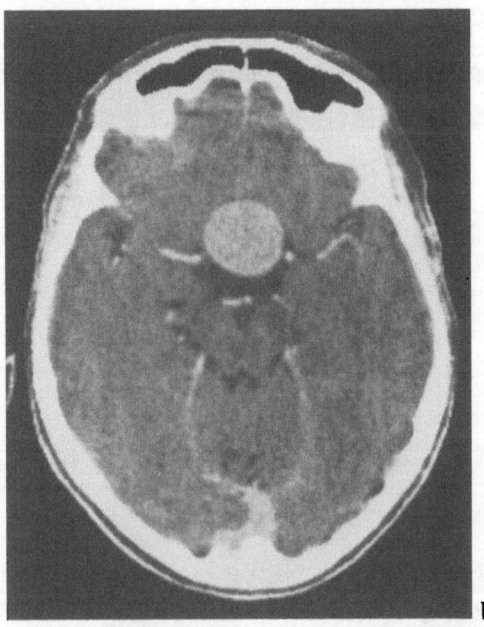

Fig. 4.13a, b. Pituitary adenomas with suprasellar extension. Axial contrast CT scans. In each case, homogeneous enhancement of the tumor which occupies the anterior part of the suprasellar cistern

Fig. 4.14a, b. Pituitary adenoma with large suprasellar ▶ posterior extension. Severe hypopituitarism in a 71-year-old man. **a** Consecutive postcontrast axial sections through the suprasellar area. **b** Midsagittal reformatted image. The posterior compartment of the tumor is extending within the interpeduncular fossa

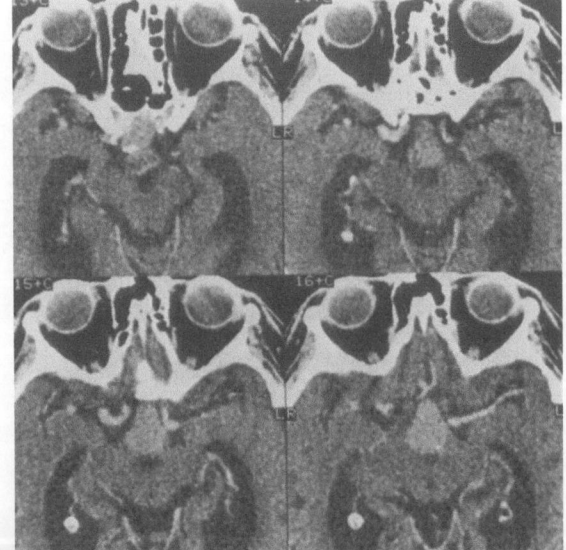

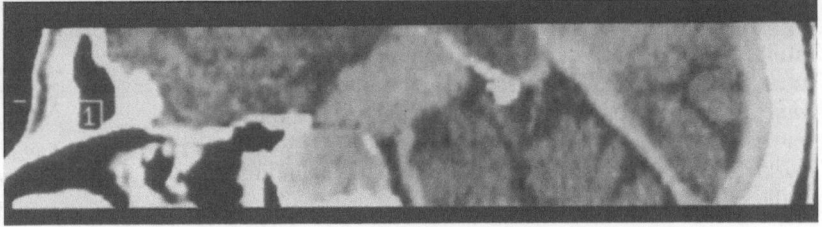

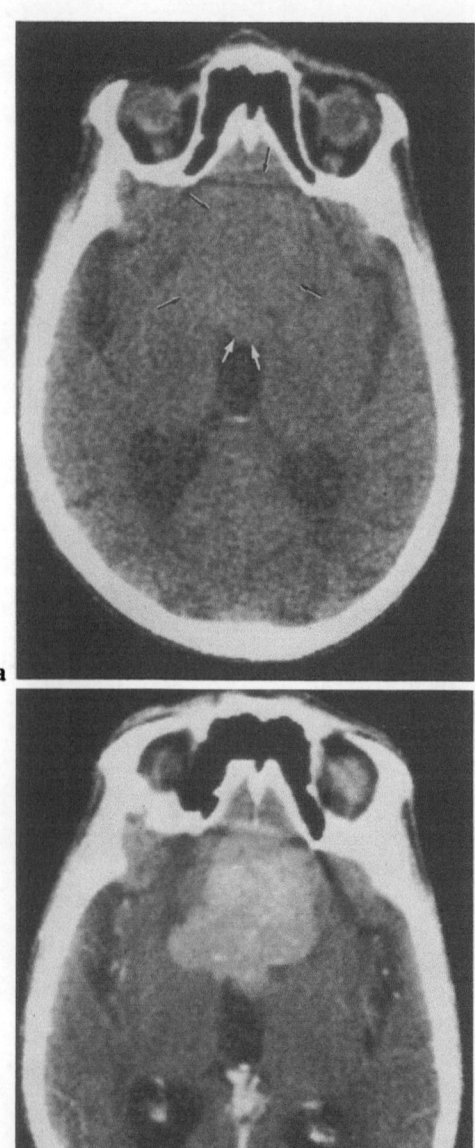

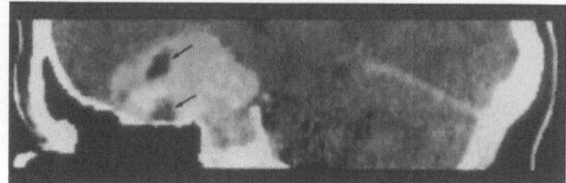

Fig. 4.16. Giant pituitary prolactinoma in an 18-year-old woman. Sagittal reformatted image after contrast. Large subfrontal expansion with hypodense necrotic areas (*arrows*)

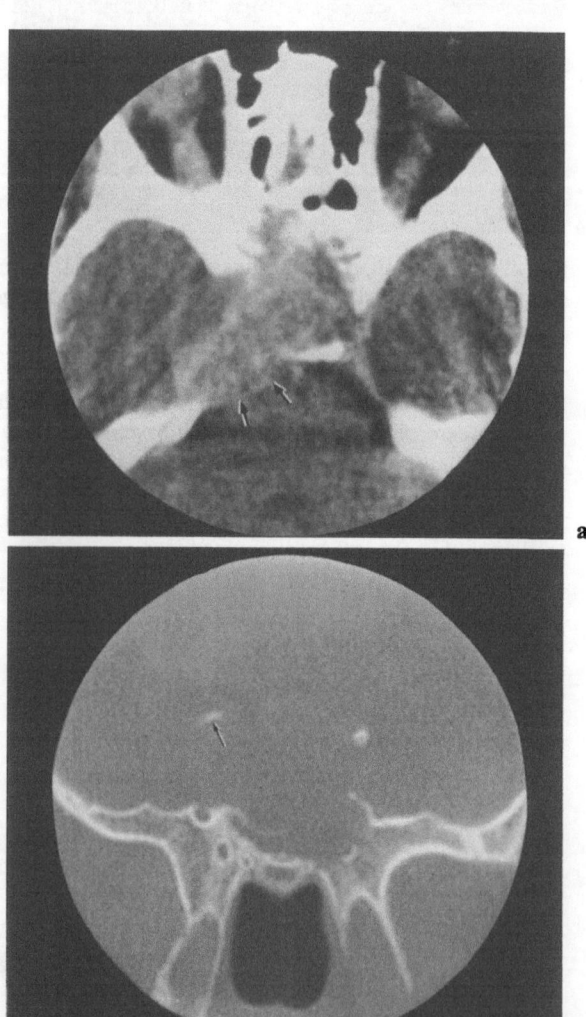

Fig. 4.15a, b. Pituitary adenoma with frontal extension. Pituitary insufficiency in an 85-year-old man. **a** Precontrast axial CT scan. Huge isodense tumor (*black* and *white arrows*) encroaching the anterior part of the third ventricle (*white arrows*). **b** Contrast axial CT scan: marked enhancement of the polycyclic mass extending in the subfrontal area

Fig. 4.17a, b. Invasive pituitary adenoma. Recurrent hyperprolactinemia 20 years after surgery and radiotherapy in a 37-year-old woman. **a** Axial contrast CT scan: large pituitary mass destroying the anterior wall of the sella and the right part of the dorsum and extending laterally towards the cavernous sinus and petrous apex (*arrows*). **b** Bone review image in coronal section. Destruction of the sellar floor and erosive changes of the right anterior clinoid process (*arrow*)

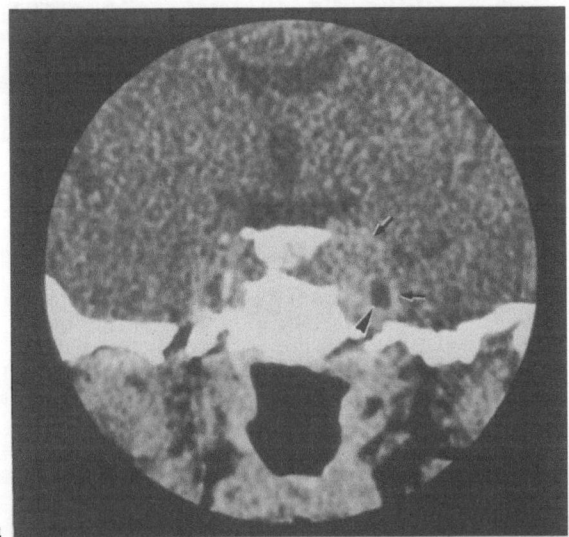

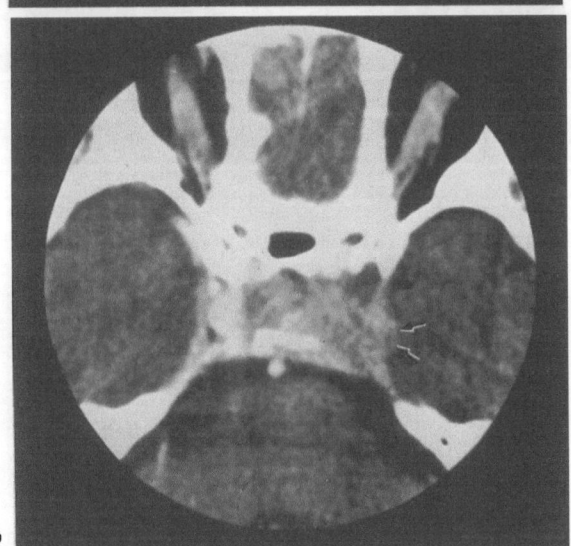

Fig. 4.18a, b. Pituitary adenoma with posterolateral extension; incidental discovery in a 69-year-old man. **a** Coronal contrast CT scan. Enlargement of the left cavernous sinus (*arrows*) and lateral displacement of the Gasser ganglion (*arrowhead*). **b** Axial contrast CT scan: erosion of the left part of the dorsum. Localized bulging of the lateral limit of the left cavernous sinus (*arrows*)

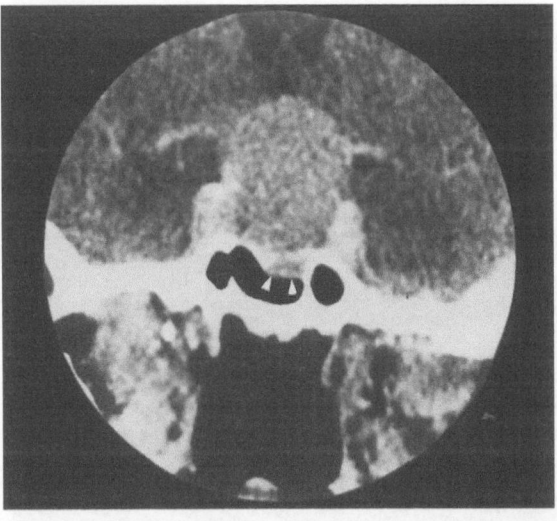

Fig. 4.19. Pituitary adenoma with suprasellar expansion. Large visual field defect in a 31-year-old man. Contrast coronal CT scan. Homogeneous enhancement of the tumor. Small extension of the mass into the upper part of the sphenoid sinus through the eroded sellar floor (*arrowheads*)

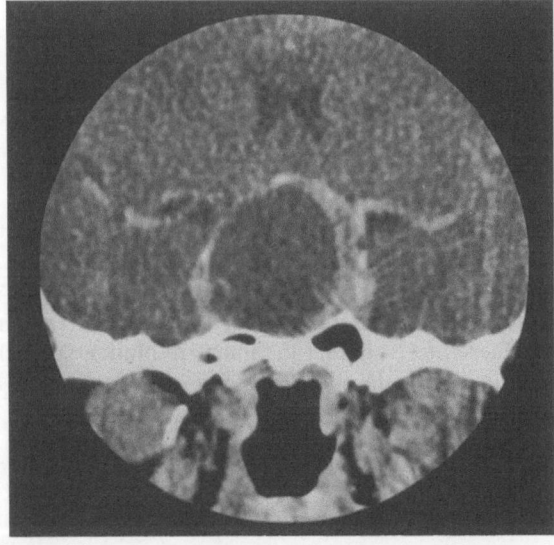

Fig. 4.20. Pituitary adenoma. Headache, bitemporal hemianopia and amenorrhea in an 18-year-old woman. Prolactinemia is 120 ng/ml. Contrast coronal CT scan. Huge hypodense adenoma with ring enhancement

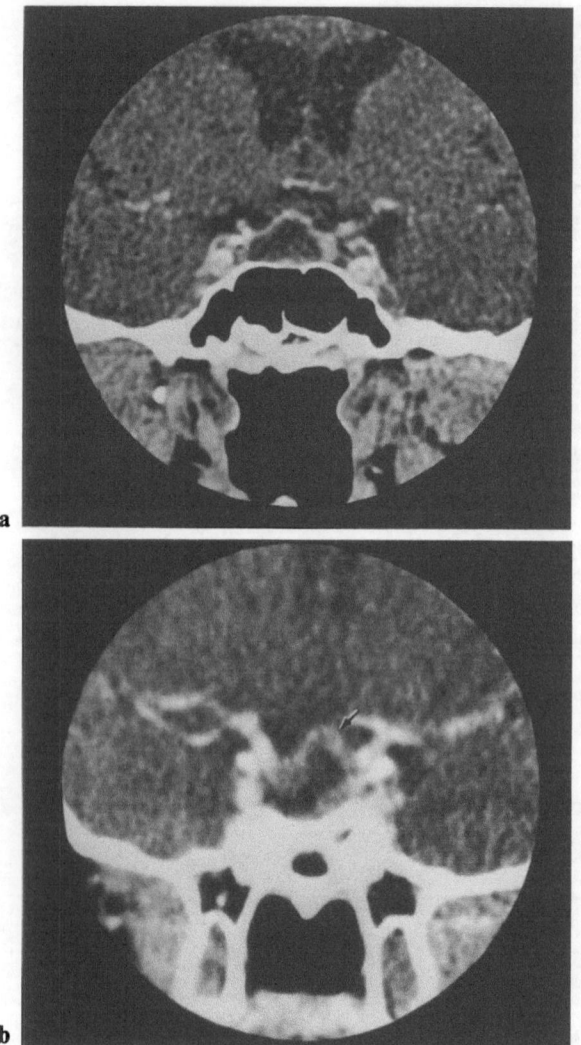

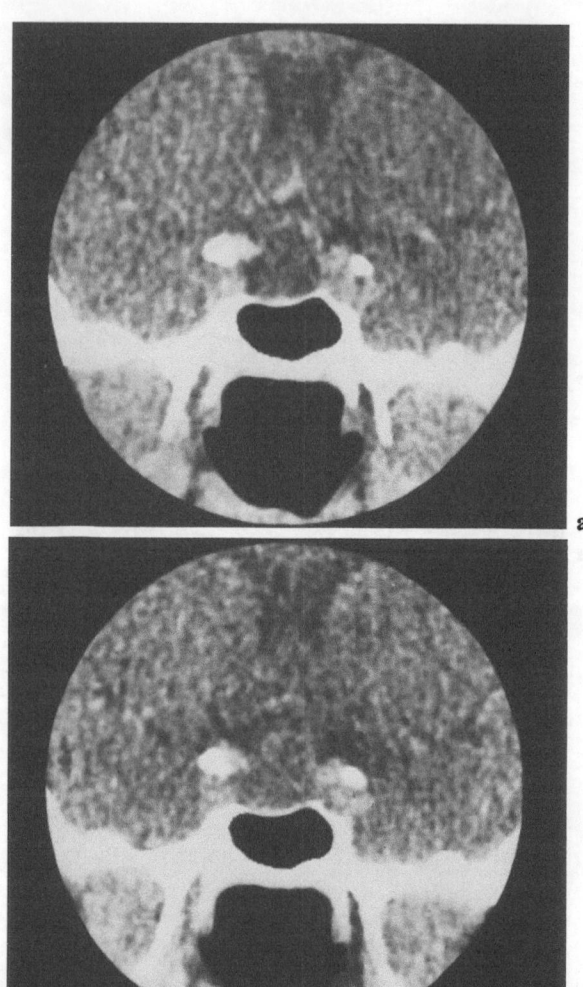

Fig. 4.21a, b. Pituitary adenomas. Contrast coronal CT scans. Different patterns of ring enhancement: **a** thin or **b** thick rim. Note that the image resolution is much better in **a** with the 9800 GE CT/T scanner than in **b** with the 8800 GE CT/T scanner

Fig. 4.22a, b. Prolactinoma. Sexual impotence in a 30-year-old man. Prolactinemia is 360 ng/ml. **a** Coronal CT scan 5 min after injection. Small suprasellar extension of the mass. Faint ring enhancement. The central part of the tumor is hypodense. **b** Coronal CT scan 1 h after contrast injection. Slight enhancement of the central part of the adenoma related to intratumoral passage of contrast material

Fig. 4.24a–c. Pituitary adenoma after tumoral hemor- ▶ rhage Bilateral sudden visual loss in a 55-year-old man. **a** Sella turcica magnified lateral view: no deepening of an open sella. The cortical bone of the sellar floor is not thinner than in the presellar region. In this case, a pituitary adenoma with large suprasellar extension is difficult to predict. **b** Coronal and **c** sagittal reformatted images after contrast injection. Pituitary adenoma with regular suprasellar extension. The low-density just above the sellar floor (*arrows*) is related to a recent hemorrhage confirmed at surgery

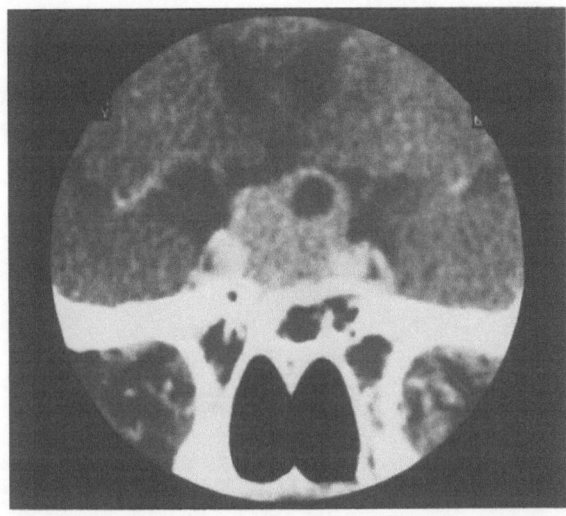

Fig. 4.23. LH-secreting pituitary adenoma Impotence and visual field defect in a 61-year-old man. Contrast coronal CT scan. Asymmetrical suprasellar development of a tumoral mass with necrotic area on the left

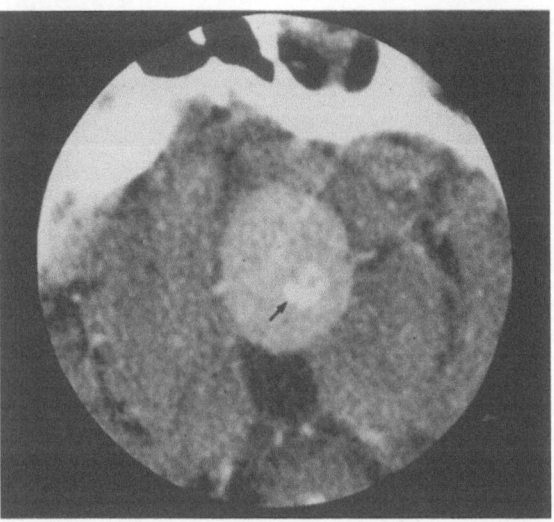

Fig. 4.25. Huge calcified pituitary adenoma. Hypopituitarism in an 80-year-old man. Axial contrast CT scan: regular dense pituitary adenoma encroaching the anterior part of the third ventricle. Intratumoral granular calcification (*arrow*)

Fig. 4.26. Suprasellar cystic formation after surgery. Visual field defect 6 years after transsphenoidal adenomectomy in a 38-year-old woman. Coronal CT cisternography. Central defect in the chiasmatic cistern corresponding to a cystic lesion found at surgery. Encroachment of the optic chiasm (*arrows*)

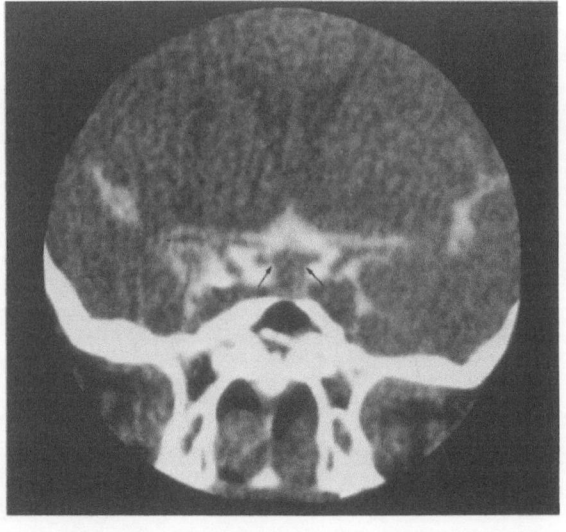

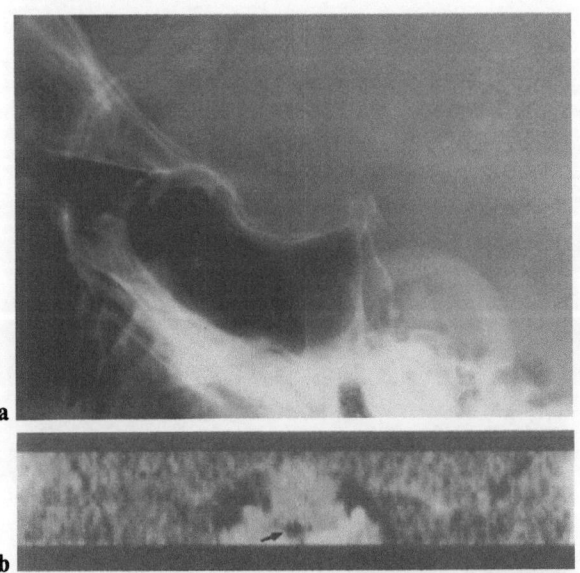

Chapter 5
Prolactin-Secreting Pituitary Adenomas

Prolactin-secreting pituitary adenomas are the most common pituitary adenomas. They occur in women in 80%–90% of cases (Derome, Domingue). In women, hyperprolactinemia is manifest by secondary amenorrhea with anovulation, in 50% of cases following a variably long period of oligomenorrhea. Galactorrhea is frequently limited, or can be manifest solely by pressure on the nipple. Increase in body hair and a decrease in libido can complete the clinical profile. The tumoral syndrome is rarely the primary disorder; however, this can occur, for example, in pregnancy or in the presence of an undiagnosed adenoma. Induction of ovulation should thus never be carried out before verification of prolactin levels and plain films of the sella turcica; if there is any doubt, a CT scan should be performed.

In men, headache and visual disturbances are the primary clinical symptoms of the adenoma which is generally discovered only after suprasellar expansion. The common symptom of impotency is frequently neglected and not spontaneously reported due to loss of libido.

CT scan is unnecessary when iatrogenic origin of hyperprolactinemia can be proven: we have shown that pituitary volume and contrast enhancement are unmodified in iatrogenic hyperprolactinemias due to treatment with neuroleptics or renal dialysis.

CT of Intrasellar Prolactinomas

Computed tomographic demonstration of intrasellar prolactinomas has been essential for the understanding of hyperprolactinemias; this was a determining factor for modification of therapeutic approach, demonstrating that tumoral volume generally decreases with medical treatment (Bonneville). Diagnosis of intrasellar prolactinomas has become reliable only with the development of high-resolution CT scanners.

General Characteristics

Before discussing semiology of intrasellar prolactinomas, we would like to list seven fundamental points:

1. The limits of sella turcica plain films are currently well known: an adenoma of less than 4–5 mm in size rarely deforms the sellar walls, and the presence of a normal sella turcica does not eliminate the possibility of microadenoma (Fig. 4.1). In the presence of hyperprolactinemia, a CT scan will be performed depending on clinical and laboratory arguments, plain films being normal or not. Indeed, X-ray studies of the sella turcica are today carried out essentially to orient the CT exam (choice of projection, thickness of sections, etc.). In a workup for hyperprolactinemia, we currently limit the plain films to a magnified lateral view using an electron gun.

2. Tomography, even with complex motion, virtually never provides supplementary information, and we have completely abandoned this technique. There is frequently no topographical correlation between an abnormality of the sellar floor and the site of the microadenoma (Rhoton, Muhr, Swanson, Burrow, Turski, Nachtigall). "Thinning" of the sellar walls, frequently described in tomography, often corresponds to localized, physiological variability in bone thickness. A good understanding of morphological variants of the sella turcica is useful (Bonneville), and personal experience of the radiologist is of fundamental importance here.

3. Only high-resolution CT scanners of the last generation yielding precise sections of 1- or 1.5 mm thickness allow reliable diagnosis of intrasellar adenomas. The first CT descriptions of microadenomas were hindered by partial volume effect (Gardeur).

4. Direct coronal sections are virtually always of quality superior to reformatted images from axial sections, and should always constitute the first line of approach. Concerning intrasellar adenomas, axial complementary sections are necessary only when plain films suggest the presence of a very anterior or very posterior adenoma, or when coronal sections are not diagnostic. In our experience with microprolactinomas, coronal sections are successful in 95% of cases.

5. Intrasellar prolactin-secreting adenomas are virtually always less enhanced by the iodinated contrast medium than is the normal pituitary. In our experience with 2300 pituitary CT scans, we have seen only two intrasellar prolactinomas denser than the normal pituitary gland. Compression of pituitary cellular cords at the periphery of the adenoma should explain this lesser enhancement.

6. Dynamic CT scan in coronal sections is essential for demonstration of small adenomas; indeed, this allows: (a) visualization of the delay in enhancement of the adenoma by comparison with the healthy pituitary tissue, and (b) visualization of the contralateral displacement of the secondary pituitary capillary bed.

7. When CT scan is carried out after prolonged treatment with a prolactin agonist, the adenoma per se may no longer be directly visible due to the antitumoral action of the dopaminergic agent. Dynamic scan is here again essential to find the remnant of the tumor.

Diagnostic Signs

For diagnosis of intrasellar adenomas direct thin coronal sections and dynamic scan have to be emphasized; there is less data provided from the axial sections and reformatting, and these data are less reliable.

Nine signs should be discussed:
1. Size of the pituitary
2. Morphology of the superior pole of the pituitary
3. Appearance of the sella turcica walls
4. General appearance of the gland before injection of contrast medium
5. Morphology and topography of the secondary pituitary capillary bed
6. Attenuation value measurement of the pituitary during dynamic CT
7. Attenuation value measurement of the pituitary at some distance of bolus injection
8. Appearance of the posterior pituitary
9. Direction of the pituitary stalk

Diagnosis of an intrasellar pituitary adenoma is increasingly reliable when a large number of these signs are found. Data from dynamic CT are essential.

Size of the Pituitary

A pituitary gland, whose height is more than 8 mm, is virtually always abnormal. When isolated, this sign is nevertheless insufficient to allow a definitive diagnosis of intrasellar adenoma.

Indeed, besides cases of pituitary adenoma, a pituitary height greater than 8 mm can be seen in pregnancy (see Chap. 7) and in hyperplasia due to dysfunction of a target organ (see Chap. 12). Furthermore, Swartz reported a mean height of 7.1 mm in 50 young women of childbearing age. While we ourselves have not confirmed these data, we have seen a pituitary gland higher than 9 mm in the absence of any endocrine pathology or pregnancy in some cases where the pituitary gland width was less than 10 mm (see Chap. 2). Finally, one must always bear in mind that an intrasellar adenoma can be present in a gland the volume of which is not increased, or with a maximal height not exceeding 2 mm (Fig. 5.1). When the gland is small, asymmetry between height of the right and left sides is not necessarily pathological. A gland height greater than 10 mm above a nondepressed floor indicates slight suprasellar expansion. These intermediate forms between intrasellar and suprasellar adenomas are seen frequently (see Chap. 4).

Morphology of the Superior Pole of the Pituitary

The superior pole of the pituitary is normally horizontal; it is frequently concave upwards when the gland is small. Swartz reported that the intrasellar content was convex upwards in 44% of women of childbearing age; however, an identical study carried out in France did not confirm this report. We find that upward convexity always requires attention, even if it alone is an insufficient criterion to decide for pituitary lesion.

The concept of pituitary hyperplasia is not clearly defined. From the radiologic standpoint, we reserve the term hyperplasia for cases where the pituitary is increased in volume and presents a regular convexity of its superior pole without any detectable adenomas (Fig. 5.2). In hyperplasia, the increase in pituitary volume usually regresses after treatment with dopaminergic agents.

Asymmetrical bulging of the intrasellar contents on direct coronal sections is highly indicative of a microadenoma. On the other hand, a triangular or rounded median rise at the pituitary stalk passage is not pathological (see Chap. 2). If coronal sections are precluded or if their quality is poor, the superior pole of the pituitary can be visualized by reformatting from axial sections. Asymmetry of the upper limit of intrasellar contents in right and left paramedian sagittal reconstructions can have some diagnostic value (Fig. 5.3). Convexity of the superior pole of the pituitary can be marked in forms intermediate with suprasellar adenomas (Figs. 4.6 and 4.7). The superior pole of the pituitary can remain entirely flat when the adenoma is of very small size (Figs. 5.4 and 5.5), if the CT scan is done several months after the beginning of treatment with dopaminergic agents (Fig. 6.7), or where extension is exclusively inferior, towards the sphenoid sinus (Fig. 5.6).

Appearance of the Walls of the Sella Turcica

Study of the appearance of the walls of the sella turcica is frequently disappointing, and rarely of absolute diagnostic value in intrasellar prolactinomas. When the adenoma is smaller than 5 mm in diameter, the floor of the sella can be perfectly normal (Fig. 5.7). Asymmetry, inclination, angulation, and differences in thickness of the sellar floor can be seen in the absence of microadenomas (see Chap. 2). In adenomas larger than 5 mm in diameter, localized thinning, especially when associated with a depression of the floor, is a supplementary diagnostic sign (Figs. 5.6, 5.8, 5.9, 5.25, and 5.30). Osseous thinning can be generalized over the entire width of the sellar floor in cases of intrasellar adenomas (Fig. 5.27). Differential diagnosis between pathological thinning and a normally thin sellar floor is not always easy. Diagnostic superiority of the CT scan with the bone reconstruction algorithm as compared with conventional CT scan with bone window setting is not self-evident (Fig. 1.6). Study of the sellar floor is interesting in adenomas treated with dopaminergic agents; a major osseous abnormality combined with an adenoma of very small size can suggest that the adenoma was far larger before treatment, and there is thus a risk of major expansion upon withdrawal of dopaminergic agents.

In axial sections, a pronounced posterior concavity of the anterior wall of the sella is highly characteristic of intrasellar adenoma (Fig. 5.28); however, this appearance is also seen in empty sella turcica. In holosellar adenomas, bulging of the anterior wall of the sella is frequently accompanied by a convexity of the medial limit of the cavernous sinus, pushing the internal carotid arteries outwards. In posteriorly located adenomas, asymmetrical erosion of the dorsum sellae is a useful supplementary sign (Fig. 5.10); however, physiological asymmetry of the dorsum sellae is not exceptional.

General Appearance of the Gland Before Injection of Contrast Medium

Data supplied by native coronal CT before injection of contrast medium should not be neglected. These coronal sections, in particular with the most recent systems, allow a rough identification of the gland, allowing determination of height and morphology of the superior pole. The relationship between the sellar contents and the chiasm are readily visible in the suprasellar cistern (see Chap. 2). Of course, the native scan also allows study of the osseous sellar walls. In our experience, it can also suggest the presence of a microadenoma if a rounded hypodense area is demonstrated within the gland (Fig. 5.11). Native scans allow identification of thin tumoral calcifications, particularly in prolactin-secreting adenomas. (Figs. 5.12 and 5.13); the frequency of such calcifications as visualized by CT scan is limited, approximately 2%, and is less than 7% of the histological series. A massive calcification of an intrasellar prolactinoma is highly uncommon (Fig. 5.14).

Morphology and Topography of the Secondary Pituitary Capillary Bed

Demonstration of the secondary pituitary capillary bed by dynamic coronal CT scan is essential if the adenoma is of a size too small to modify the sellar floor or the superior pole of the pituitary. Displacement of the pituitary capillary bed (Figs. 3.16, 5.15, 5.16, 5.17, 5.18, and 5.19), or its compression (Fig. 3.18) provide formal evidence for an intrasellar expansive process (Bonneville). This can be the sole CT sign in the rare cases of prolactinomas presenting the same attenuation value as the healthy pituitary (Fig. 3.17). In the unusual cases where the capillary bed is not nodular but rather linear, parallel, and in contact with the superior pole, localized defects in filling of the capillary bed take on particular significance (Fig. 6.12g). These modifications in the capillary bed can be absent or minimal in cases where microadenomas are on the midline or posteriorly located. Of course, the capillary bed may not be visualized if the projection is improperly chosen or if the patient moved between the scoutview and the injection of contrast medium. Finally, when the adenoma is larger than 8–10 mm in diameter, the secondary pituitary bed will generally not be seen due to its compression (Fig. 3.15).

Attenuation Value of the Pituitary Gland During Dynamic CT

Contrast enhancement of the pituitary gland normally begins about 10 s after optimal opacification of the secondary pituitary capillary bed (see Chap. 3). In case of adenoma, enhancement of the parenchyma on the side of the lesion is delayed by comparison with the opposite side (Figs. 3.19, 3.20, and 5.20).

Attenuation Value of the Pituitary Gland at Some Distance of Bolus Injection

At the end of dynamic CT, when the plasma iodine level reaches a plateau, intrasellar microprolactinomas virtually always appear as rounded or oval zones, less dense than the normal pituitary (Figs. 5.21 and 5.22) (Bonneville). In frequently, intrasellar adenomas present a rectilinear border; this can be the case after treatment with dopaminergic agents (Fig. 5.23). An adenoma with a rectilinear border will be differentiated from a streak artifact going beyond the limits of the sella (Fig. 1.12). Hypodensity of the lesion by comparison with the healthy pituitary does not necessarily imply that the adenoma is necrotic except where density is approximately 0 Hounsfield units. Heterogeneity of the adenoma with greater density of the lower part of the tumor can be seen in the absence of any hemorrhagic process (Fig. 5.24). Microprolactinomas denser than the normal pituitary are exceptional; we have seen only two cases in more than 2300 pituitary CT scans (Fig. 5.25).

Demonstration of prolactinomas is far less reliable in axial sections, even when thin. Only lesions greater than 4–5 mm in diameter are readily identifiable in this projection, and these should be differentiated from:

1. The normal posterior pituitary (see Chap. 2)
2. A partial intrasellar expansion of the subarachnoid spaces (see Chap. 13)
3. Artifacts of reformatting (streak artifacts resulting from inclusion of the petrous bones in the section) (see Chap. 1).

Frontal and sagittal reformatting can also be useful. In general, these are not considered to be highly reliable for diagnosis of intrasellar prolactinomas.

Prolactin microadenomas generally have an anterolateral topography and less frequently posterolateral, due to the presence of prolactin cells in the lateral part of the pituitary. In actual fact, adenomas can develop anywhere in the gland, including on the midline.

Holosellar adenomas present a usually homogeneous enhancement; density of the sellar content is generally less than that of the normal pituitary; however, it is sometimes very similar: in such cases, increase in volume of the gland, convexity of the superior pole and, especially, osseous deformation can constitute the sole diagnostic elements (Fig. 5.9).

Appearance of the Posterior Pituitary

Under normal conditions, in 50% of the cases the posterior pituitary appears on thin axical sections as an oval zone, less dense than the anterior pituitary and in direct contact with the anterior surface of the dorsum sellae (Bonneville). This characteristic pattern can be deformed by pressure of an intrasellar adenoma (Fig. 5.26).

Direction of the Pituitary Stalk

The pituitary stalk is frequently inclined from the normal vertical axis, even in cases of strictly intrasellar adenomas, due to its close relationship with the secondary pituitary capillary bed (Fig. 5.17). Nevertheless, it must be borne in mind that the pituitary stalk can be inclined in the absence of any pituitary or hypothalamic disorder (see Chap. 15).

Particular CT Signs of Intrasellar Prolactinomas with Anterior, Posteroinferior or Inferior Development

Prolactinomas with anterior, posteroinferior, or inferior development may not be seen due to partial volume effect. Greater significance will thus be given to osseous deformations (Fig. 5.27). Hollowing and thinning of the anterior wall are sometimes better seen in axial sections (Fig. 5.28) or by reformatting, rather than in coronal sections (Fig. 5.29). Deformation of the dorsum sellae is invariably better visualized in axial sections. In some cases, the only sign of small adenomas with inferior extension is the presence of a rounded imprint occupying the top of the sphenoid sinus in axial section (Fig. 5.30). If the adenoma is situated within the sphenoid sinus, coronal sections generally do not allow identification of the hypodensity which usually characterizes intrasellar prolactinomas due to partial volume effect. Where hyperprolactinemia has been treated with dopaminergic agents, it is very difficult to determine whether the adenoma is still present above the osseous deformation.

Particular CT Signs of Intrasellar Prolactinomas with Lateral Development

Lateral extension into the cavernous sinus is infrequent in prolactinomas. Visualization is facilitated by dynamic CT and axial sections, along with the usual coronal sections. One must differentiate between:

1. Simple lateral extension of intrasellar adenomas displacing both cavernous sinuses (Fig. 5.31)

2. Pattern of false enlargement of the cavernous sinus, due to poorly delimited adenomatous growth within the sella along the medial wall of the cavernous sinus (Fig. 5.32); dynamic CT scan is essential here to distinguish the pituitary gland from vascular structures of the cavernous sinus

3. Postsurgical adenomatous residues attached to the medial wall of the cavernous sinus and yielding an image of asymmetric empty sella (Fig. 14.5)

4. True invasion of the cavernous sinus indicating an invasive adenoma. This causes true enlargement of the cavernous sinus with convexity of the lateral wall. Dynamic CT allows identification of the intracavernous carotid arteries within the enhanced tumoral tissue which infiltrates between structures of the cavernous sinus. In these cases, visualization of the intracavernous nerves may be blurred as is that of the intracavernous venous spaces (Figs. 4.12 and 5.33). In such cases, ocular signs may be seen due to involvement of the intracavernous nerves.

CT Signs of Prolactinomas with Suprasellar Development

Prolactinomas with suprasellar development are seen as frequently in men as in women. Hyperprolactinemia is here of secondary importance, while the tumoral syndrome and sometimes hypopituitarism tend to be the presenting signs. CT signs of suprasellar prolactinomas are not different from those of nonsecreting pituitary adenomas. The old designation "chromophobe adenoma" included both prolactinomas and nonsecreting adenomas with suprasellar extension. These signs have been discussed in Chap. 4.

References

Andersen AN, Starup J, Tabor A, Jensen HK, Westergaard JG (1983) The possible prognostic value of serum prolactin increment during pregnancy in hyperprolactinaemic patients. Acta Endocrinol 102:1–5

Annegers JF, Coulam CB, Abboud CF, Laws ER, Kurland LT (1978) Pituitary adenoma in Olmsted Country, Minnesota, 1935–1977. A report of an increasing incidence of diagnosis in women of childbearing age. Mayo Clin Proc 53:641–643

Antunes JL, Housepian EM, Frante AG, Holub DA, Hui R, Carmel PW, Quest DO (1977) Prolactin-secreting pituitary tumors. Ann Neurol 2:148

Arafah BM (1981) Cure of hypogonadism after removal of prolactin secreting adenoma in men. J Clin Endocrinol 52:91–94

Aubourg PR, Derome PJ, Peillon F, Jedynak CP, Visot A, Le Gentil P, Balagura S, Guiot G (1980) Endocrine outcome after trans-sphenoidal adenomectomy for prolactinoma: prolactin levels and tumor size as predicting factors. Surg Neurology 14:141–143

Azar Kia B, Palacios E, Churchill RJ (1977) Diagnosis of sellar and parasellar lesions by CT and other diagnostic modalities. CT 1:249–256

Bamberger C, Rosier J (1980) Examen tomodensitométrique des lésions sellaires et parasellaires. Feuillets de radiologie 20:197–208

Banna M (1980) CT arteriography of microadenomas. J Comp Ass Tomog 4:690–692

Banna M, Nicholas W, McLachlan M (1978) The borderline pituitary fossa in patients with amenorrhea and/or galactorrhea. Neuroradiology 16:440–442

Banna M, Baker HL Jr, Houser OW (1980) Pituitary and parapituitary tumours on CT. Br J Radiol 53:1123–1143

Barbarino A, De Marvis L, Menini E, Anile C, Maira G (1979) Prolactin secreting pituitary adenomas: prolactin dynamics before and after transsphenoidal surgery. Acta Endocrinol 91:397–400

Belloni G, Baciocco A, Borrelli R, Sagui G, Di Rocco C, Maira G (1978) The value of CT for the diagnosis of pituitary microadenoma in children. Neuroradiology 15:179–181

Besser GM (1976) The pituitary fossa: normal or abnormal. Br J Radiol 49:652–653

Blanchi SD, Gatti G, Andreis M, Ghigo E (1982) Personal experience with highresolution CT in the diagnosis of microprolactinoma. In: Molinatti GM (ed) A clinical problem: microprolactinoma. Diagnosis and treatment. Excerpta Medica, Amsterdam-Oxford-Princeton, p 69

Bonafe A, Sobel D, Manelfe C (1981) Relative value of CT and hypocycloidal tomography in the diagnosis of pituitary microadenoma. A radio-surgical correlative study. Neuroradiology 22:133–137

Bonneville JF, Dietemann JL (1981) Radiology of the sella turcica. Springer Verlag, Berlin Heidelberg New York

Bonneville JF, Poulignot D, Cattin F, Couturier M, Mollet E, Dietemann JL (1982) Apport des méthodes nouvelles dans l'exploration morphologique des tumeurs hypophysaires. Ann Endocrinol (Paris) 43:303–308

Bonneville JF, Poulignot D, Cattin F, Couturier M, Mollet E, Dietemann JL (1982) Computed tomographic demonstration of the effects of bromocriptine on pituitary microadenoma size. Radiology 143:451–455

Bonneville JF, Poulignot D, Coche G, Portha C, Cattin F, Bacha M (1982) Radiological techniques in the diagnosis of microprolactinoma. In: Molinatti GM (ed) A clinical problem: microprolactinoma. Diagnosis and treatment. Excerpta Medica, Amsterdam-Oxford-Princeton, p 57

Bonneville JF, Cattin F, Moussa Bacha K, Portha C (1983) Un plus dans l'exploration de l'hypophyse: l'angioscan. Presse Méd 12:1669

Bonneville JF, Cattin F, Poulignot D, Dietemann JL, Couturier M (1983) Value of high-resolution CT in the follow-up of prolactinomas treated with bromocriptine. In: Tolis G (ed) "Human Prolactin". Raven Press, New York

Bonneville JF, Cattin F, Moussa-Bacha K, Portha C (1983) Dynamic computed tomography of the pituitary gland: the "tuft sign". Radiology 149:145–148

Bonneville JF, Cattin F, Dietemann JL (to be published) The convex pituitary gland. 24th meeting of the American Society of Neuroradiology, San Diego, 22 Jan 1986

Bruneton JN, Drouillard JR, Sabatier JC, Elie GP, Tavernier JF (1979) Normal variants of the sella turcica. Radiology 131:99–104

Burrow GN, Wortzman B, Rewcastle NB, Holgate RC, Kovacs K (1981) Microadenomas of the pituitary and abnormal sellar tomograms in an unselected autopsy series. N Engl J Med 304:156–158

Buvat J, Buvat-Herbaut M (1982) Prolactine, bromocriptine et fonction gonadique de la femme: données récentes. I. Physiologie de la prolactine, physiopathologie et diagnostic des hyperprolactinémies. J Gyn Obst Biol Repr 11:341–353

Cabanis EA, Van Effenterre R, Iba-Zizen MT (1979) CT in parasellar space-occupying lesions and therapeutic decision. Acta Neurochirurgica (suppl 28) 28:329–333

Campagnoli C, Belforte L, Lotano M, Sandri A, Camanni F (1982) Prognosis of microprolactinoma and hyperprolactinemia in relation to pregnancy. In: Molinatti GM (ed) A clinical problem: Microprolactinoma. Diagnosis – treatment. Excerpta Medica, Amsterdam

Carter JN, Tyson TE, Tolis G, Van Vliet S, Faiman C, Friesen HG (1979) Prolactin-secreting tumors and hypogonadism in 22 men. N Engl J Med 32:483–485

Chambers EF, Turski PA, La Masters D, Newton TH (1982) Regions of low density in the contrast-enhanced pituitary gland: normal and pathologic processes. Radiology 144:109–113

Chang JR, Keye WR, Young JR, Wilson CB, Jaffe RB (1977) Detection, evaluation and treatment of pituitary microadenoma in patients with galactorrhea and amenorrhea. Am J Obstet Gynecol 128:350

Citrin CM, Davis DO (1977) CT in the evaluation of pituitary adenomas. Invest Radiol 12:27–35

Costello RT (1936) Subclinical adenoma of the pituitary gland. Am J Pathol 12:205–215

Coulam CB, Annegers JF, Abboud CF, Laws ER Jr, Kurland LT (1979) Pituitary adenoma and oral contraceptives: a case-control study. Fertil Steril 31:25

Coulon G, Fellmann D, Arbez-Gindre F, Pageaut G (1983) Les adénomes hypophysaires latents. Etude autopsique. Sem Hôp Paris 59:2747–2750

Cusick JF, Haughton VH, Hagen TC (1980) Radiological assessment of intrasellar prolactin-secreting tumors. Neurosurgery 6:376–379

Daniels DL, Williams AL, Thornton RS, Meyer GA, Cusick JF, Haughton VM (1981) Differential diagnosis of intrasellar tumors by CT. Radiology 141:697–701

Davis PC, Hoffman JC, Tindall GT, Braun IF (1984) Prolactin-secreting pituitary microadenomas: inaccuracy of high-resolution CT imaging. AJNR 5:721–726

Derome PJ (1982) Les adénomes hypophysaires. Encycl Med Chir 17340 A 10

Derome PJ, Jedynak CP, Peillon F (1980) Pituitary adenomas. Biology, physiopathology and treatment. Asclepios Publishers, Paris

Derome PJ, Peillon F, Bara RM, Jedynak CP, Racadot J, Guiot G (1979) Adénomes à prolactine, résultats du traitement chirurgical. Nouv Presse Méd (Paris) 8:577–583

De Villiers JFK, Bam WV (1984) Diagnosis of pituitary micro-adenomas by computed tomography. South African Medical Journal 57:564–565

Di Chiro G, Nelson KB (1962) The volume of the sella turcica. AJR 87:989–1007

Dietemann JL, Bonneville JF (1985) Radiological Diagnosis of Pituitary Diseases. In: Imura H (ed) The pituitary gland. Raven Press, New York, pp 341–361

Domingue JN, Wing SD, Wilson CB (1978) Coexisting pituitary adenomas and partially empty sella. J Neurosurg 48:23–28

Domingue JN, Richmond IL, Wilson CB (1980) Results or surgery in 114 patients with prolactin-secreting pituitary adenomas. Am J Obst Gynecol 137:102–108

Drayer BP, Kattah J, Rosenbaum A, Kennerdell J, Maroon J (1979) Diagnostic approaches to pituitary adenomas. Neurology 29:161–169

Dubois PJ, Orr DP, Hoy RJ, Herbert DL, Heinz ER (1979) Normal sella variations in frontal tomograms. Radiology 131:105–110

Dupuy M, Derome PJ, Peillon F, Jedynak CP, Visot A, Racadot J, Guiot G (1984) L'adénome à prolactine chez l'homme: étude pré- et post-opératoire de 80 cas. Sem Hôp Paris 60:2943–2954

Earnest FIV, McCullough EC, Frank DA (1981) Fact or artifact: an analysis of artifact in high-resolution CT scanning of the sella. Radiology 140:109–113

Eresue J, Drouillard J, Philippe JC, Guibert JL, Poux P, Tavernier J (1982) L'exploration des adénomes hypophysaires par scanographie à haute résolution et angioscanographie. Ann Radiol 25:509–517

Faglia G, Giovanelli MA, MacLeod RM (1980) Pituitary microadenomas. Academic Press, London

Fargason RD, Jacques S, Rand RW, Shelden CH, McCann GD, Linn P (1981) Visualization and three-dimensional reconstruction of pituitary microadenomas from CT data: a technical report. Surg Neurology 15:450–454

Faria MA, Tindall GT (1982) Transsphenoidal microsurgery for prolactin-secreting pituitary adenomas. J Neurosurg 56:33–43

Flerko B (1980) The hypophysial portal circulation today. Neuroendocrinology 30:56–63

Gardeur D, Metzger J (1982) Adénomes hypophysaires intra-sellaires. In: Tomodensitométrie intra-crânienne. Livre V: Pathologie sellaire. Ellipses, Paris, p 41

Gardeur D, Nachanakian A, Kulesza E, Metzger J (1979) La tomodensitométrie dans les adénomes hypophysaires. Ann Radiol 22:489–499

Gardeur D, Naidich TP, Metzger J (1981) CT analysis of intrasellar pituitary adenomas with emphasis on patterns of contrast enhancement. Neuroradiology 20:241–247

Geehr RB, Allen WE, Rothman SGL (1978) Pluridirectional tomography in the evaluation of pituitary tumors. AJR 130:105–109

Ghoshhajra K (1981) High-resolution metrizamide CT cisternography in sellar and suprasellar abnormalities. J Neurosurg 54:232–239

Godin D, Stevenaert A, Thibault A (1981) Reliability of the CT scan for the diagnosis of microadenoma in a normal sized sella turcica. Neuroradiology 20:261

Gold EB (1981) Epidemiology of pituitary adenomas. Epidemiol Rev 3:163–183

Gonzalez ER (1982) To treat or not to treat: hyperprolactinemia. JAMA 248:513–514

Grisoli F, Vincentelli F, Jaquet P, Guibout M, Hassoun J, Farnarier P (1980) Prolactin secreting adenoma in 22 men. Surg Neurol 13:241–247

Guibout M, Jaquet P, Grisoli F, Tapias PL, Mouly A (1982) Effet antitumoral de la bromocriptine sur cinq prolactinomes envahissants. Ann Med Int 133:48–50

Hardy J, Mohr G (1981) Prolactinoma: surgical aspects. In: Hardy J (ed) Prolactinoma. Masson, Paris New York

Hemminghytt S, Kalkhoff RK, Daniels DL, Williams AL, Grogan JP, Haughton VM (1983) Computed tomographic study of hormone-secreting microadenomas. Radiology 146:65–69

Hershon KS, Kelly WA, Shaw CM, Schwartz R, Bierman EL (1983) Prolactinomas as part of the multiple endocrine neoplastic syndrome type 1. Am J Med 74:713–729

Keye WR, Chang RJ, Monroe SE, Wilson CB, Jaffe RB (1979) Prolactin-secreting pituitary adenomas in women. Am J Obstet Gynecol 134:360

Keye WR Jr, Chang RJ, Wilson CB, Jaffe RB (1980) Prolactin-secreting pituitary adenomas. III. Frequency and diagnosis in amenorrhea-galactorrhea. JAMA 244:1329–1332

Kishmore PRS, Kaufman AB, Melichar FA (1979) Intrasellar carotid anastomosis simulating pituitary microadenoma. Radiology 132:381–383

Kleinberg DL, Noel GL, Frantz AG (1977) Galactorrhéa: a study of 235 cases including 48 with pituitary tumors. N Engl J Med 296:589–600

Klibanski A, Neer RM, Beitins IZ, Ridgway EC, Zervas NT, McArthur JW (1980) Decreased bone density in hyperprolactinaemic women. N Engl J Med 303:1511–1514

Kovacs K, Ryan N, Howath E, Singer W, Exrin C (1980) Pituitary adenomas in old age. J Gerontol 35:16–22

Kricheff II (1979) The radiologic diagnosis of pituitary adenoma. Radiology 131:263–265

Kuuliala I (1981) CT of pituitary adenomas. Clinical Radiology 32:259–264

Lemaître G, Linquette M, Fossati P, Cappoen JR (1982) Détection des adénomes hypophysaires sécrétants par tomodensitométrie. Ann Med Int 133:33–34

Linfoot JA (1979) Recent advances in the diagnosis and treatment of pituitary tumors. Raven Press, New York

Linquette M, Buvat J, Gauthier A, Gasnault JP, Pagniez T, Decoulx M, Laine E (1977) Apoplexie hypophysaire révélatrice d'un adénome à cellules prolactiniques au cours d'une grossesse permise par la bromocriptine. Nouv Presse Med 6:3525–3531

Linquette M, Fossati P (1980) Les adénomes hypophysaires sécrétants. Rev Prat, Paris 30:3017–3040

Lipman JK, Marshal W (1982) Practical errors in measurement of the pituitary at CT (letter). AJNR 3:87

Lundberg PO, Osterman PO, Wide L (1981) Serum prolactin in patients with hypothalamus and pituitary disorders. J Neurosurg 55:194–199

March CM, Kletzky OA, Davajan V, Teal J, Weiss M, Apuzzo MLJ, Marrs RP, Mishell DR Jr (1981) Longitudinal evaluation of patients with untreated prolactin-secreting pituitary adenomas. Am J Obstet Gynecol 139:835–839

Marrs RP, Kletzky DA, Teal J, Davajan V, March C, Mishell DR Jr (1979) Comparison of serum prolactin, plain radiography and hypocycloidal tomography of the sella turcica in patients with galactorrhea. Am J Obstet Gynecol 135:467–469

Mass S, Norman D, Newton TH (1978) Coronal computed tomography: indications and accurancy. AJR 131:875–879

McGregor A, Scanlon MF, Hall R (1982) Les adénomes hypophysaires à prolactine. Explorations et traitements. Ann Med Int 133:51–57

Mealna S, Barbuti D, Borrelli P, Fariello G, Carnevale E, Parri C, Gugiannini P (1981) Present value of conventional X-ray examination in pituitary microadenomas in childhood: relation with CT. Ann Radiol (Paris) 24:109–115

Metzger J, Gardeur D, Houlbert D, Thibierge M (1982) Surveillance neuro-radiologique des adénomes hypophysaires sécrétants. Ann Med Int 133:29–32

Mindel JS, Sachdev VP, Kline LB, Sivak MA, Bergman DA, Yang WC, Choi IS, Huang YP (1983) Bilateral intracavernous carotid aneurysm mimicking a prolactin secreting pituitary tumor. Surg Neurol 19:163–167

Mohr G, Hardy J (1982) Hemorrhage, necrosis, and apoplexy in pituitary adenomas. Surg Neurol 18:181–189

Molitch ME, Schwartz J, Mukherji B (1981) Is prolactin secreted ectopically? Am J Med 70:803–807

Moore TJ (1979) Prolactinomas – our raised consciousness. Arch Inter Med 139:1223–1224

Muhr C, Bergström K, Grimelius L, Larsson SG (1981) A parallel study of the roentgen anatomy of the sella turcica and the histopathology of the pituitary gland in 205 autopsy specimens. Neuroradiology 21:55–65

Nachtigall RD, Monroe SE, Wilson CG, Jaffe RB (1981) Prolactin-secreting pituitary adenomas in women. VI. Absence of demonstrated adenomas in patients with altered menstrual function and abnormal sellar polytomography. Am J Obstet Gynecol 140:303–308

Nagagawa Y, Matsumoto K, Fukami T, Takase K (1984) Exploration of the pituitary stalk and gland by high-resolution computed tomography. Comparative study of normal subjects and cases with microadenoma. Neuroradiology 26:473–478

Naidich TP, Pinto RS, Kushner MJ, Lin JP, Kricheff II, Leeds NE, Chase NE (1976) Evaluation of sellar and parasellar masses by computed tomography. Radiology 120:91–99

Peyster RG, Hoover ED (1984) CT of the abnormal pituitary stalk. AJNR 5:49–52

Pituitary adenoma study group (1982) Variation in assessing sella turcica tomograms for pituitary microadenomas. Am J Obstet Gynecol 60:700–704

Pituitary adenoma study group (1983) Pituitary adenomas and oral contraceptives: a multicenter case-control study. Fertility and Sterility 39:753–759

Popa G, Fielding U (1930) A portal circulation from the pituitary to the hypothalamic region. J Anat 65:88–91

Powers SK, Wilson CB (1981) Simultaneously occurring prolactinomas. Case report. J Neurosurg 55:124–126

Prosser PR (1979) Prolactin secreting pituitary adenoma in multiple endocrine adenomatosis type I. Ann Intern Med 91:41–44

Racadot J (1980) Adénomes de l'antéhypophyse: histoire naturelle, classification et histopathologie. Rev Prat (Paris) 30:2981–3002

Raji MR, Kishore PRS, Becker AP (1981) Pituitary microadenoma: a radiological surgical correlative study. Radiology 139:95–99

Reichlin S (1979) The prolactinoma problem. N Engl J Med 300:313–315

Rhoton AL Jr, Hardy DG, Chambers SM (1979) Microsurgical anatomy and dissection of the sphenoid bone, cavernous sinus and sellar region. Surg Neurol 12:63–104

Rhoton AL, Harris FS, Renn WH (1977) Microsurgical anatomy of the sellar region and cavernous sinus. Clin Neurosurg 24:54–85

Richmond IL, Newton TH, Wilson CB (1980) Indications for angiography in the preoperative evaluation of patients with prolactin secreting pituitary adenomas. J Neurosurg 52:378–380

Richmond IL, Newton TH, Wilson CB (1980) Prolactin-secreting pituitary adenomas: correlation of radiographic and surgical findings. AJR 134:707–710

Richmond IL, Wilson CB (1978) Pituitary adenomas in childhood and adolescence. J Neurosurg 49:163–168

Rilliet B, Mohr G, Robert F, Hardy J (1981) Calcifications in pituitary adenomas. Surg Neurol 15:249–255

Robertson HJ, Rose A, Ehmi B, England G, Meriweather R (1981) Trends in the radiological study of pituitary adenoma. Neuroradiology 21:75–78

Robertson WD, Newton TH (1978) Radiologic assessment of pituitary microadenomas. AJR 131:489–492

Roppolo HMN, Latchaw RE (1983) Normal pituitary gland: 2. Macroscopic anatomy CT correlation. AJNR 4:937–944

Ross LS (1981) Routine hormonal screening of infertile men: is it worthwhile? J Urol 126:756–758

Sakoda K, Mukada K, Yonezawa M, Matsumura S, Yoshimoto H, Mori S, Uozumi T (1981) CT scan of pituitary adenomas. Neuroradiology 20:249–253

Segal S, Yaffe H, Laufer N, Ben David M (1979) Male hyperprolactinemia: effects on fertility. Fertil Steril 32:556

Sherman BM, Schlechte J, Halmi N, Chapler FK, Harris CE, Duello TM, Van Gilder J, Granner DK (1978)

Pathogenesis of prolactin-secreting pituitary adenomas. Lancet 2:1019–1021

Shucart WA (1980) Complications of very high serum prolactin levels associated with pituitary tumors. J Neurosurg 52:226–228

Spark RF, O'Reilly G, Wills CA, Ransil BJ, Bergland R (1982) Hyperprolactinemia in males with and without pituitary macroadenomas. Lancet 2:129–131

Speroff L, Levin RM, Haning RV Jr, Kase NG (1979) A practical approach for the evaluation of women with abnormal polytomography or elevated prolactin levels. Am J Obstet Gynecol 135:896–906

Swanson HA, Du Boulay G (1975) Borderline variants of the normal pituitary fossa. Br J Radiol 48:366–369

Swanson JA, Sherman BM, Van Gilder JC, Chapler FK (1979) Coexistent empty sella and prolactin-secreting microadenoma. Am J Obstet Gynecol 53:258

Swartz JD, Russel KB, Basile BA, O'Donnell PC, Popky GL (1983) High-resolution CT appearance of the intrasellar contents in women of childbearing age. Radiology 147:115–117

Syvertsen A, Haughton VM, Williams AL, Cusick JF (1979) The computed tomographic appearance of the normal pituitary gland and pituitary microadenomas. Radiology 133:385–391

Taylor CR, Jaffe CC (1983) Methodological problems in clinical radiology research: pituitary microadenoma detection as paradigm. Radiology 147:279–283

Taylor S (1982) High resolution computed tomography of the sella. Radiologic clinics of North America 20:207–236

Teasdale E, McPherson P, Teasdale G (1981) The reliability of radiology in detecting prolactin-secreting pituitary microadenomas. Br J Radiol 54:566–571

Tenner MS, Weitzner I Jr (1980) Pitfalls in the diagnosis of erosive changes in expanding lesions of the pituitary fossa. Radiology 137:393–396

Thibaut A, Rausin L, Stenevaert A (1981) Correlative study of the sella turcica with the site and size of 82 secreting microadenomas. Neuroradiology 20:281–285

Tindall GT, McLanahan CS (1980) Hyperfunctional pituitary tumors: pre- and postoperative management considerations. Clin Neurosurg 27:48–82

Tourniaire J, Trouillas J, Maillet P, David L, Pallo D, Tran-Minh V, Bressot C (1976) Polyadénomatose endocrinienne associant un adénome hypophysaire à prolactine et un adénome parathyroîdien intra-thyroîdien. Ann Endocrinol 37:465

Tucker HS, Grubb SR, Wigand JP, Taylon A, Lankford HV, Blackard WG, Becker DP (1981) Galactorrhea-amenorrhea syndrome: follow-up of forty-five patients after pituitary tumor removal. Ann Int Med 94:302–307

Turski PA, Newton TH, Horten BH (1981) Sellar contour: anatomic-polytomographic correlation. AJR 137:213–216

Valenta LJ, Sostrin RD, Eisenberg H, Tamkin JA, Elias AN (1982) Diagnosis of pituitary tumors by hormone assays and computerized tomography. Am J Med 72:861–873

Vermeulen C, Bouallouche A, Brerault JL, Cathelineau G (1982) Les hyperprolactinémies pathologiques. II. Evolution et problèmes diagnostiques et thérapeutiques. Sem Hôp Paris 58:2085–2095

Vertosick FT Jr (1985) Role of defective dopaminergic inhibition of prolactin secretion in the pathogenesis of prolactinoma. Neurosurgery 16:261–266

Vezina JL, Sutton TJ (1974) Prolactin-secreting pituitary microadenomas. AJR 120:46–54

Virapongse C, Bhimani S, Sarwar M, Greenberg A, Kim J (1984) Prolactin-secreting pituitary adenomas: CT appearance in diffuse invasion. Radiology 152:447–451

Von Werder K (1983) Prolactinomas. An overview. In: 3rd European Workshop on pituitary adenomas. Amsterdam, p 50

Warner BA, Santen RJ, Page RB (1982) Growth hormone and prolactin secretion by a tumor of the pharyngeal pituitary. Ann Int Med 96:65–66

Wass JA, Moult PJ, Thorner MO (1979) Reduction of pituitary-tumour size in patients with prolactinomas and acromegaly treated with bromocriptine with or without radiotherapy. Lancet 2:66–69

Weiss MH (1981) Medical and surgical management of functional pituitary tumors. Clin Neurosurg 28:374–383

Weiss MH, Teal J, Gott P, Wycoff R, Yadley R, Apuzzo MLJ, Giannotta SL, Kletzky O, March C (1983) Natural history of microprolactinomas: six-year follow-up. Neurosurgery 12:180–183

Wilson CB, Dempsey LC (1978) Transsphenoidal microsurgical removal of 250 pituitary adenomas. J Neurosurg 48:13–22

Wingrawe SJ, Kay CR, Vessey MP (1980) Oral contraceptives and pituitary adenomas. Br Med J 280:685–686

Wolpert SM, Post KD, Biller BJ, Molitch ME (1979) The value of computed tomography in evaluating patients with prolactinomas. Radiology 131:117–119

Wortzman G, Rewcastle NB (1982) Tomographic abnormalities simulating pituitary microadenomas. AJNR 3:505–512

Xuereb GB, Prichard MML, Daniel PM (1954) The hypophysial portal system of vessels in man. Q J Exp Physiol 39:219–230

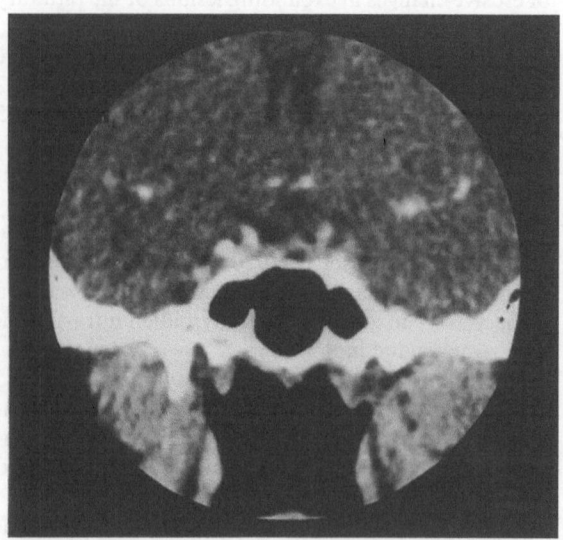

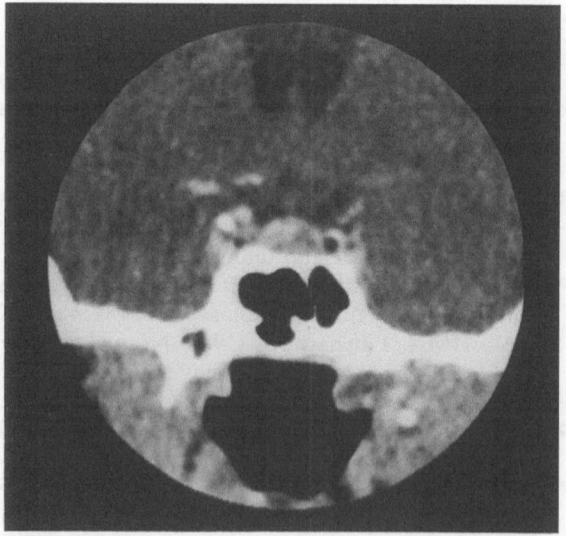

Fig. 5.1. Microprolactinoma. Amenorrhea-galactorrhea in a 20-year-old woman. Prolactinemia is 92 ng/ml. Contrast coronal CT scan: small median defect within a nonenlarged pituitary. Bromocriptine was not well accepted. At surgery, a 3-mm microprolactinoma is found

Fig. 5.2. Pituitary "hyperplasia." Galactorrhea in a 31-year-old woman. Prolactinemia is 30 ng/ml. Contrast coronal CT scan: Upward regular convexity of the superior aspect of the gland. Inhomogeneous enhancement without localized well-defined pituitary adenoma

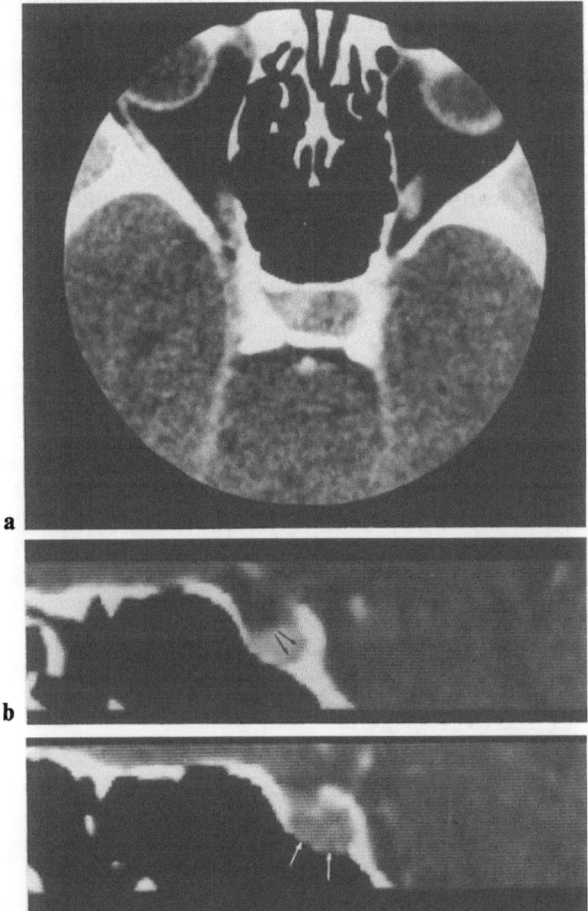

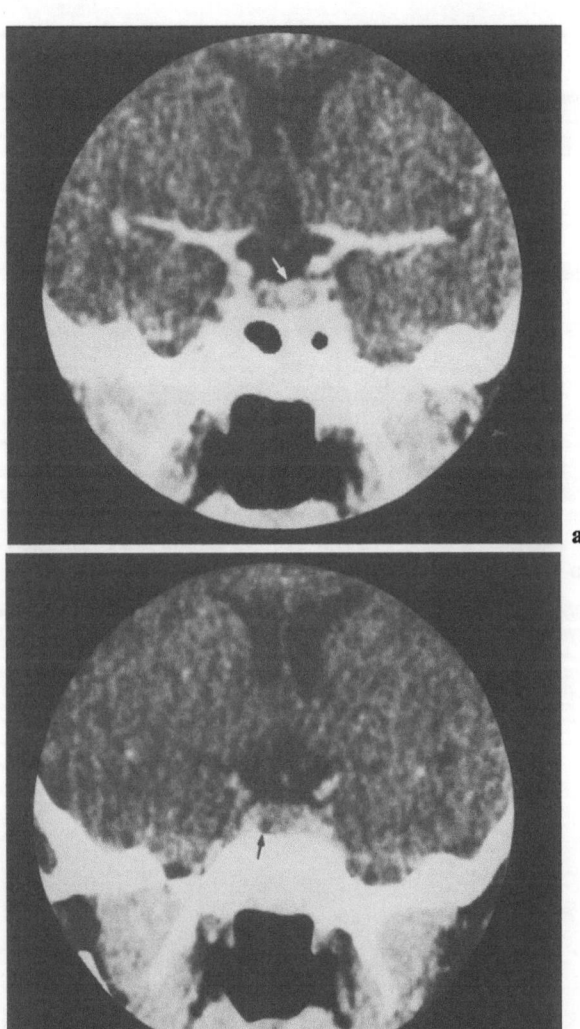

Fig. 5.3 a–c. Prolactinoma. Amenorrhea-galactorrhea in a 38-year-old woman. Hyperprolactinemia (180 ng/ml). **a** Axial contrast CT scan: The left part of the pituitary is less enhanced. Normal enhancement on the right. **b** Midsagittal reformatted image. Normal pattern of the enhanced pituitary except in its most posterior part (*arrows*). **c** Para-midsagittal reformatted image through the left part of the sella. Erosion of the sellar floor (*arrows*). The hypodense prolactinoma is well shown

Fig. 5.4 a, b. Microprolactinoma. Secondary amenorrhea in a 30-year-old woman. Prolactinemia is 140 ng/ml. **a** Dynamic coronal CT scan: Displacement to the left of the secondary capillary bed ("tuft sign") (*arrow*). **b** Contrast coronal CT 1 min later: a 2-mm hypodense rounded lesion is demonstrated on the right (*arrow*). Sellar floor, superior aspect of the gland, and pituitary stalk are not modified

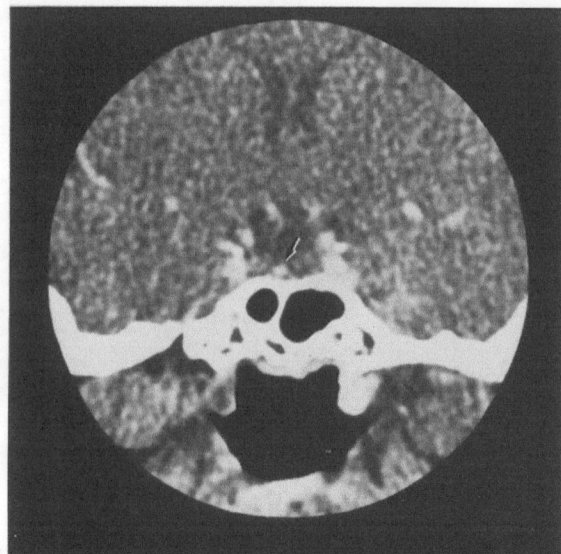

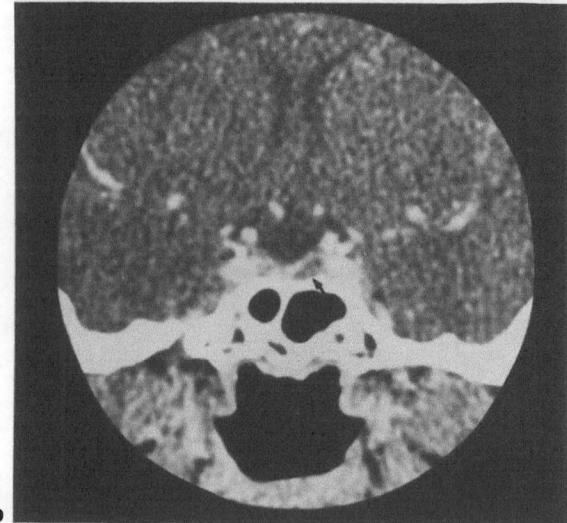

Fig. 5.5a, b. Microprolactinoma. Galactorrhea in a 34-year-old woman. Prolactinemia is only 27 ng/ml. **a** Dynamic coronal CT scan: Discrete displacement to the right of the capillary bed ("tuft sign") (*arrow*). **b** Coronal CT scan: A 3-mm oval hypodense microprolactinoma is demonstrated on the left (*arrow*). No upward bulging of the gland

Fig. 5.6a–d. Prolactinoma with inferior extension. ▶ Amenorrhea and infertility in a 32-year-old woman. Prolactinemia is 125 ng/ml. **a** Sella turcica magnified lateral view: Depression and thinning of the sellar floor (*arrows*). **b** Axial CT scan through the upper part of the sphenoid sinus. A central rounded defect represents the adenoma imprint into the sphenoid sinus (*arrows*). **c** Contrast coronal CT scan: Left-sided 7-mm pituitary adenoma. The upper surface of the gland remains flat. The pituitary stalk is not displaced. **d** Bone algorithm image: Depression and thinning of the sellar floor

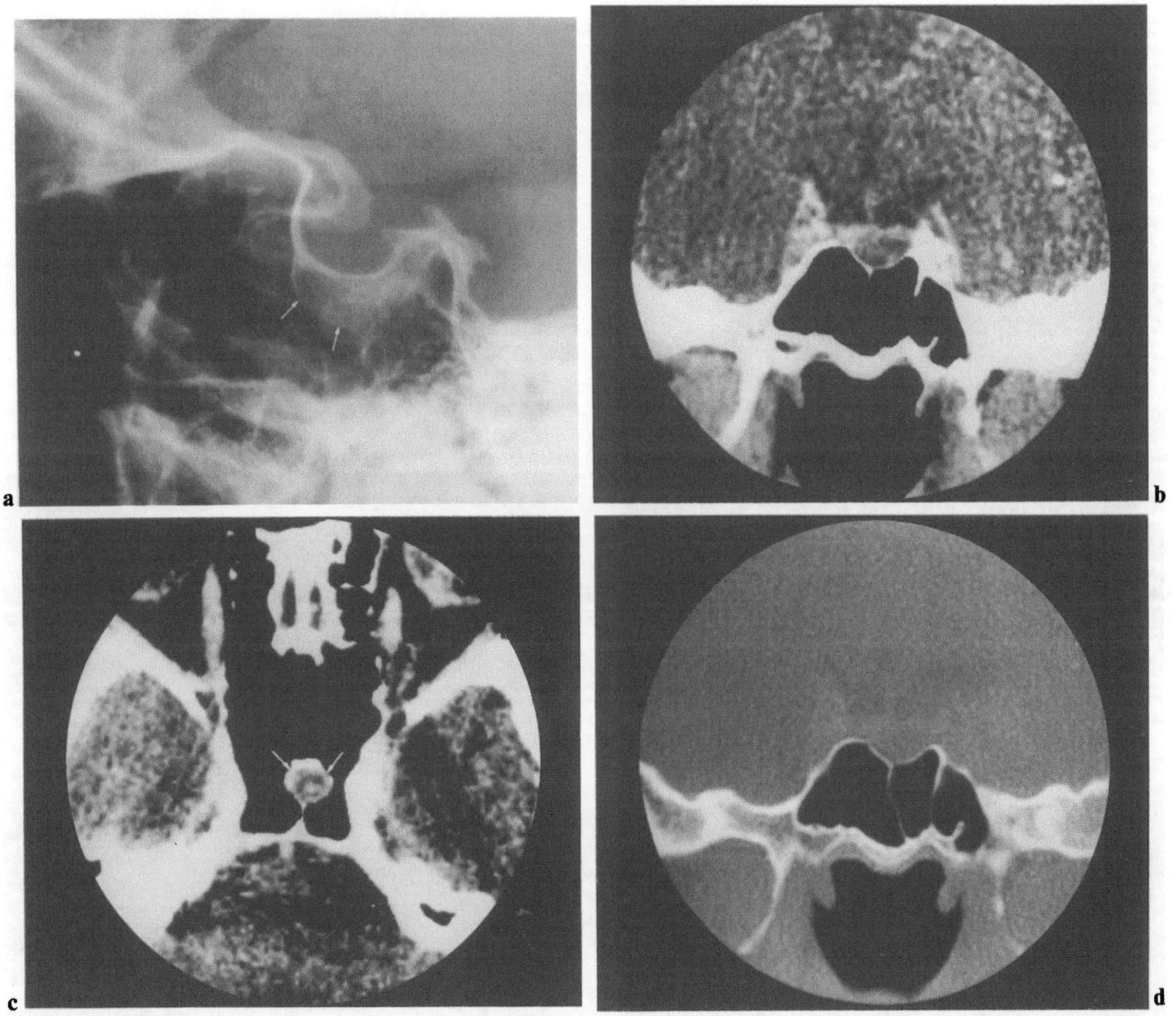

Fig. 5.7. Microprolactinoma. Post-pill amenorrhea and ▶ galactorrhea with anovulation in a 30-year-old woman. Hyperprolactinemia (138 ng/ml). Contrast coronal CT scan. Median 3-mm microadenoma. Slight convexity of the upper surface of the pituitary above the lesion (*arrow*). The sellar floor is normal

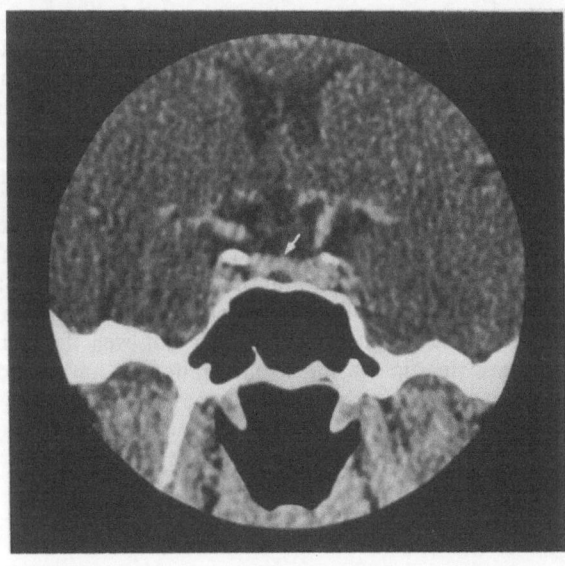

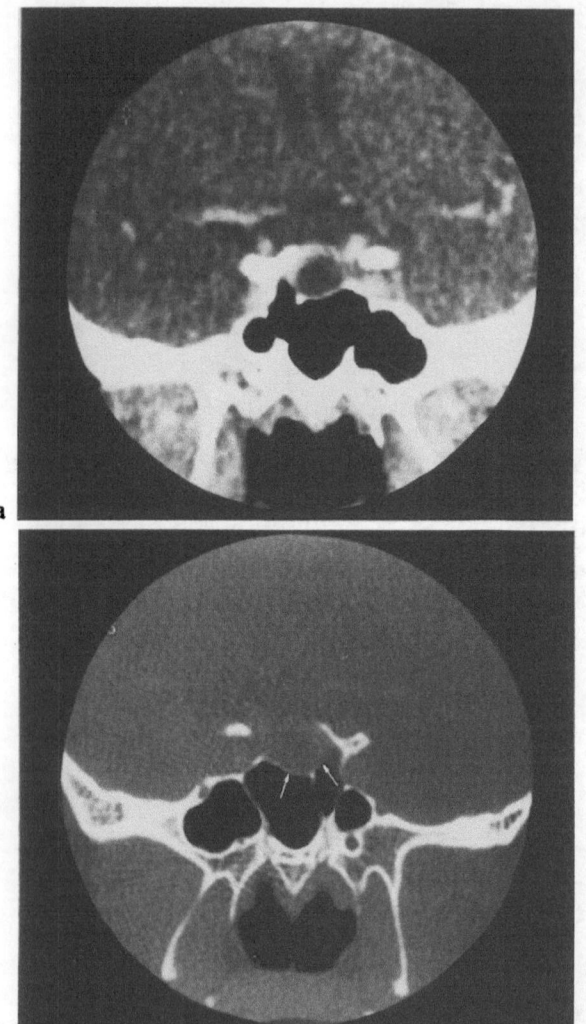

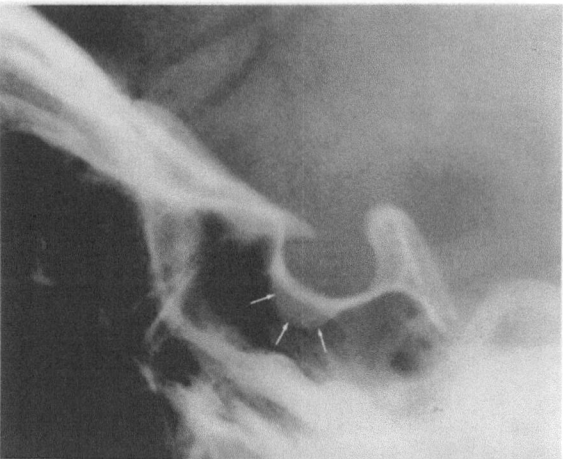

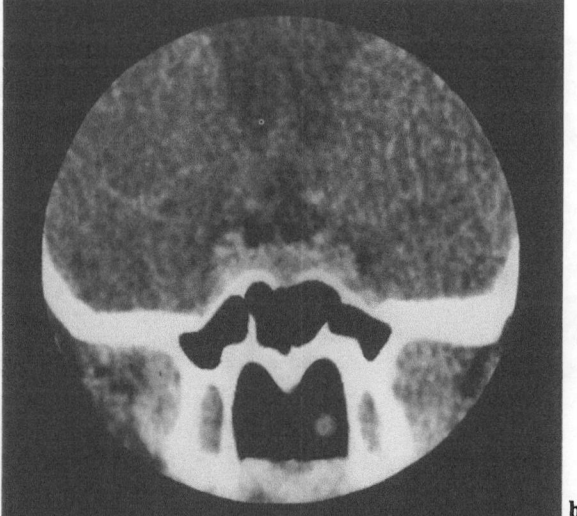

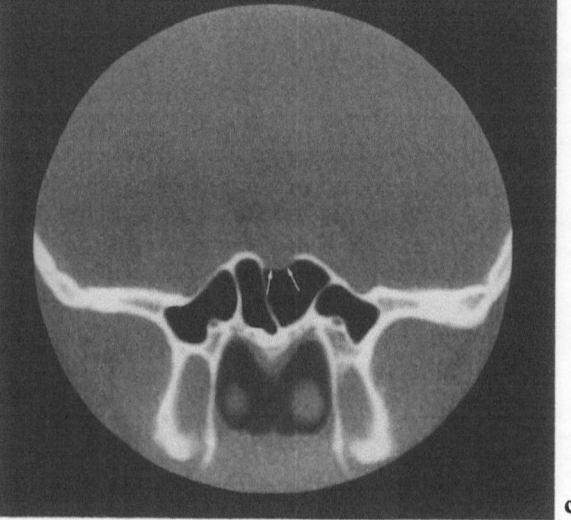

Fig. 5.8a, b. Intrasellar prolactinoma. Amenorrhea-galactorrhea for 3 years in a 30-year-old woman. Prolactinemia was 73 ng/ml. Bromocriptine was given for 2 years. Prolactin is now 23 ng/ml. Regular ovulatory menses. **a** Contrast coronal CT scan. Very hypodense 9-mm pituitary adenoma. Slight bulging of the sellar diaphragm. **b** Bone review image: Depressed and thinned sellar floor (*arrows*)

Fig. 5.9a–c. Microprolactinoma. Oligomenorrhea and ▶ hyperprolactinemia (79 ng/ml) in a 20-year-old woman. **a** Magnified lateral view of the sella. Anteroinferior bulging of the sellar floor (*arrows*). **b** Contrast coronal CT scan. The pituitary adenoma is not well defined. **c** Bone review image for demonstration of an eroded sellar floor is here essential for diagnosis (*arrows*)

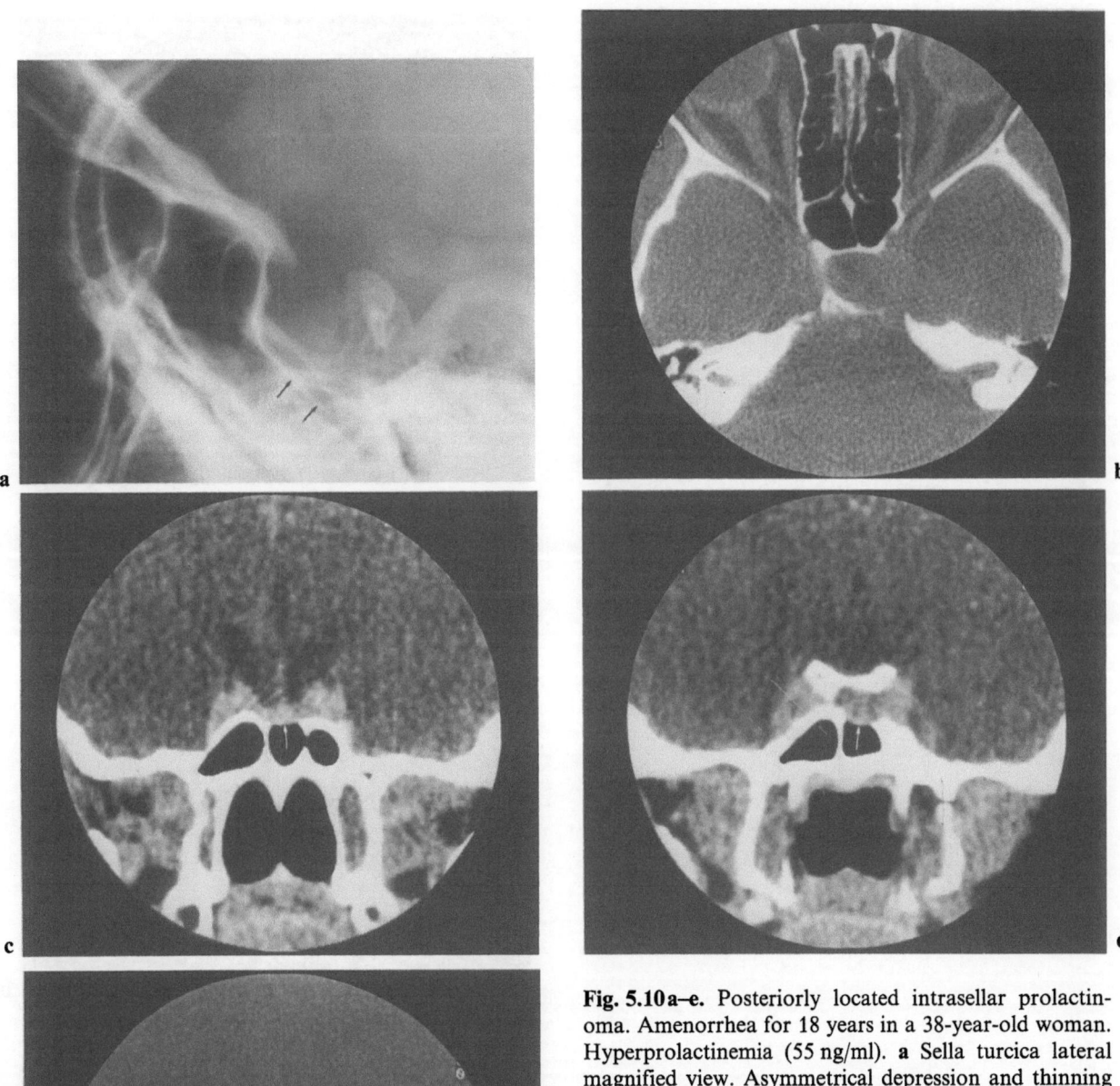

Fig. 5.10a–e. Posteriorly located intrasellar prolactinoma. Amenorrhea for 18 years in a 38-year-old woman. Hyperprolactinemia (55 ng/ml). **a** Sella turcica lateral magnified view. Asymmetrical depression and thinning of the posterior part of the sellar floor (*arrows*). **b** Axial bone review image. Erosion of the left part of the dorsum sellae. **c** Contrast coronal CT scan. Above the midportion of the sellar floor, the pituitary gland appears almost normal. **d** Contrast coronal CT scan just in front of the dorsum. Localized depression and thinning of the left posterior portion of the floor (*arrow*) better shown in **e** (*arrows*). Volume averaging compromises direct demonstration of the adenoma when posterior in location

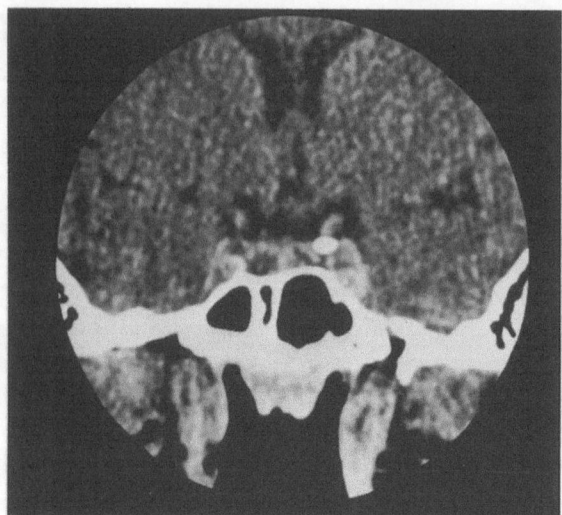

Fig. 5.11. Microprolactinoma. Anovulatory menses and galactorrhea in a 28-year-old woman. Prolactinemia is 33 ng/ml. Native coronal CT scan (GE 9800 CT/T). A rounded right-sided 4-mm pituitary adenoma is well shown. In such cases, contrast injection can be spared

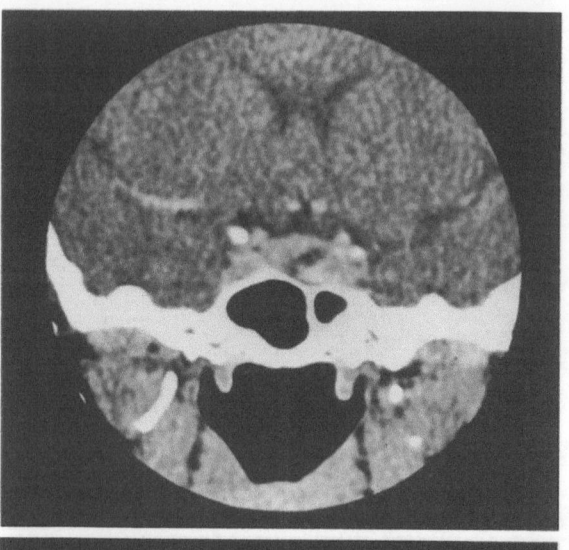

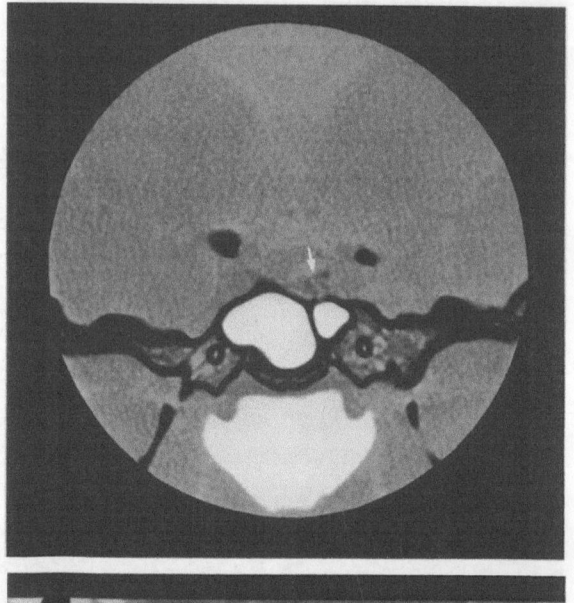

Fig. 5.12a–d. Calcified intrasellar prolactinoma. Amenorrhea and galactorrhea for 5 years in a 31-year-old woman. Prolactinemia is 150 ng/ml. **a** Contrast coronal CT. Depressed sellar floor on the left. Bulging of the sellar diaphragm. **b** Hypodense area representing the adenoma with calcification better demonstrated on bone review image (**b**) (*arrow*) and sagittal reformatted images (**c**, **d**) (*arrow*)

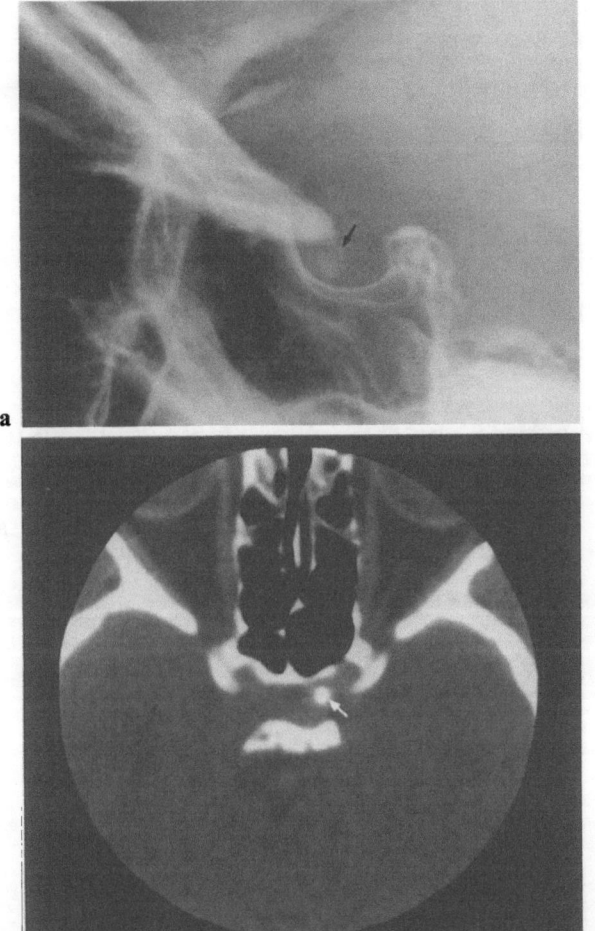

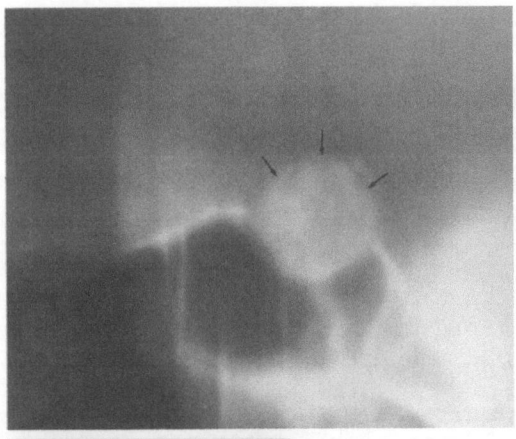

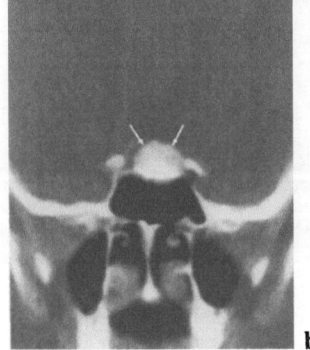

Fig. 5.13a, b. Calcified intrasellar prolactinoma. Oligomenorrhea and galactorrhea in a 30-year-old woman. Prolactinemia is 50 ng/ml. **a** Lateral magnified view of the sella. Intrasellar anterior calcification (*arrow*). **b** Axial CT scan, bone window. The calcified microprolactinoma is found on the left (*arrow*)

Fig. 5.14a, b. Calcified prolactinoma. Galactorrhea and hyperprolactinemia (800 ng/ml) in a 26-year-old woman. **a** Sagittal sellar tomogram. Homogeneous calcification occupying the enlarged sella turcica (*arrows*). **b** Coronal CT scan. The calcification occupies the entire enlarged pituitary fossa (*arrows*). A calcified pituitary adenoma with psammoma bodies was found at surgery (courtesy of CM Carapella, Roma)

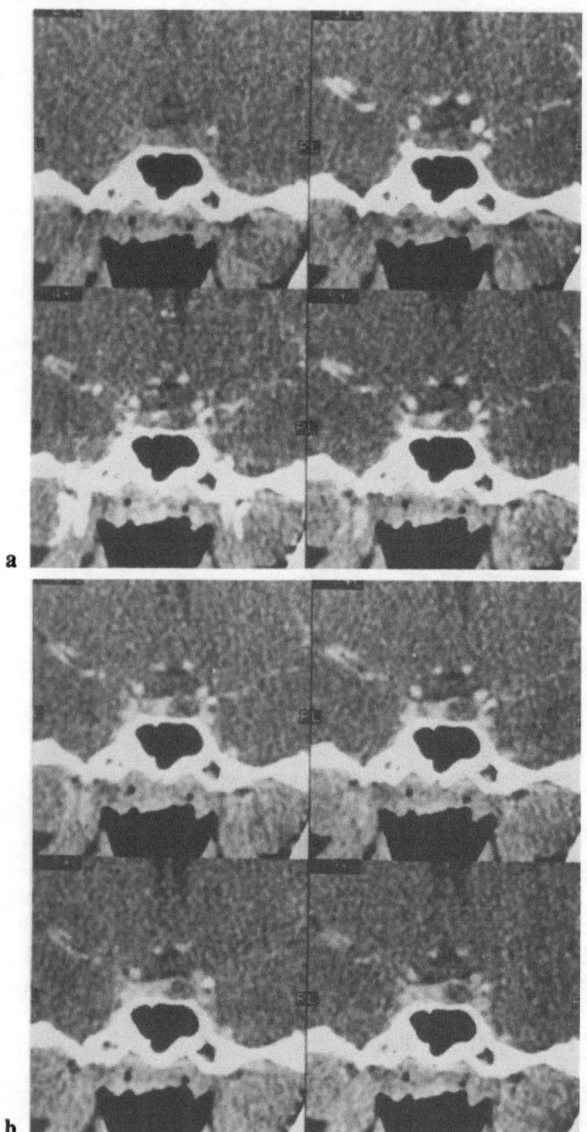

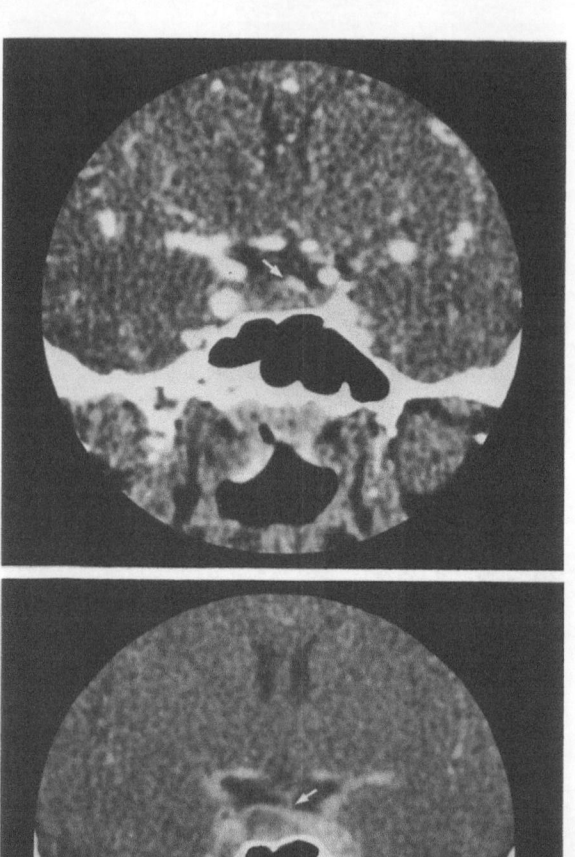

Fig. 5.15a, b. Prolactinoma. Irregular menses and galactorrhea in a 19-year-old woman. Prolactinemia is 100 ng/ml. Dynamic coronal CT. Pituitary enhancement is delayed on one side. Displacement of a compressed capillary bed. An 8-mm intrasellar prolactinoma is well demonstrated

Fig. 5.16a, b. Intrasellar prolactinoma. Amenorrhea and galactorrhea for 4 years in a 38-year-old woman. Prolactinemia is 90 ng/ml. **a** Dynamic CT scan. Displacement to the left of the compressed pituitary bed (*arrow*). **b** Coronal contrast CT scan. Hypodense area on the right representing a 6-mm prolactinoma. Displacement of the pituitary stalk (*arrow*)

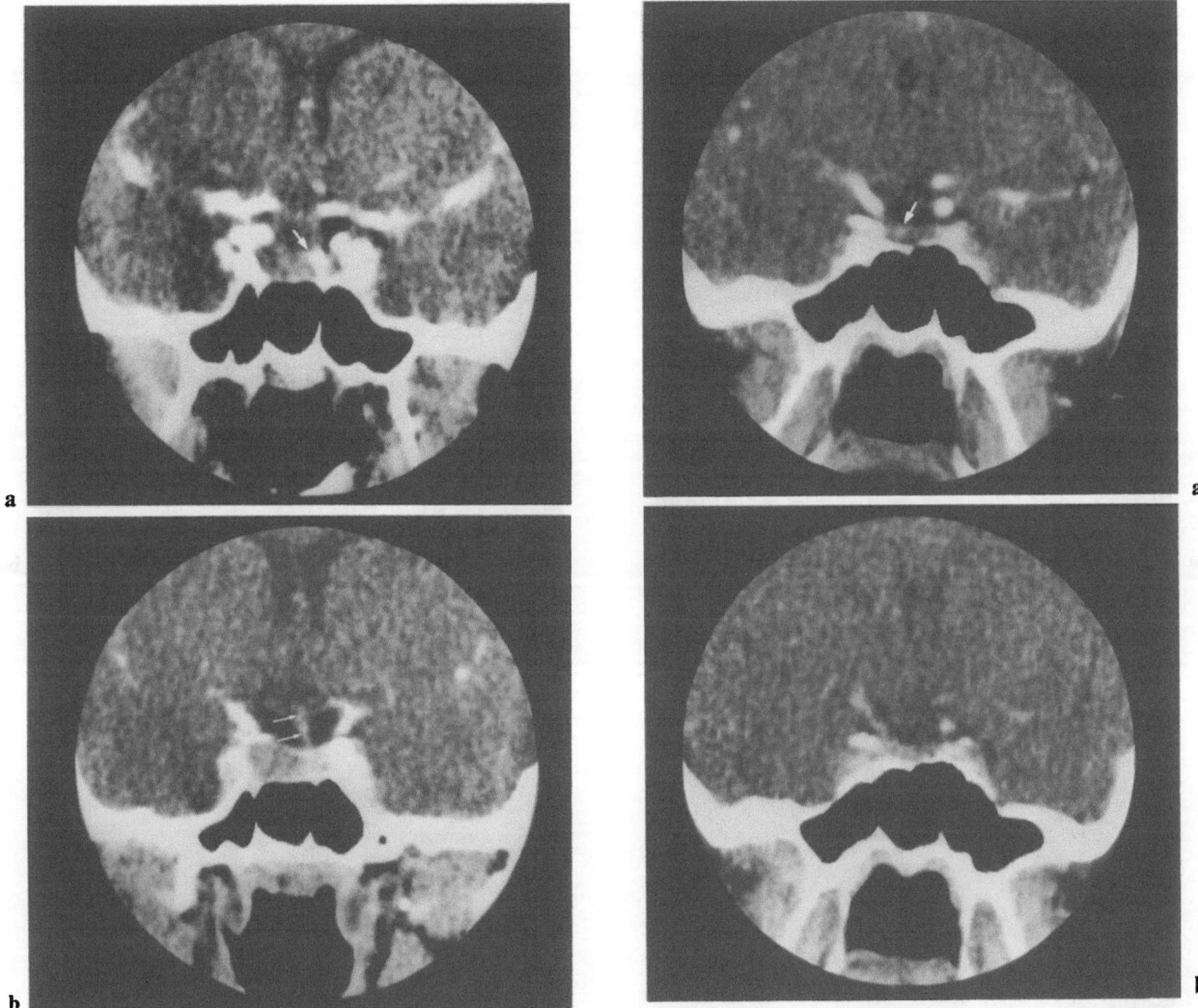

a

b

a

b

Fig. 5.17 a, b. Intrasellar prolactinoma. Amenorrhea and galactorrhea in a 21-year-old woman. Hyperprolactinemia is 310 ng/ml. **a** Dynamic coronal CT. Displacement to the left of the pituitary bed ("tuft sign") (*arrow*). **b** Coronal CT scan. Right-sided less-enhanced pituitary adenoma. The pituitary stalk is also pushed to the left (*arrows*)

Fig. 5.18 a, b. Microprolactinoma. Amenorrhea and galactorrhea in a 29-year-old woman. Prolactinemia is 80 ng/ml. **a** Dynamic CT scan. Displacement to the right of the pituitary tuft (*arrow*). **b** Coronal CT scan 1 min after injection. Questionable low dense area on the left. Dynamic CT is here once more essential

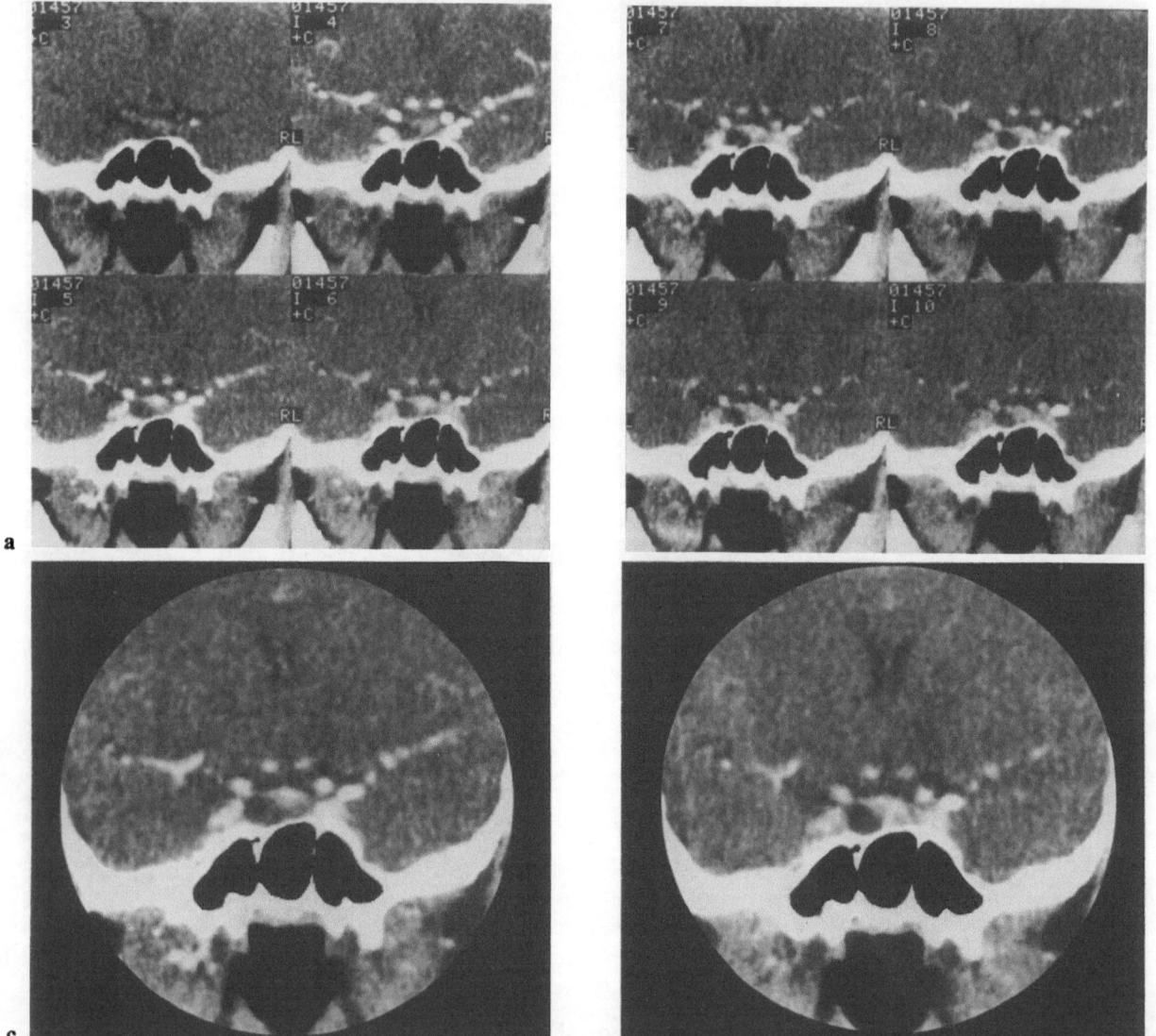

Fig. 5.19a–d. Prolactinoma. Amenorrhea and galactorrhea in an 18-year-old woman. Hyperprolactinemia (150 ng/ml). **a, b** Dynamic CT scan. Before contrast injection (*upper left*), low dense area within the pituitary. After bolus injection, displacement of the capillary bed (*lower left*) and delayed enhancement of the pituitary adenoma. The tuft sign is better demonstrated on a magnified view (**c**) as is the pituitary adenoma (**d**). The sellar floor is slightly depressed. Upward bulging of the sellar diaphragm

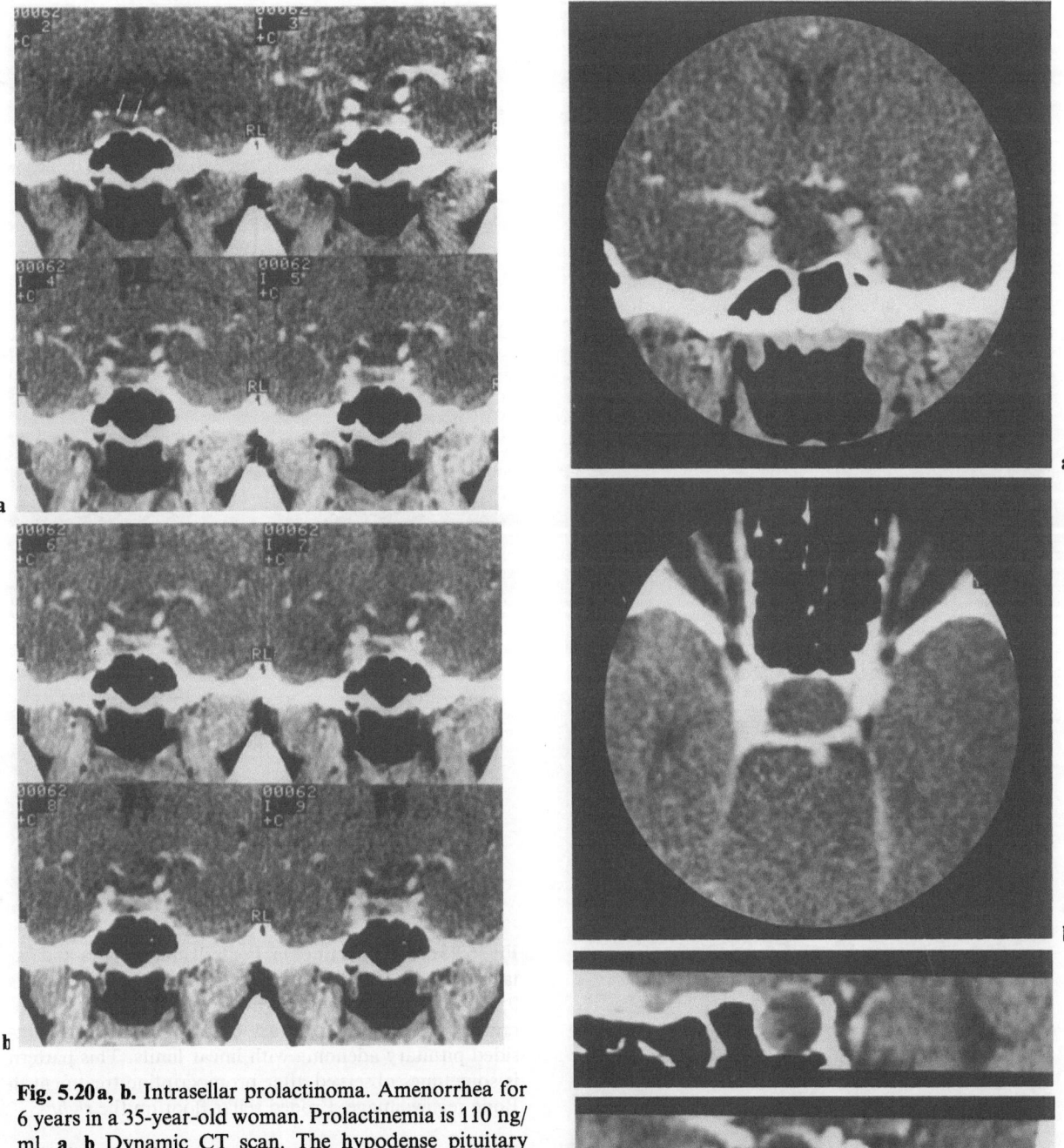

Fig. 5.20a, b. Intrasellar prolactinoma. Amenorrhea for 6 years in a 35-year-old woman. Prolactinemia is 110 ng/ml. **a, b** Dynamic CT scan. The hypodense pituitary microadenoma is blurred by a streak artifact (*arrows*) on the native scan (*upper left*). After bolus injection, delayed enhancement better delineates the prolactinoma

Fig. 5.21a–c. Prolactin-secreting pituitary adenoma. Irregular menses, then postpartum amenorrhea and galactorrhea in a 21-year-old woman. Prolactinemia is 2500 ng/ml. **a** Dynamic CT scan. Large hypodense adenoma with depression of the sellar floor and bulging of the sellar diaphragm. **b** Axial contrast CT. Hypodense lesion occupying the entire enlarged pituitary fossa. **c** Midsagittal and coronal reformatted images

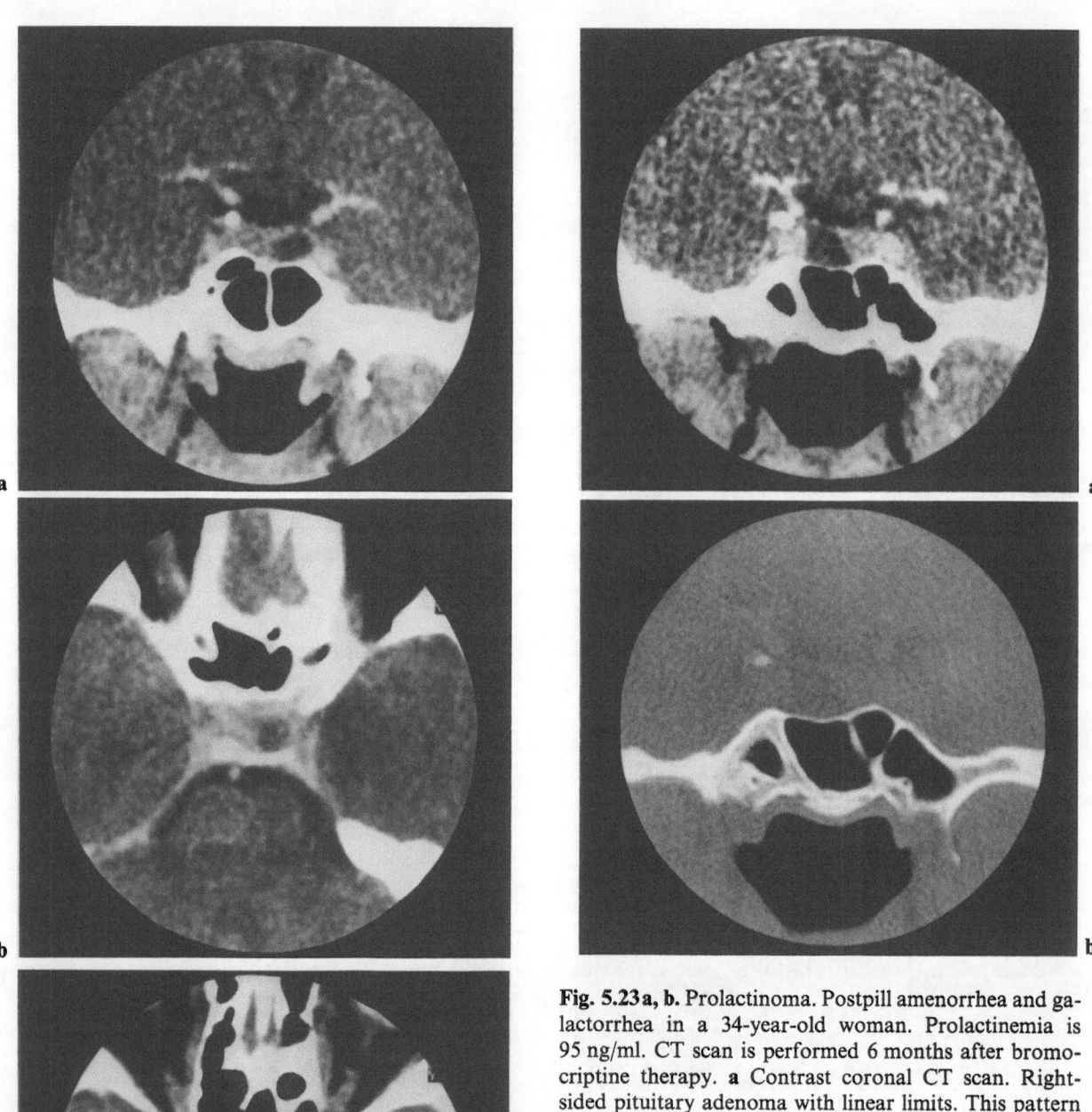

Fig. 5.23a, b. Prolactinoma. Postpill amenorrhea and galactorrhea in a 34-year-old woman. Prolactinemia is 95 ng/ml. CT scan is performed 6 months after bromocriptine therapy. **a** Contrast coronal CT scan. Right-sided pituitary adenoma with linear limits. This pattern is sometimes observed after bromocriptine treatment. **b** Bone review image. Eroded sellar floor on the *right*

◀ **Fig. 5.22a–c.** Prolactinoma. Infertility. Amenorrhea and galactorrhea in a 30-year-old woman. Prolactinemia is 140 ng/ml. **a** Contrast coronal CT scan. Rounded well-defined pituitary adenoma on the left. Discrete erosion of the left part of the sellar floor. The sellar diaphragm is stretched over the tumor. **b** Axial contrast CT. The pituitary adenoma appears here clearly on the left. **c** Axial contrast CT 1.5 mm below **b**. The posterior pituitary is not displaced (*arrows*)

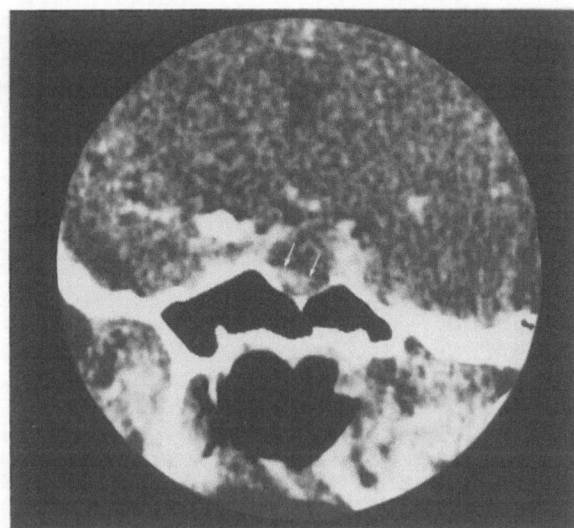

Fig. 5.24. Prolactinoma. Amenorrhea and galactorrhea for 3 years in a 24-year-old woman. Prolactinemia is 124 ng/ml. Contrast coronal CT scan. Inhomogeneous intrasellar adenoma with higher density of the inferior part of the lesion (*arrows*) above an eroded sellar floor

Fig. 5.25a, b. Hyperdense mixed prolactin- and GH-secreting microadenoma. Hyperprolactinemia (52 ng/ml) and elevated GH level (18 ng/ml) in a 20-year-old woman. **a** Contrast-enhanced coronal CT. Rounded midline lesion hyperdense relative to the surrounding pituitary (*arrows*). **b** Bone review image. Focal ballooning and erosion of the sellar floor

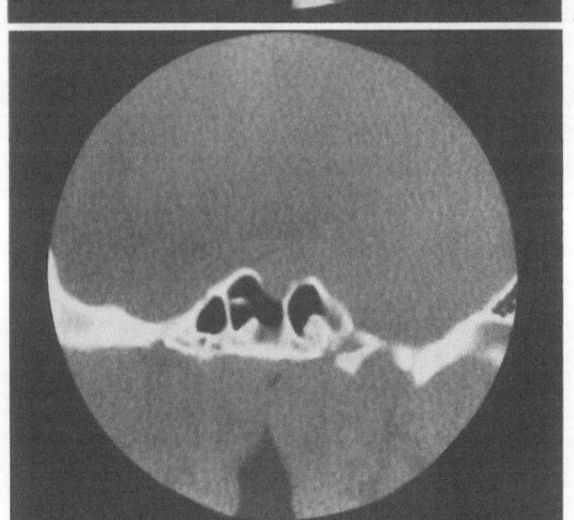

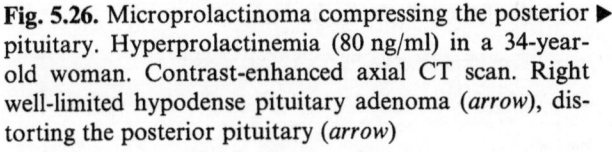

Fig. 5.26. Microprolactinoma compressing the posterior ▶ pituitary. Hyperprolactinemia (80 ng/ml) in a 34-year-old woman. Contrast-enhanced axial CT scan. Right well-limited hypodense pituitary adenoma (*arrow*), distorting the posterior pituitary (*arrow*)

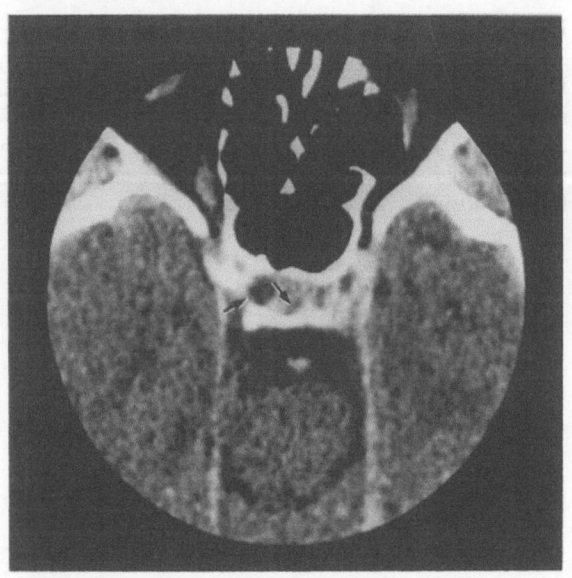

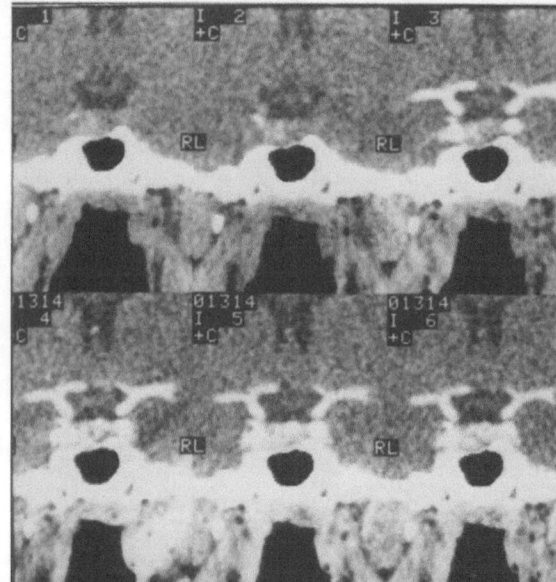

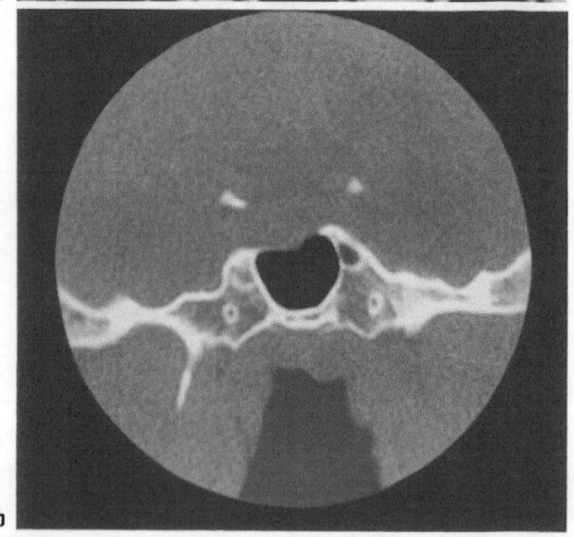

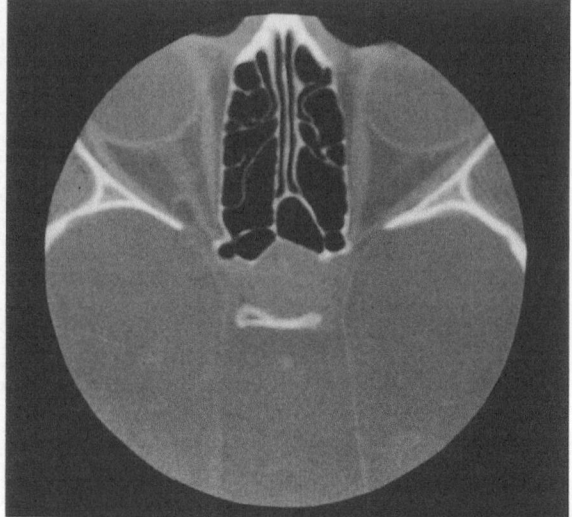

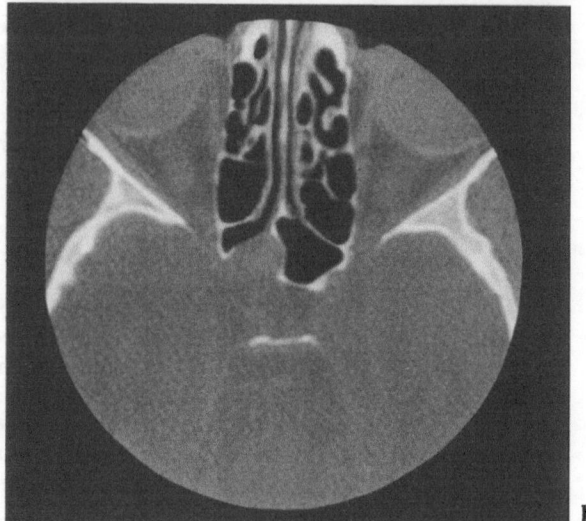

Fig. 5.27a, b. Prolactinoma. Primary amenorrhea in a 23-year-old hyperprolactinemic woman (320 ng/ml). **a** Dynamic coronal CT scan. Slight increase of the pituitary content height above a sloped sellar floor. Mottled appearance of the gland during dynamic CT. **b** Bone review image. Bone changes are in this case essential for diagnosis

Fig. 5.28a, b. Anteriorly located prolactinomas. Bone review axial images. Typical bone changes of the anterior wall of the sella in each case

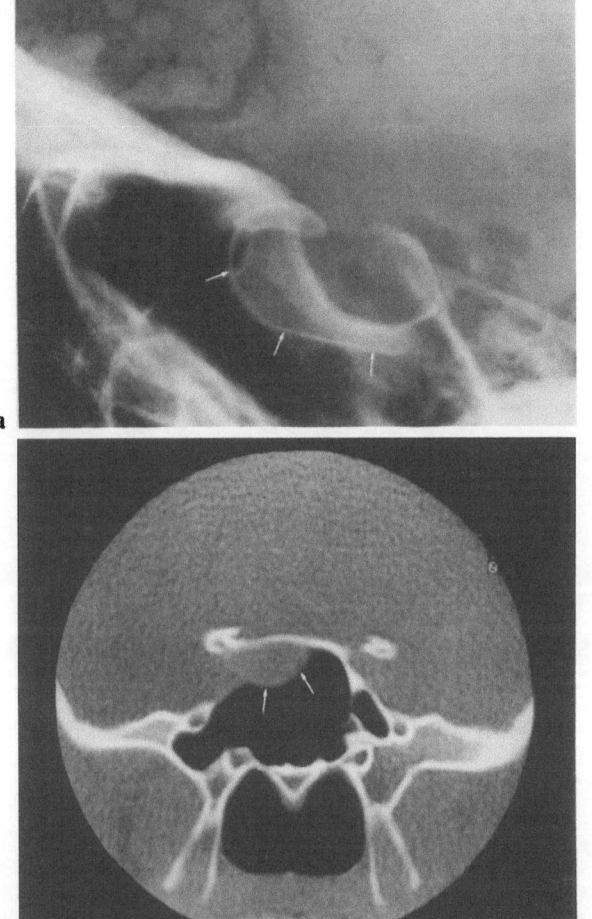

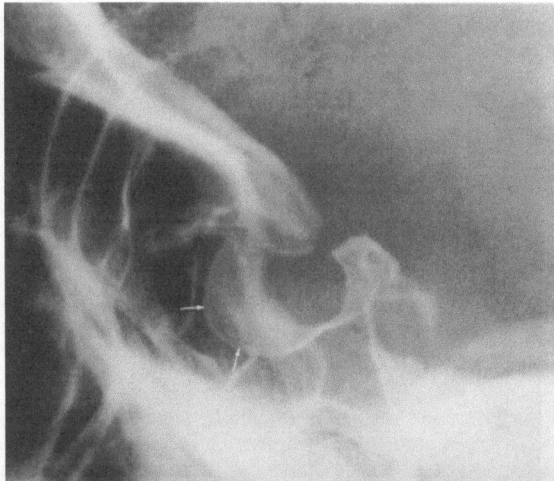

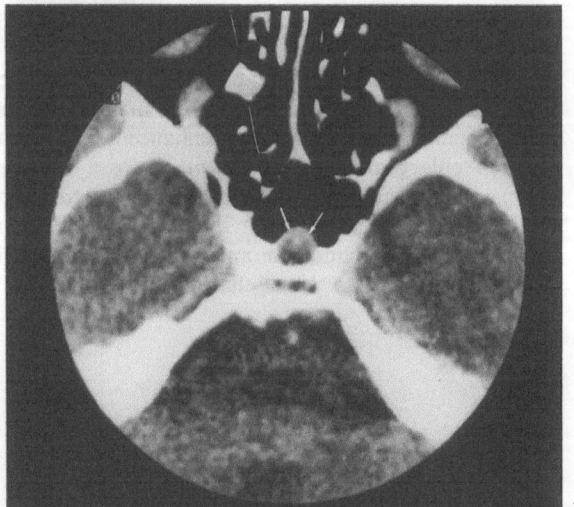

Fig. 5.29 a, b. Anteriorly located prolactinoma. **a** Magnified lateral view. Asymmetric ballooning and thinning of the anterior wall and of the floor of the sella (*arrows*). **b** Bone review image. Anterior extension of the pituitary tumor beneath the planum sphenoidale (*arrows*)

Fig. 5.30 a–c. Prolactinoma with anteroinferior develop- ▶ ment. Amenorrhea for 7 years and spontaneous galactorrhea in a 32-year-old woman. Hyperprolactinemia. **a** Magnified lateral view. Marked bulging of the anterior limit of the sella (*arrows*). **b** Axial contrast CT scan. Imprint of the adenoma within the sphenoid sinus (*arrows*). **c** Contrast coronal CT scan. Thinning of the midpart of the sellar floor (*arrow*). The adenoma itself is poorly defined

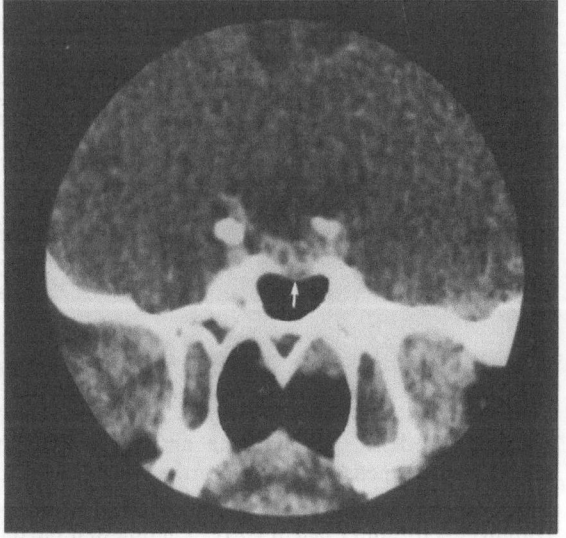

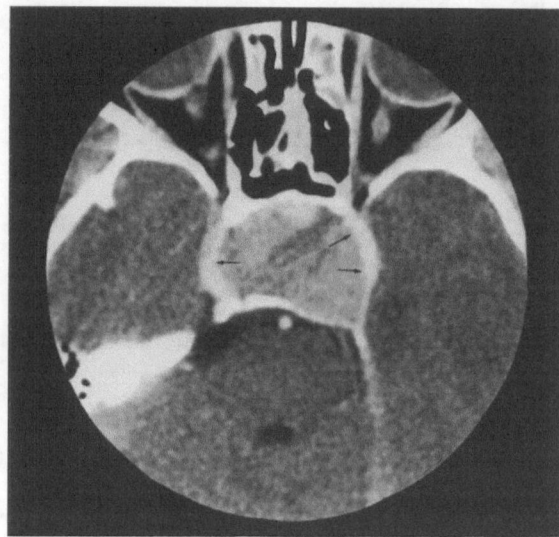

Fig. 5.31. Holosellar prolactinoma with bilateral extension. Sexual deficiency in a 40-year-old man with hyperprolactinemia. Axial contrast-enhanced CT scan. Marked enlargement of the pituitary fossa. The adenoma pushes the carotid arteries laterally (*arrows*), but the cavernous sinuses are not invaded. A small dense nodule is visible behind the anterior wall of the sella. A streak artifact crosses the sellar area

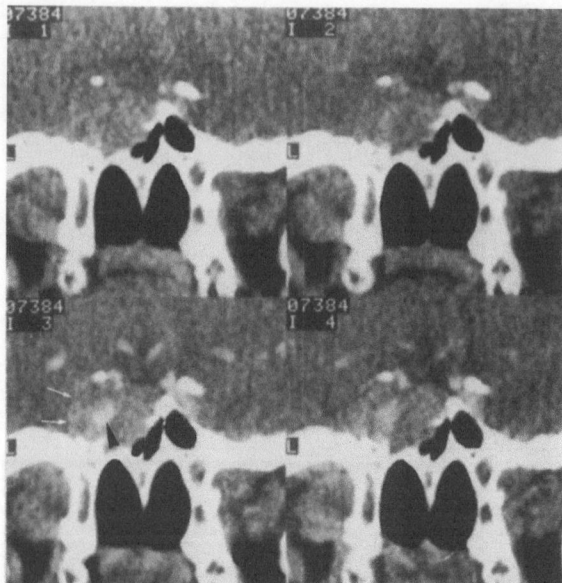

a

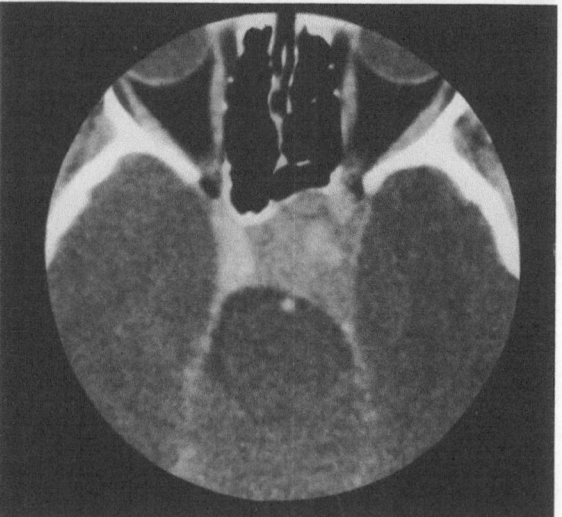

b

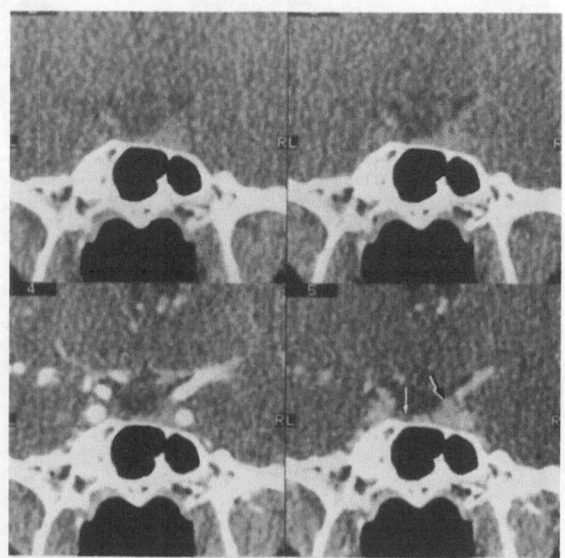

Fig. 5.32. Isodense prolactinoma and partially empty sella. Irregular menses and galactorrhea in a 43-year-old woman. Prolactin level is 120 ng/ml. Dynamic CT scan. Diminished pituitary height corresponding to a partially empty sella (*white arrow*). Isodense adenomatous tissue against the medial aspect of the cavernous sinus (*black* and *white arrow*). The ipsilateral carotid artery is not displaced

Fig. 5.33a, b. Prolactinoma with intracavernous invasion. Infertility and secondary amenorrhea with hyperprolactinemia in a 47-year-old woman. **a** Dynamic CT scan. Eroded sellar floor with soft tissue density within the sphenoid sinus. Abnormal convexity of the lateral aspect of the cavernous sinus (*arrows*). Increased size of the cavernous sinus between its lateral aspect and the carotid artery (*black arrowhead*). Convexity of the sellar diaphragm. Elevation of the ipsilateral clinoid process. **b** Axial contrast CT. Lateral and posterior extent of the tumor surrounding the carotid artery

Chapter 6
Prolactinomas and Dopamine Agonists

For several years dopaminergic agents have been used as a promising alternative to surgical treatment of prolactin-secreting adenomas. In the majority of centers, use of these drugs has led to transsphenoidal removal of adenomas to be relegated to the rare cases of tumors refractory to medical treatment.

In *macroadenomas*, we now know that surgery alone does not allow satisfactory normalization of hyperprolactinemia; it also presents a risk of anterior pituitary insufficiency, particularly when the surgeon is not highly experienced in the technique.

X-ray therapy rarely normalizes hyperprolactinemia; it can contribute to a reduction in tumoral volume in large adenomas, but it also poses the risk of anterior pituitary insufficiency. In view of the limited results of surgery for macroprolactinomas, even when supplemented by X-ray therapy, medical treatment with dopaminergic agents now constitutes a major alternative in the therapeutic arsenal. Demonstration that bromocriptine and other dopamine agonists can very rapidly reduce tumoral volume has greatly restricted the indications for surgery.

Under the influence of skilled neurosurgeons following J. Hardy, selective transsphenoidal removal of *microadenomas* was until recently considered to be the ideal treatment, immediately restoring gonadal function in more than 85% of patients. Nevertheless, in view of its simplicity and safety, bromocriptine treatment is now of interest to many specialists. Recent publications by the J. Hardy group (Serri 1983) showing 50% recurrence of hyperprolactinemia over the years following selective removal of microprolactinomas has been very discouraging for the advocates of surgical treatment. These latter now brandish economic arguments (cost of lifetime treatment with bromocriptine – but must

such treatment be continued indefinitely?) or technical problems (technical difficulties and less good results when surgery is carried out after treatment with dopaminergic agonists – but why operate if medical treatment is satisfactory?). For microprolactinomas, radiologic demonstration of antitumoral efficacy of bromocriptine (Bonneville 1982) has probably contributed to the demise of the surgical approach. The prophecy enunciated by Harvey Cushing in 1926 is now becoming reality: "the surgery of the hypophysis at the present time is practically in the stone age of its development. The time will come, before long, perhaps, when the biochemists will have shown us how to cure most of the common functional adenomas of this gland."

CT Follow-up of Macroprolactinomas Treated with Bromocriptine and Other Dopaminergic Drugs

The antitumoral effect of bromocriptine on prolactin adenomas with suprasellar extension was evaluated in terms of regression of visual field abnormalities (Vaidya), radiologic and tomographic changes in the sellar floor (Sobrinho), and pneumoencephalography or metrizamide cisternography (Corenblum). High-resolution CT scanners now allow precise evaluation of the volume of tumors following medical treatment. Corenblum studied 16 macroadenomas treated with bromocriptine for periods of between 5 and 9 years. In five cases, there was no modification in volume of the adenoma and in 11 there was a reduction in tumoral volume; no tumor increased in size during medical treatment.

CT scans have also allowed clarification of mechanisms involved in shrinkage of prolactinomas; it has been possible to show that the action of bromocriptine was not identical in all types of pituitary adenomas, and that it was thus not a vascular phenomenon. Tumoral infarction due to decreased blood flow is rarely involved: necrosis should yield a CT pattern of a hypodense central zone which is rarely seen in practice (Rengachary). Furthermore, a rapid increase in tumoral volume is most frequently seen upon withdrawal of treatment, and this does not suggest tumoral necrosis (Thorner). Shrinkage of tumors appears to be essentially a function of a decrease in size of prolactin-secreting cells as has been shown in the rat. In the largest macroprolactinomas, inefficacy of bromocriptine for reduction in volume of tumoral tissue outside the sellar region can be accounted for by the fact that this ectopic adenomatous tissue is out of contact with the hypothalamo-pituitary portal system, and is thus not susceptible to the effect of dopamine agonists (Spark). This failure of bromocriptine to reduce the volume of extrasellar tumors is nevertheless not universally seen (Wollesen). With the exception of cases of major frontal or temporal extension, efficacy of bromocriptine to reduce tumoral volume can be demonstrated by CT in 70%–75% of cases (Figs. 6.1, 6.2, 6.3, and 6.4).

These considerations would suggest that a trial of medical treatment with dopamine agonists is today desirable for all macroprolactinomas. The action of bromocriptine is very rapid and clinical and CT signs of decreased tumoral volume can be seen within 2–3 days (Thorner). On the other hand, interruption of treatment can cause a rapid recurrence of signs of chiasmatic compression; 13 days after withdrawal in one case reported by Thorner.

In bromocriptine-treated macroprolactinomas, CT monitoring is obviously essential. Scotti suggested that closely spaced serial controls be carried out initially after 7, 21, 45, and 180 days, and subsequently on an annual basis using rigorous techniques including thin axial and coronal sections, and reformatting. We prefer a less constraining schedule, coupled with evaluation of the visual field every 2 months. Since the purpose of the examination is simply

to control reduction in tumoral volume, in the majority of cases we consider that direct coronal sections or reformatting of axial sections is sufficient. In a few cases, the control examination can be further simplified, excluding use of contrast media. For large adenomas, thin (1–1.5 mm) sections are unnecessary. An acceptable compromise would be 5-mm sections with overlap to facilitate reformatting. The same methods are applicable when the examination is aimed at visualization of effects of bromocriptine on tumoral residues or tumoral recurrences after neurosurgery (Fig. 6.5).

CT Follow-up of Microprolactinomas Treated with Bromocriptine

The antitumoral activity of bromocriptine for microprolactinomas was demonstrated relatively late; indeed, at one time it was considered that the antitumoral action of bromocriptine did not encompass intrasellar adenomas. Development of more rigorous techniques with high-resolution CT machines allowed both the detection of the smallest adenomas and the demonstration that microprolactinomas regress when treated with dopamine agonists (Bonneville). There is here a very good correlation between clinical, laboratory, and radiologic efficacy of medical treatment although antitumoral effects may be of slower onset in microprolactinomas than in macroprolactinomas; in a few cases we have seen positive effects after 6 months' treatment, while a control study carried out 2 months after the beginning of treatment failed to show any noteworthy decrease in tumoral volume. In virtually all cases, maximal reduction in tumoral volume was seen after between 6 months and 1 year of treatment; beyond 1 year, there was generally no further modification in microprolactinoma morphology.

For microprolactinomas we recommend control CT with direct coronal sections, and if possible with dynamic CT after 3 months and 1 year. Subsequently, a CT study every 2 years, sometimes without use of contrast media, is clearly sufficient if clinical and laboratory efficacy is maintained. In general (80%–85% of cases), a reduction in volume of microadenomas is seen less than 3 months after the beginning of medical treatment.

For clearly circumscribed hypodense microadenomas 5–10-mm in diameter, there is a mean 50% reduction in tumoral volume (Figs. 6.6, 6.7, 6.8, and 7.5). This is accompanied by a decrease in height of the gland, usually with disappearance of the bulging of the superior pituitary pole, with a decreased deviation or return to the midline of the pituitary stalk (Fig. 6.6) or the pituitary capillary bed. In a few cases, reduction in volume of microadenomas can be accompanied by a sharp decrease in density of the lesion, but this does not necessarily indicate necrosis of the adenoma (Figs. 6.8 and 6.9). Remineralization of the sellar floor can also be seen (see Chap. 7). For intrasellar adenomas ranging between 5 and 10 mm in size, in a few cases complete disappearance of the adenoma may be seen (Figs. 6.10 and 6.11).

In the uncommon poorly delimited intrasellar adenomas without clear hypodensity, efficacy is generally evaluated on the basis of a simple reduction in vertical height of the gland, with resolution of the bulging of the sellar diaphragm. Nevertheless, in a few cases perfectly defined hypodensities appear within the gland a few weeks after the beginning of treatment with dopaminergic agents; it would appear that bromocriptine here reveals the microadenoma, acting as a sort of "sympathetic ink" (Fig. 6.12).

In the smallest adenomas with initial diameter less than 5 mm, complete radiologic resolution of the lesion with normalization of the morphologic aspect of the pituitary is seen in 50% of cases (Figs. 6.13 and 6.14). When necessary, the disappearance of a hypodense pituitary lesion after bromocriptine allows subsequent confirmation of an initial diagnosis of microprolactinoma.

In a few cases, the remnant of the microadenoma after medical treatment can be constituted by a slight abnormality in progressive enhancement of the pituitary after injection, or by a minimal modification in the normal appearance of the pituitary capillary bed visualized during dynamic CT scan (Fig. 6.12g).

CT and "Nontumoral Hyperprolactinemia"

Is "nontumoral hyperprolactinemia." a real medical entity, or simply a microprolactinoma masked by medical treatment? Numerous hyperprolactinemias are treated with bromocriptine, with CT studies sometimes carried out only several months after the beginning of treatment: in the absence of visualization of a microadenoma, an erroneous diagnosis of nontumoral functional hyperprolactinemia is frequently made. It is in all cases clearly preferable to carry out a CT study before medical treatment, or where this is not done, at least after a sufficient interval of time following treatment with dopamine agonists. If CT is carried out during medical treatment, in an evocative clinical context, some degree of value will be attributed to a very small hypodensity or a discrete localized heterogeneity of the gland. Lesions approximately 1 mm in size can now be visualized with the best CT scanners, and the dynamic scan can also be of interest here, allowing demonstration of a deformed or asymmetrical capillary bed (see Chap. 3). Finally, if a first CT scan is carried out after prolonged treatment with bromocriptine, an absence of correlation between frank osseous deformation and discovery of a small adenoma can be seen, and it should be borne in mind that the volume of the adenoma can increase rapidly if treatment is stopped (see Chap. 11).

CT of Prolactinomas and Bromocriptine Failure

In 10%–15% of patients, bromocriptine does not reduce the tumoral volume. This morphological inefficacy does not necessarily correlate with an absence of clinical and laboratory results. Failure of bromocriptine can be affirmed only 6 months after the beginning of treatment, and in our experience, inefficacy of bromocriptine for reduction in the volume of microprolactinomas has been seen either in very hypodense adenomas with a water-like appearance (probably necrotic) or in apparently diffuse adenomas with poorly defined margins. Nevertheless, some initially very hypodense adenomas show a morphologic improvement following medical treatment (Fig. 6.10).

We have very rarely seen secondary escape phenomena (clinical, laboratory and radiologic evidence) after an initial 6-month period of successful treatment (Fig. 6.15).

CT of Prolactinomas After Withdrawal of Medical Treatment

Withdrawal of medical treatment brings about a recurrence of initial CT signs within a few weeks in 80%–85% of patients, in parallel with a new rise in prolactin levels. Thus, in 10%–15% of cases it appears that we can speak of complete recovery (Nillius), with blood prolactin levels remaining normal and CT scans showing no further evidence of adenoma. In view of these recoveries, a 2–3 months "drug holiday" should be allowed after 2 years continuous treatment with dopamine agonists. Nevertheless, Corenblum (1983), who allowed a drug holiday after only 12 months of bromocriptine, reported a consistent recurrence of hyperprolactinemia. For the majority of authors, if blood prolactin levels are definitively normalized in only 10%–15% of patients, prolactin levels following withdrawal of therapy are virtually always below initial levels in all subjects, and we would thus expect to see a decrease in volume of the lesion in radiologic studies.

Development of pregnancy during treatment requires a modification in the monitoring protocol (see Chap. 7). It would appear that pregnancy alone can bring about a definitive recovery in 10%–15% of microprolactinomas. Overall, prolonged treatment with dopamine agonists and the pregnancy which this allows both appear to benefit the clinical course of microprolactinomas, such that we can expect complete recovery following withdrawal of treatment in almost 20% of patients (Fig. 7.10). Only Corenblum (1983) appears to disagree with this finding.

The effect of dopaminergic agents other than bromocriptine (lisuride, pergolide mesylate) ap-

pears identical, and the only noteworthy differences are in toleration of the various drugs. For example, the longer duration of action of pergolide mesylate allows administration in a single evening dose; in our experience, this attenuates the side effects seen with other dopamine agonists which must be administered in two daily doses due to a shorter duration of action, in particular over the first few days of treatment. In case of clinical failure or severe intolerance, the availability of several dopamine agonists improves probability of succes in medical treatment of prolactinomas.

References

Arlot-Debry S, Jaffiol C (1983) La bromocriptine en endocrinologie. Conc Med 105:1713–1727

Aronoff SL, Daughaday W, Laws ER (1979) Bromocriptine – treatment of prolactinomas. N Engl J Med 339:1391

Barbieri RL, Ryan KJ (1983) Bromocriptine: endocrine pharmacology and therapeutic applications. Fertil Steril 39:727–741

Barrow DL, Tindall GT, Kovacs K, Thorner MO, Horvath E, Hoffman JC (1984) Clinical and pathological effects of bromocriptine on prolactin-secreting and other pituitary tumors. J Neurosurg 60:1–7

Baskin DS, Wilson CB (1981) Bromocriptine treatment of pituitary adenomas. Neurosurgery 8:7141–7144

Belforte L, Bruno M, Campagnoli C, Fessia L, Massara F, Molinatti GM (1980) Hormone pattern during bromocriptine. Induced pregnancy in hyperprolactinaemic patients. Eur J Obstet Gynec Reprod Biol 10:309–317

Bergh T, Nillius SJ, Wide L (1982) Menstrual function and serum prolactin levels after long-terme bromocriptine treatment of hyperprolactinaemic amenorrhea. Clin Endocrinol 16:587–593

Bonneville JF, Dietemann JL (1981) Radiology of the sella turcica. Springer Verlag, Berlin Heidelberg New York

Bonneville JF, Poulignot D, Cattin F, Couturier M, Mollet E, Dietemann JL (1982) Apport des méthodes nouvelles dans l'exploration morphologique des tumeurs hypophysaires. Ann Endocrinol (Paris) 43:303–308

Bonneville JF, Poulignot D, Cattin F, Couturier M, Mollet E, Dietemann JL (1982) Computed tomographic demonstration of the effects of bromocriptine on pituitary microadenoma size. Radiology 143:451–455

Bonneville JF, Cattin F, Poulignot D, Dietemann JL,

Couturier M (1983) Value of high resolution computerized tomography in the follow-up of prolactinomas treated with bromocriptine. In: Tolis G (ed) Prolactin and Prolactinomas. Raven Press, New York, p 395–402

Bonneville JF, Poulignot D, Coche G, Portha C, Cattin F, Bacha M (1982) Radiological techniques in the diagnosis of microprolactinoma. In: Molinatti GM (ed) A clinical problem: microprolactinoma. Diagnosis and treatment. Excerpta Medica, Amsterdam-Oxford-Princeton, p 57

Bricaire H (1982) Régression sous traitement médical par des adénomes hypophysaires à prolactine. Ann Med Int 133:58

Burry KA, Schiller HS, Mills R, Harris B, Heinrich L (1978) Acute visual loss during pregnancy after bromocriptine-induced ovulation. Obstet Gynecol 52:19–22

Buvat J, Buvat-Herbaut M (1982) Prolactine bromocriptine et fonction gonadique de la femme: données récentes. I. Physiologie de la prolactine, physiopathologie et diagnostic des hyperprolactinémies. J Gyn Obst Biol Repr 11:341–353

Buvat J, Buvat-Herbaut M (1982) Prolactine, bromocriptine et fonction gonadique de la femme: données récentes. II. Traitement des hyperprolactinémies féminines et autres indications de la bromocriptine. J Gyn Obst Biol Repr 11:509–521

Campagnoli C, Belforte L, Massara F, Peris C, Molnatti GM (1981) Partial remission of hyperprolactinaemic amenorrhea after bromocriptine-induced pregnancy. J Endocrinol Invest 4:85–91

Chambers EF, Turski PA, La Masters D, Newton TH (1982) Regions of low density in the contrast-enhanced pituitary gland: normal and pathologic processes. Radiology 144:109–113

Chang JR, Keye WR, Young JR, Wilson CB, Jaffe RB (1977) Detection, evaluation and treatment of pituitary microadenoma in patients with galactorrhea and amenorrhea. Am J Obstet Gynecol 128:350

Chernow B, Buck DR, Early CB, Ray J, O'Brian JT (1982) Rapid shrinkage of a prolactin-secreting pituitary tumor with bromocriptine; CT documentation. AJNR 3:442–443

Chiodini P, Liuzzi A, Cozzi R, Verdi G, Oppizzi G, Dallabonzana D, Spelta B, Silvestrini F, Borghi G, Luccarelli G, Rainer E, Horowski R (1981) Size reduction of macroprolactinomas by bromocriptine or lisuride treatment. J Clin Endocrinol Metab 53:737–743

Corenblum B, Taylor PJ (1978) Bromocriptine in pituitary tumours. Lancet 2:786

Corenblum B, Taylor PJ (1983) Long-term follow-up of hyperprolactinaemic women treated with bromocriptine. Fertil Steril 40:596–598

Corenblum B, Webster BR, Mortimer CB, Ezrin C (1978) Bromocriptine in pituitary tumours. Lancet 2:786

Cowden EA, Thomson JA (1979) Resolution of hyperprolactinemia after bromocriptine-induced pregnancy. Lancet 1:613

Cusick JF, Haughton VH, Hagen TC (1980) Radiological assessment of intrasellar prolactin-secreting tumors. Neurosurgery 6:376–379

Del Pozo E, Gerber L, Hunziker S (1983) Response to bromocriptine therapy in 115 prolactinoma cases. In: Tolis G (ed) Prolactin and prolactinomas. Raven Press, New York, p 403–414

Dietemann JL, Bonneville JF (1985) Radiological Diagnosis of Pituitary Diseases. In: Imura H (ed) The pituitary gland. Raven Press, New York, pp 341–361

Domingue JN, Richmond IL, Wilson CB (1980) Results of surgery in 114 patients with prolactin-secreting pituitary adenomas. Am J Obst Gynecol 137:102–108

Edwards CRW, Feek CM (1981) Prolactinomas: a question of rational treatment (Editorial). Br Med J 283:1561–1562

Eversmann T, Farlbusch R, Rjosk HK, Von Werder K (1979) Persisting suppression of prolactin-secretion after long-term treatment with bromocriptine in patients with prolactinomas. Acta Endocrinol (Copenh) 92:413–427

Faglia G, Nissim M, Giannattasio G, Moriondo P, Travaglini P, Ambrosi B, Bernasconi V, Giovanelli MA, Vaccari V (1982) Medical therapy of prolactinomas: effects of bromocriptine on serum PRL levels, and tumor size and morphology. In: Molinatti GM (ed) A clinical problem: microprolactinoma. Diagnosis and therapy. Excerpta Medica, Amsterdam-Oxford-Princeton

Ferrari C, Mattei A, Rampini P, Benco R, Caccamo A, Zavaglia C, Crosignani PG (1982) Long-term effects of drug treatment on hyperprolactinaemic disorders: a study after discontinuation of bromocriptine and metergoline. In: Molinatti GM (ed) A clinical problem: microprolactinoma. Diagnosis and therapy. Excerpta Medica, Amsterdam-Oxford-Princeton

Francks S, Horrocks PM, Lynch SS, Butt WR, London DR (1981) Treatment of hyperprolactinemia with pergolide mesylate: acute effects and preliminary evaluation of long-term treatment. Lancet 2:659–661

Galvan G, Frick J, Irnberger T (1981) Bromocriptine-induced cystic tumour regression in advanced prolactinomas. Dtsch Med Wochenschr 106:637–642

George SR, Burrow GN, Zinman B, Ezrin C (1979) Regression of pituitary tumors: a possible effect of bromoergocryptine. Am J Med 66:697–702

Gonzalez ER (1982) To treat or not to treat: hyperprolactinemia. JAMA 248:513–514

Grimson BS, Bowman ZL (1983) Rapid decompression of anterior intracranial visual pathways with bromocriptine. Arch Ophthalmol 101:604–606

Guibout M, Jaquet P, Grisoli F, Tapias PL, Mouly A (1982) Effet antitumoral de la bromocriptine sur cinq prolactinomes envahissants. Ann Med Int 133:48–50

Hall K, Johnston DG, Kendall-Taylor P, Patrick D (1983) The effect of stopping long-term bromocriptine therapy upon serum prolactin levels and the size of large prolactinomas. In: 3rd European Workshop on pituitary adenomas. Amsterdam, p 15

Hancock KW, Scott JS, Lamb JT, Gibson RM, Chapman C (1980) Conservative management of pituitary prolactinomas; evidence for bromocriptine-induced regression. Brit J Obstet Gynaecol 87:523–529

Howards P, Woltman SS, MacLeod RM (1980) Rapid regression of pituitary prolactinomas during bromocriptine treatment. J Clin Endocrinol Metab 51:438–445

Isaac AJ (1979) Resolution of hyperprolactinemia after bromocriptine-induced pregnancy. Lancet 1:784–785

Kellett J, Friesen HG (1979) Bromocriptine and pituitary disorders. Ann Intern Med 90:980–982

Kendall-Taylor P, Hall K, Johnson DG, Prescott KWG (1982) Reduction in size of prolactin-secreting tumours in men treated with pergolide. Br Med J 285:465–467

Kleinberg DL, Boyd AE, Wardlaw S, Frantz AG, George A, Bryan N, Hilal S, Greising J, Hamilton D, Seltzer T, Sommers CJ (1983) Pergolide for the treatment of pituitary tumors secreting prolactin or growth hormone. N Engl J Med 22:704–709

Klibanski A, Neer RM, Beitins IZ, Ridgway EC, Zervas NT, McArthur JW (1980) Decreased bone density in hyperprolactinaemic women. N Engl J Med 303:1511–1514

Lancranjan I (1981) The endocrine profile of bromocriptine: its application in endocrine diseases. J Neural Transmission 51:61–82

Landolt AM, Wüthrich R, Fellmann H (1979) Regression of pituitary prolactinoma after treatment with bromocriptine. Lancet 1:1082–1083

Le Pogamp C, Grall JY, Massart C, Ramee A, Toulouse R (1979) Poussée évolutive d'adénome à prolactine au cours d'une grossesse après stérilitée traitée. 2 observations. Nouv Presse Med 8:2009–2011

Linquette M, Buvat J, Gauthier A, Gasnault JP, Pagniez T, Decoulx M, Laine E (1977) Apoplexie hypophysaire révélatrice d'un adénome à cellules prolactiniques au cours d'une grossesse permise par la bromocriptine. Nouv Presse Med 6:3525–3531

March CM, Kletzky OA, Davajan V, Teal J, Weiss M, Apuzzo MLJ, Marrs RP, Mishell DR Jr (1981) Longitudinal evaluation of patients with untreated prolactin-secreting pituitary adenomas. Am J Obstet Gynecol 139:835–839

McGregor AM, Scanlon MF, Hall K, Cook DB, Hall R (1979) Reduction in size of a pituitary tumor by bromocriptine therapy. N Engl J Med 300:291–293

McGregor AM, Scanlon MF, Hall R, Hall K (1979) Effects of bromocriptine on pituitary tumour size. Brit Med J 2:700–703

McGregor AM, Scanlon MF, Hall R (1982) Les adénomes hypophysaires à prolactine. Explorations et traitements. Ann Med Int 133:51–57

Mohr G, Hardy J (1982) Hemorrhage, necrosis, and apoplexy in pituitary adenomas. Surg Neurol 18:181–189

Moore TJ (1979) Prolactinomas – our raised consciousness. Arch Inter Med 139:1223–1224

Mornex R, Orgiazzi J, Hugues B, Gagnaire JP, Claustrat B (1978) Normal pregnancies after treatment of hyperprolactinemia with bromoergocryptine, despite suspected pituitary tumors. J Clin Endocrinol Metab 47:290–295

Nelson PB, Robinson AG, Archer DF, Maroon JC (1978) Symptomatic pituitary tumor enlargement after induced pregnancy: case report. J Neurosurg 49:283–287

Nillius SJ, Bergh T, Lundberg PO, Stahle J, Wide L (1978) Regression of a prolactin-secreting pituitary tumor during long-term treatment with bromocriptine. Fertil Steril 30:710–712

Parkes D (1979) Bromocriptine. N Engl J Med 301:873–878

Pawlikowski M, Kunert-Radek J, Stepien H (1978) Direct antiproliferative effect of dopamine agonist on the anterior pituitary gland in organ culture. J Endocrinol 79:245–246

Pearson KC (1981) Mental disorders from low dose bromocriptine. N Engl J Med 305:173

Peillon F, Cesselin F, Bression D, Zygelman N, Brandl AM, Nousbaum A, Mauborgne A (1979) In vitro effect of dopamine and L-dopa on prolactin and growth hormone release from tumor pituitary adenomas. J Clin Endocrinol Metab 49:737–741

Perryman RL, Rogol AD, Kaiser DL, MacLeod RM, Thorner MO (1981) Pergolide mesylate: its effects on circulating anterior pituitary hormones in man. J Clin Endocrinol Metab 53:772–778

Reichlin S (1979) The prolactinoma problem. N Engl J Med 300:313–315

Rengachary SS, Tomita T, Jefferies BF, Watanabe T (1982) Structural changes in human pituitary tumor after bromocriptine therapy. Neurosurgery 10:242–251

Scotti G, Scialfa G, Pieralli S, Chiodini PG, Spelta B, Dallabonzana D (1982) Macroprolactinomas: CT evaluation of reduction of tumor size after medical treatment. Neuroradiology 23:123–126

Sobrinho LG, Nunes MC, Calhaz-Jorge C, Mauricio JE, Santos MA (1981) Effect of treatment with bromocriptine on the size and activity of prolactin producing pituitary tumours. Acta Endocrinol (Copenh) 96:24–29

Sobrinho LG, Nunes MCR, Santos MA, Mauricio JC (1978) Radiological evidence for regression of prolactinoma after treatment with bromocriptine. Lancet 2:257–258

Spark RF, Baker R, Biewfang DC, Bergland R (1982) Bromocriptine reduces pituitary tumor size and hypersecretion: requiem for pituitary surgery? JAMA 247:311–316

Thorner MO, Flückiger E, Calne DB (1980) Bromocriptine: a clinical and pharmacological review. Raven Press, New York

Thorner MO, Martin VH, Rogol AD, Morris JL, Perryman RL, Conway BP, Howards SS, Wolfman MG, MacLeod RM (1980) Rapid regression of pituitary prolactinomas during bromocriptine treatment. J Clin Endocrinol Metab 51:438–445

Thorner MO, Perryman L, Rogol AD, Conway BP,

MacLeod RM, Login TS, Morris JL (1981) Rapid changes of prolactinoma volume after withdrawal and reinstitution of bromocriptine. J Clin Endocrinol Metab 53:480–483

Tindall GT, Kovacs K, Horvath E, Thorner MO (1982) Human prolactin-producing adenomas and bromocriptine: a histological, immunocytochemical, ultrastructural, and morphometric study. J Clin Endocrinol Metab 55:1178–1183

Vaidya RA, Aloorkar SD, Rege NR, Maskati BT, Jahangir RP, Sheth AR, Pandya SK (1978) Normalization of visual fields following bromocriptine treatment in hyperprolactinemic patients with visual field constriction. Fertil Steril 29:632–636

Van Dalen JW, Greve EL (1977) Rapid deterioration of visual fields during bromocriptine-induced pregnancy in a patient with a pituitary adenoma. Br J Ophthalmol 61:728–733

Van Roon E, Van der Vijver JCM, Gerretsen G, Hekster REM, Wattendorff RA (1981) Rapid regression of a suprasellar extending prolactinoma after bromocriptine treatment during pregnancy. Fertil Steril 36:173–177

Velentzas C, Carras D, Vassilouthis J (1981) Regression of pituitary prolactinoma with bromocriptine administration. JAMA 245:1149–1150

Vertosick FT Jr (1985) Role of defective dopaminergic inhibition of prolactin secretion in the pathogenesis of prolactinoma. Neurosurgery 16:261–266

Völker W, Gehring WG, Berning R, Schmidt RC, Schneider J, Von Zur Mühlen A (1982) Impaired pituitary response to bromocriptine suppression: reversal after bromocriptine plus tamoxifen. Acta Endocrinol 101:491–500

Von Werder K, Brendel C, Eversmann T (1980) Medical therapy of hyperprolactinemia and Cushing's disease associated with pituitary adenomas. In: Faglia G, Giovanelli MA, MacLeod RM (ed) Pituitary microadenomas. Academic Press, New York, p 383–397

Von Werder K, Eversmann T, Fahlbusch R, Rjosk HK (1980) Medical treatment of prolactinomas: persisting suppression after bromocriptine withdrawal. In: Derome PJ, Jedynak CP, Peillon F (eds) Pituitary adenomas – Biology, physiopathology and treatment. 2rd European Workshop. La Pitié Salpétrière – Asclepios Publ

Wass JA, Moult PJ, Thorner MO (1979) Reduction of pituitary-tumor size in patients with prolactinomas and acromegaly treated with bromocriptine with or without radiotherapy. Lancet 2:66–69

Wass JA, Williams J, Charlesworth M (1982) Bromocriptine in management of large pituitary tumours. Br Med J 284:1908–1911

Weiss MH, Wycoff RR, Yadley R, Gott P, Feldon S (1983) Bromocriptine treatment of prolactin-secreting tumors: surgical implications. Neurosurgery 12:640–642

Winkelmann W, Allolio B, Deuss U, Heesen D, Kaulen D, Wilcke O (1983) Persistent normoprolactinaemia after withdrawal of bromocriptine therapy in patients with prolactinomas. In: 3rd European Workshop on pituitary adenomas. Amsterdam, p 16

Wollesen F, Andersen T, Karle A (1982) Size reduction of extrasellar pituitary tumors during bromocriptine treatment. Ann Intern Med 96:281–286

Woodhouse NJY, Khouqueer F, Sieck JO (1981) Prolactinomas and optic nerve compression: disappearance of a suprasellar extension and visual recovery after two week bromocriptine treatment. Horm Res 14:141–147

Yamaji T, Ishibaski M, Kosaka K, Fukushima T, Hori T, Manaka S, Sano K (1981) Pituitary apoplexy in acromegaly during bromocriptine therapy. Acta Endocrinol 98:171–177

Yuen BH (1978) Bromocriptine, pituitary tumours, and pregnancy. Lancet (Letter):1314

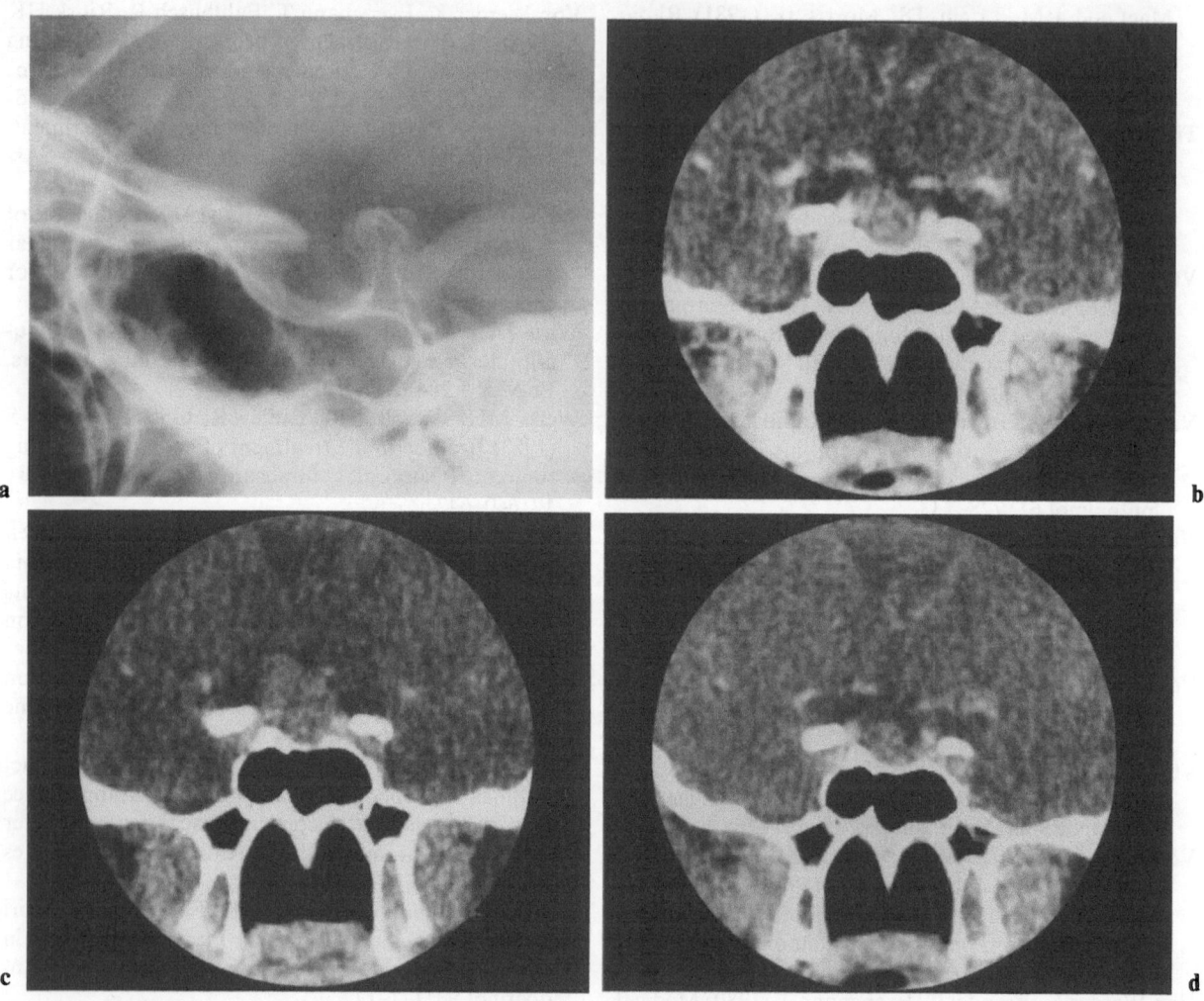

Fig. 6.1 a–d. Prolactin-secreting pituitary adenoma with suprasellar extension. Evolution with medical treatment. Primary amenorrhea and galactorrhea in a 19-year-old woman. Hyperprolactinemia (1330 ng/ml). **a** Sella turcica lateral magnified view: thin and depressed anterior wall and sellar floor. A suprasellar extension cannot be suspected from this image. **b** Contrast coronal CT. Pituitary adenoma with large asymmetrical suprasellar extent. **c** CT follow-up after bromocriptine (15 mg per day for 3 months). Prolactinemia is now 100 ng/ml. Shrinkage of the suprasellar extension. The sellar floor is thicker. **d** CT follow-up 1 year after initial CT scan. Bromocriptine is always given. Further regression of the tumor volume. Prolactinemia is 80 ng/ml and amenorrhea is still present

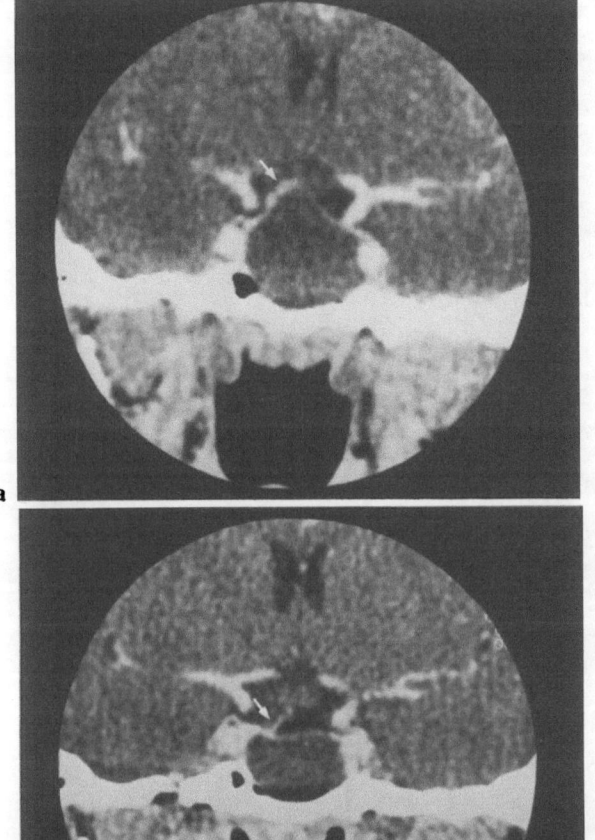

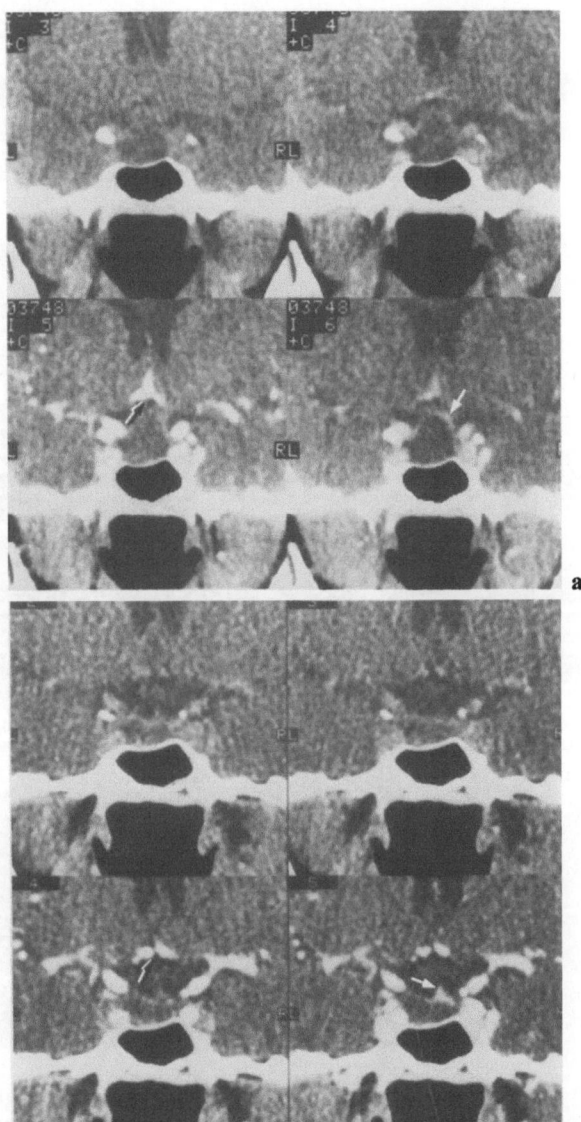

Fig. 6.2 a, b. Prolactin-secreting pituitary adenoma. Evolution with medical treatment. Amenorrhea and hyperprolactinemia in a 22-year-old woman. **a** Dynamic coronal CT scan. Homogeneous hypodense adenoma with suprasellar extension. The suprasellar extension is underlined by a thin rim enhancement and by the compressed pituitary stalk (*arrow*). The sellar floor is no longer visible. **b** Dynamic CT scan follow-up 6 months after pergolide therapy (100 μg per day). Prolactin levels are still elevated. No menses. The suprasellar expansion is no longer present. Persistent deviation of the pituitary stalk (*arrow*)

Fig. 6.3 a, b. Prolactinoma. Evolution with medical treatment. Infertility and hyperprolactinemia (364 ng/ml) in a 30-year-old man. No loss of libido. **a** Dynamic coronal CT scan. Hypodense pituitary adenoma with small suprasellar extension. Rim enhancement (*arrow*). Anterior communicating artery is raised (*black* and *white arrows*). **b** Dynamic coronal CT 7 months after pergolide therapy (100 μg per day). Normoprolactinemia. Important change of the adenoma volume. The anterior cerebral arteries are no longer compressed (*black* and *white arrow*). Persistent deviation of the pituitary stalk (*arrow*). Note that the sellar floor is now thicker

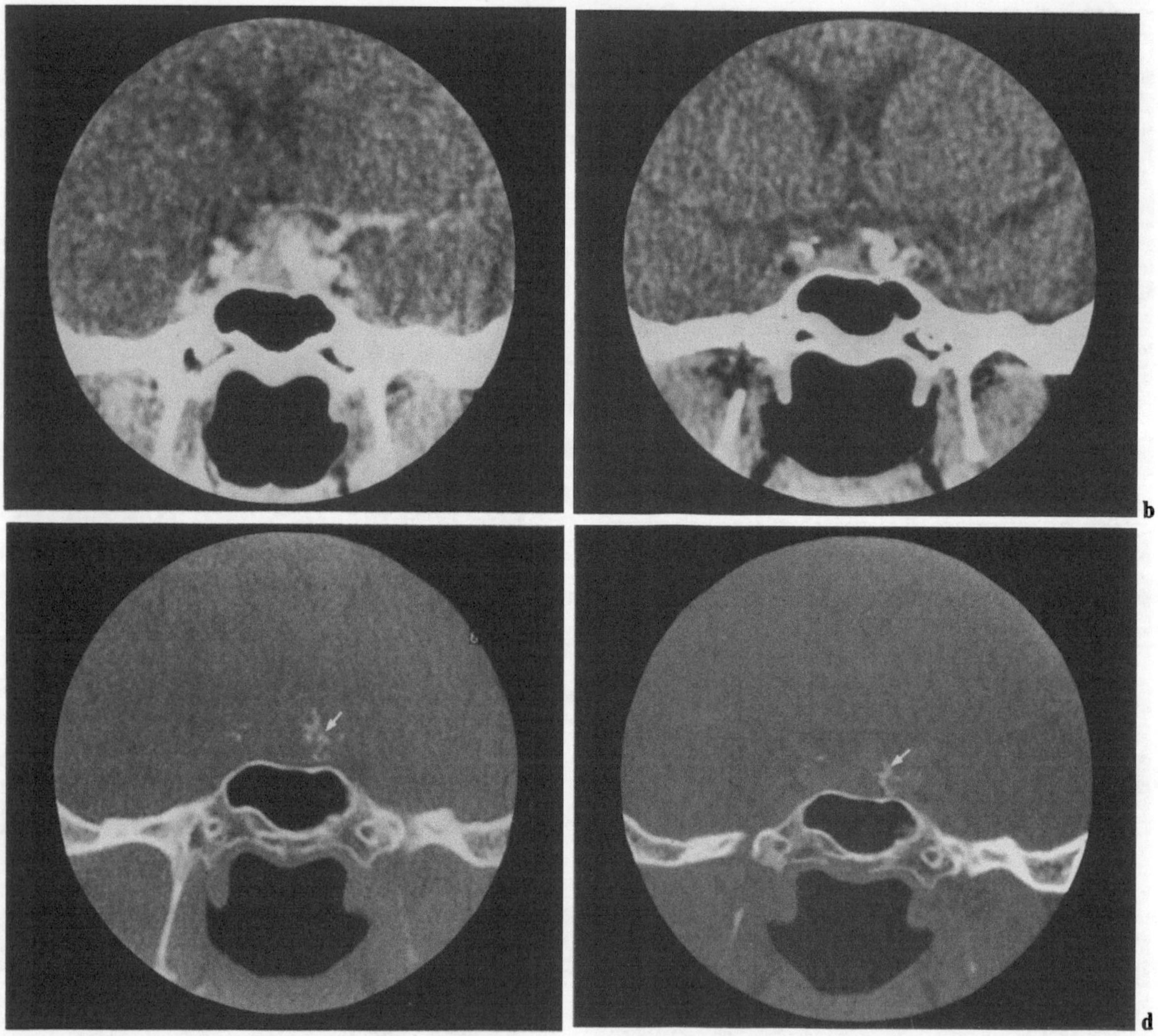

Fig. 6.4a–d. Calcified prolactinoma. Evolution with medical treatment. Secondary amenorrhea for 30 years in a 49-year-old woman. Skull X-ray for headache reveals an intrasellar calcification. Hyperprolactinemia (900 ng/ml). **a** Contrast coronal CT scan. Dense pituitary adenoma with suprasellar extension. Large calcification on the left. **b** CT scan with algorithm for bone detail for better demonstration of calcification (*arrow*). **c** Contrast coronal CT scan 6 months after pergolide therapy (50 µg per day). **d** Prolactinemia is 25 ng/ml. Shrinkage of the suprasellar extent and "coiling" of the calcification better demonstrated (*arrow*)

Fig. 6.5a, b. Prolactin-secreting pituitary adenoma after ▶ surgery in a 59-year-old woman. Residual cyst-like lesion within the sella. Evolution with bromocriptine. **a** Reformatted sagittal image. Large cystic image at the posterior part of the sella. Residual adenomatous tissue behind the modified anterior wall. **b** Reformatted sagittal image 6 months after bromocriptine (5 mg per day). Diminution of size of the sellar content and depression of its superior aspect (*arrows*)

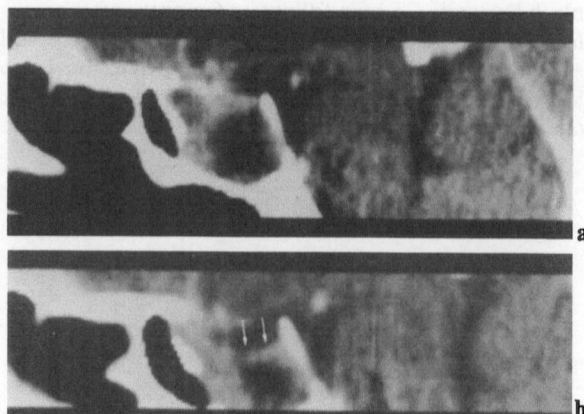

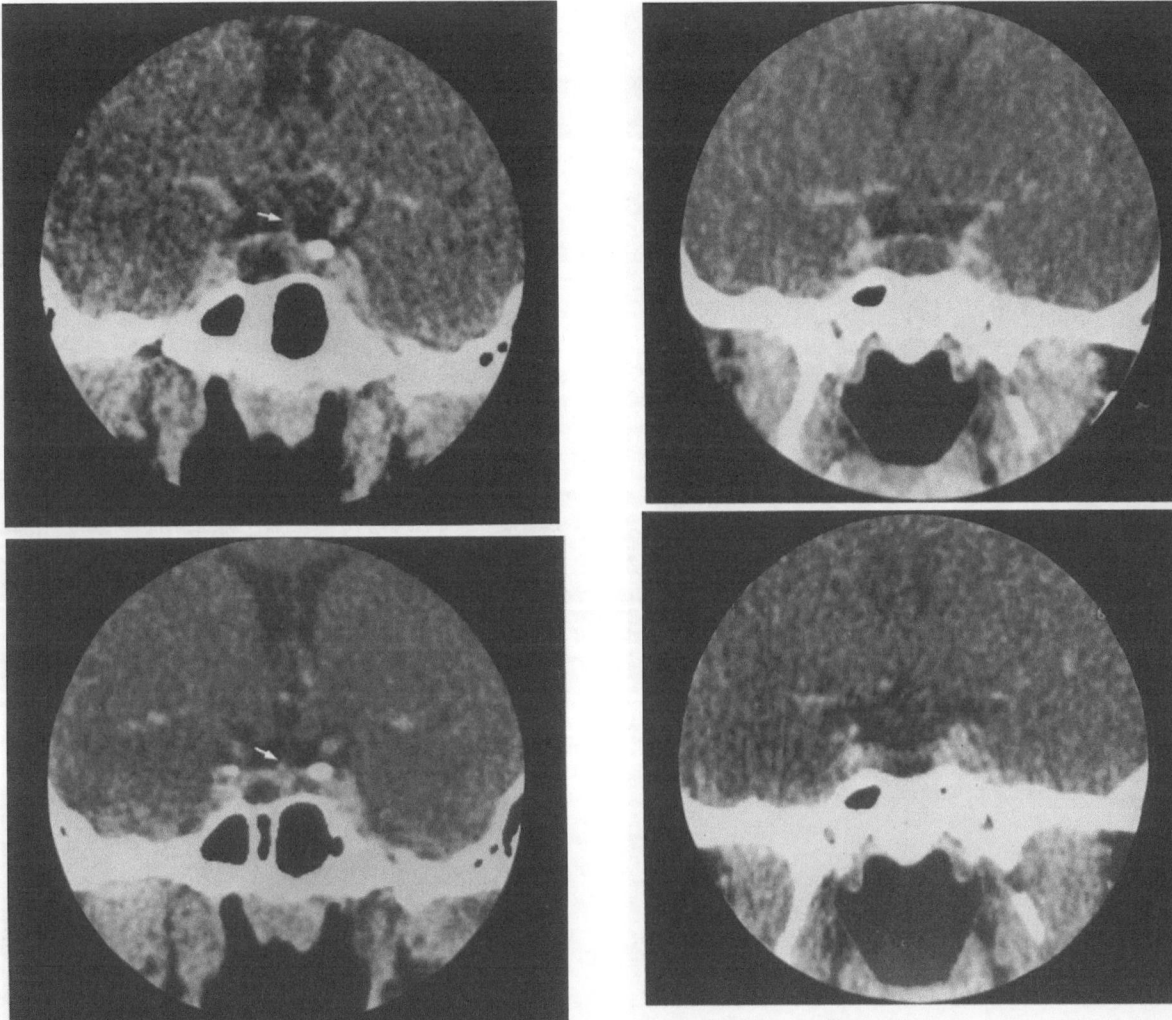

Fig. 6.6a, b. Intrasellar prolactinoma. Spaniomenorrhea and galactorrhea for 10 years in a 28-year-old woman. Prolactinemia is 33 ng/ml. **a** Contrast coronal CT scan. Well-defined hypodense adenoma pushing the pituitary stalk contralaterally (*arrow*). **b** CT follow-up 8 months after medical treatment. Normal menses and normoprolactinemia. Frank decrease of the adenoma size. The pituitary stalk is almost on the midline (*arrow*)

Fig. 6.7a, b. Prolactin-secreting pituitary adenoma. Evolution with bromocriptine. Postpartum amenorrhea and persistent galactorrhea in a 27-year-old woman. Prolactinemia is 70 ng/ml. **a** Contrast coronal CT scan. Hypodense homogeneous pituitary adenoma with bulging of the sellar content. **b** CT follow-up after 4 months of medical treatment. Restoration of normal menses. Prolactinemia is now 32 ng/ml. Decrease of the height of the sellar content

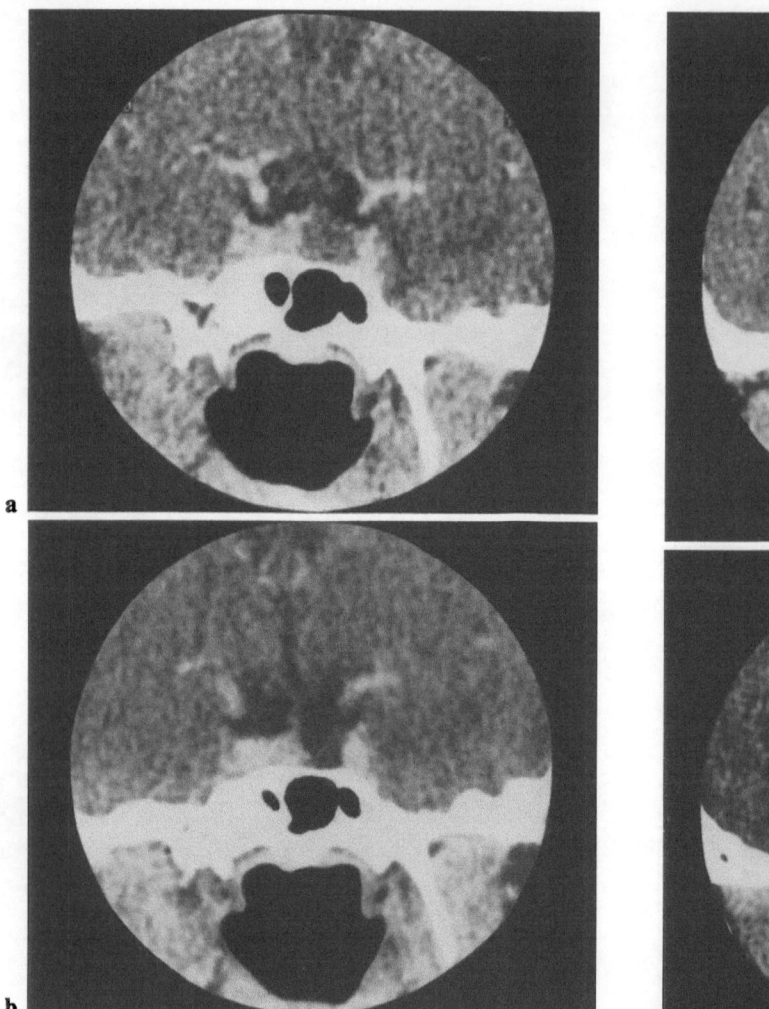

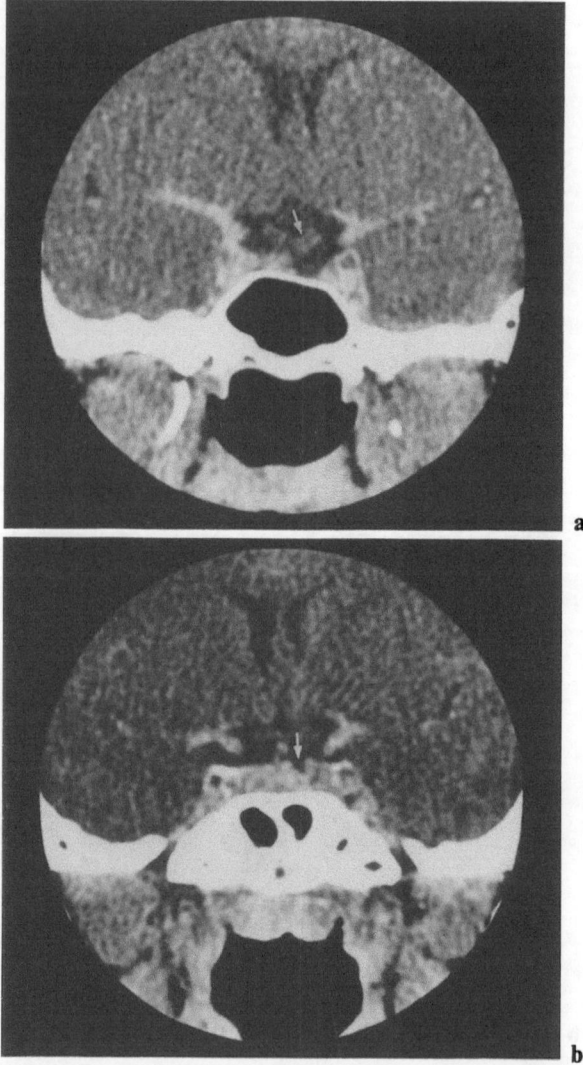

Fig. 6.8a, b. Intrasellar prolactinoma. Postpartum amenorrhea and galactorrhea in a 27-year-old woman. Prolactinemia is 29 ng/ml. **a** Contrast coronal CT scan. Left-sided pituitary adenoma and bulging of the sellar content. **b** CT follow-up 2 months after pergolide therapy (50 µg per day). Restoration of normal menses and normoprolactinemia. Shrinkage of the left part of the sellar content. The residual microadenoma is less dense than prior therapy

Fig. 6.9a, b. Prolactinomas after medical treatment. **a** Prolactinoma with small suprasellar extension in a 40-year-old woman (not shown). After bromocriptine treatment, a pregnancy occurs. Medical treatment is reintroduced after delivery and a CT scan is performed 18 months after delivery. No adenoma is identified. Partial empty sella. Note that the optic chiasm has a low position (*arrow*). **b** Prolactinoma with suprasellar extension (not shown). Bromocriptine is given for 6 months. Coronal CT scan: The pituitary adenoma is no more identified. Abnormal triangular defect is only noted at the sellar diaphragm level (*arrow*)

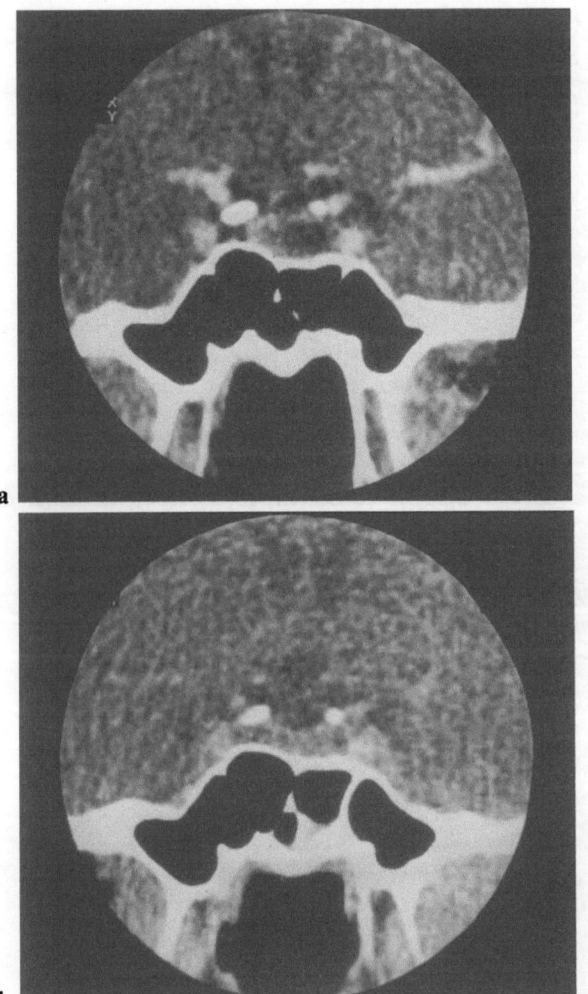

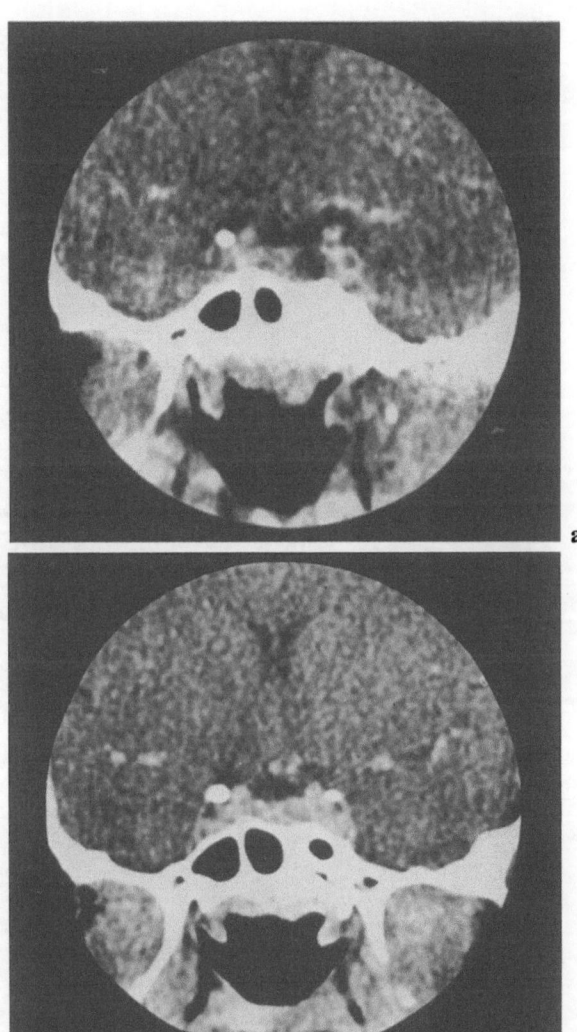

Fig. 6.10a, b. Intrasellar prolactinoma. Sexual deficiency in a 29-year-old man. Hyperprolactinemia (178 ng/ml). **a** Contrast coronal CT scan. Low-dense left-sided pituitary adenoma deviating the pituitary stalk contralaterally. The sellar floor is intact. **b** CT follow-up 10 months after bromocriptine (5 mg per day). Restoration of normal gonadal function. Prolactinemia is now 27 ng/ml. "Cleaning" of the sellar content. The adenoma is no longer visible

Fig. 6.11a, b. Intrasellar prolactinoma. Amenorrhea and galactorrhea for 10 years in a 33-year-old woman. Prolactinemia is 40 ng/ml. **a** Contrast coronal CT scan. Left-sided rounded pituitary adenoma. **b** CT follow-up 1 year after medical treatment (bromocriptine, 5 mg per day). Normal menses; no galactorrhea. Normoprolactinemia. Pituitary pattern appears now almost normal. Slight bulging of the sellar diaphragm on the left

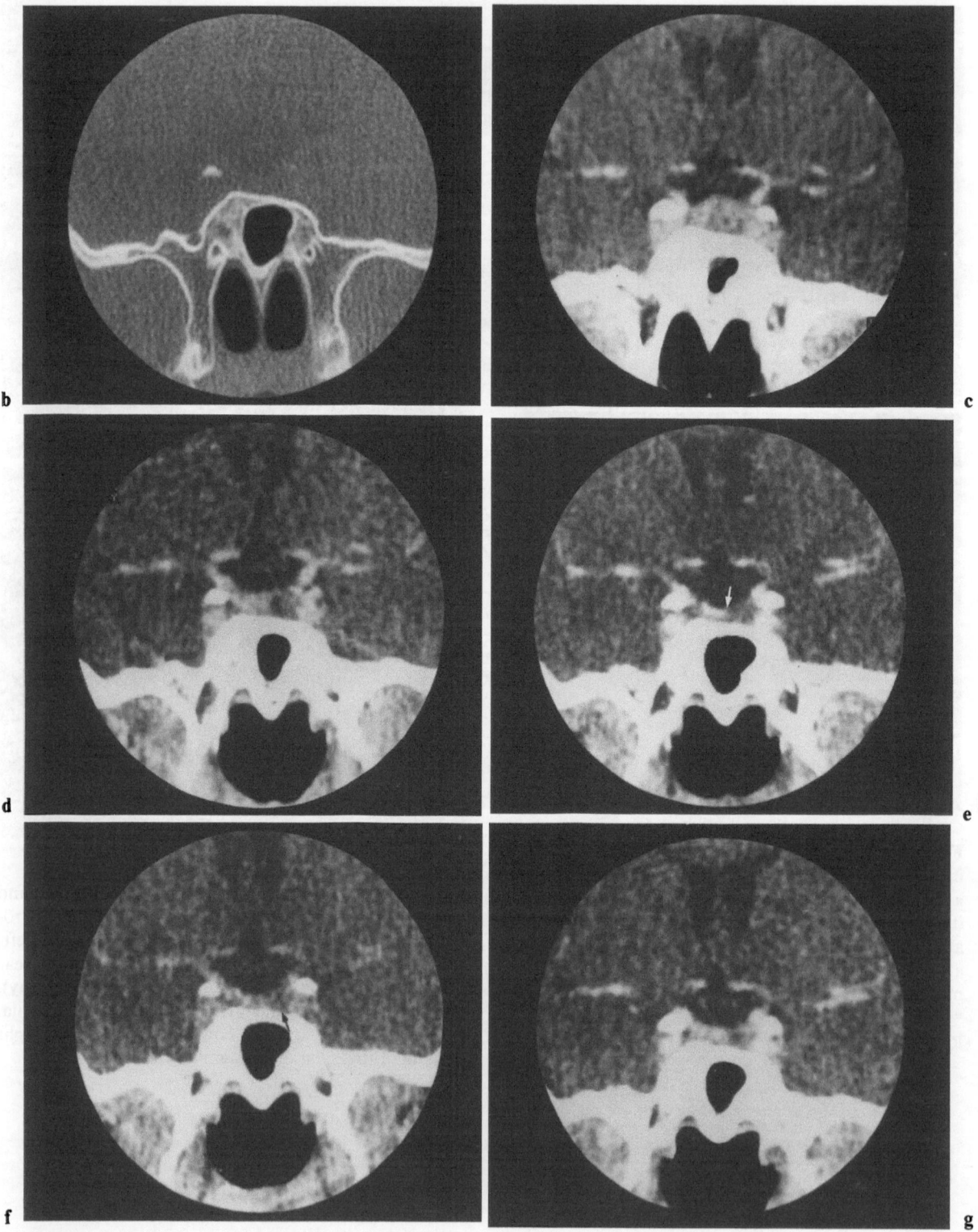

Fig. 6.12b–g

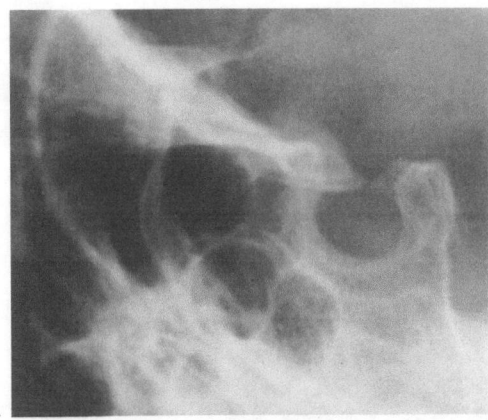

a

Fig. 6.12a–g. Prolactin-secreting pituitary adenoma. Amenorrhea-galactorrhea for 8 months in a 15-year-old girl. Prolactinemia is 260 ng/ml. **a** (see above) Sella turcica magnified lateral view. Partial lack of pneumatization of the infrasellar sphenoid bone so that unilateral depression of the sellar floor is difficult to assert. **b** Bone algorithm of high-resolution coronal CT. Slanted sellar floor without visible cortical thinning. **c** Contrast coronal CT scan. Bulging of an inhomogeneous sellar content without well-defined pituitary microadenoma. **d** CT follow-up 8 months after bromocriptine therapy (5 mg per day). Normal menses. Galactorrhea disappeared. Prolactinemia is now 40 ng/ml. Decrease of the pituitary height. A well-defined 7-mm microadenoma is now well shown on the left. **e** New CT follow-up 6 months after **d**. Further regression of the pituitary microadenoma. The size of the lesion is now 3 mm. **f** CT follow-up 7 months after **e**. Prolactinemia is now normal (7 ng/ml). The microprolactinoma is no longer well demonstrated: Faint loss of enhancement of the left part of the gland (*arrow*). **g** Dynamic coronal CT on the same day. Cutting-off of the left half of the band-like capillary bed (*arrow*) only represents the memory of the prolactinoma

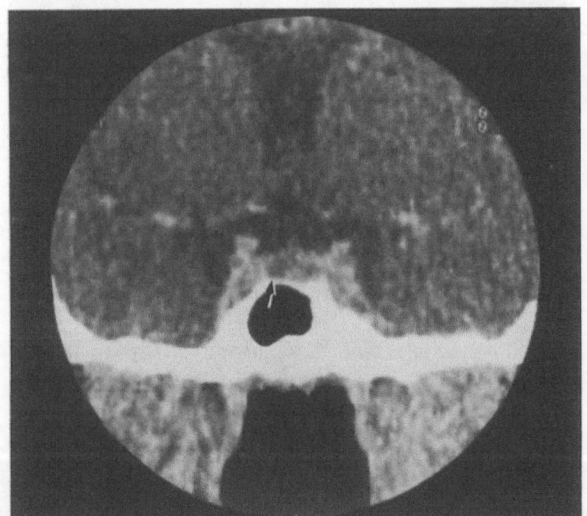

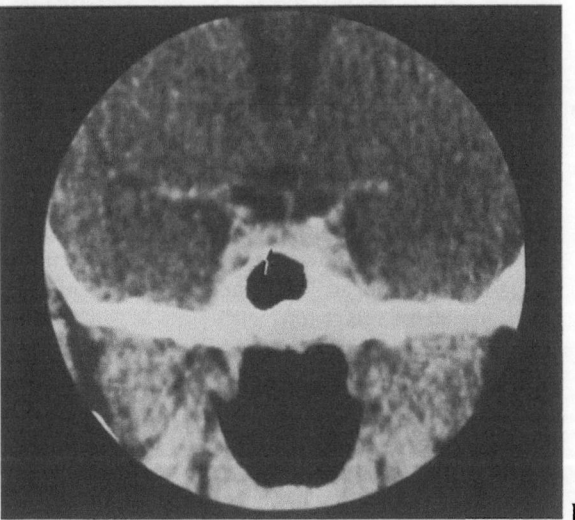

a   b

Fig. 6.13a, b. Microprolactinoma. Oligomenorrhea in a 25-year-old woman. Prolactinemia is 40 ng/ml. **a** Contrast coronal CT scan. Right-sided well-defined 3-mm pituitary microadenoma (*arrow*). **b** CT follow-up after 3 months of bromocriptine treatment (5 mg per day). Normal menses and normoprolactinemia. Diminution of the adenoma size. A 1-mm low dense area only is still visible (*arrow*)

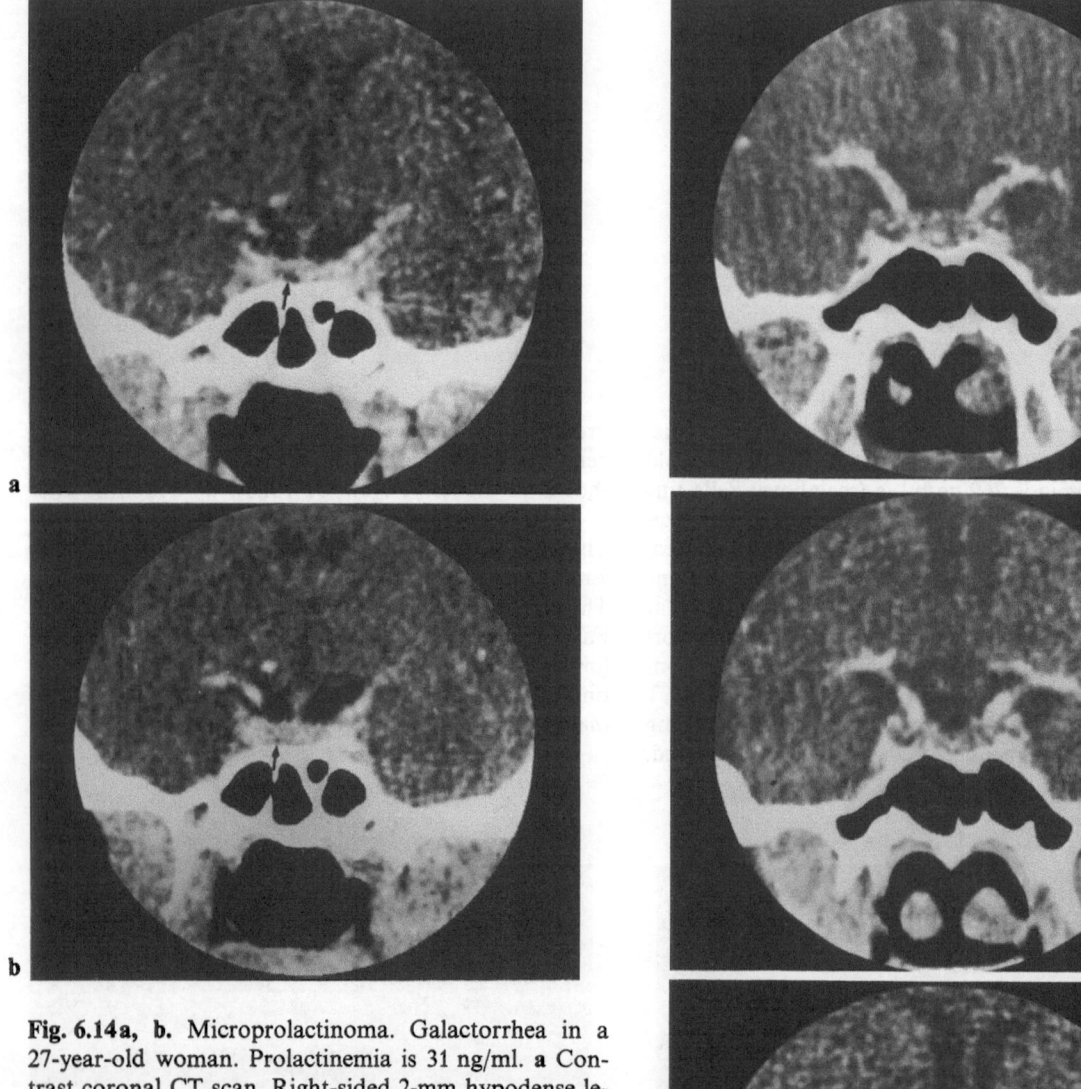

Fig. 6.14a, b. Microprolactinoma. Galactorrhea in a 27-year-old woman. Prolactinemia is 31 ng/ml. **a** Contrast coronal CT scan. Right-sided 2-mm hypodense lesion (*arrow*). **b** CT follow-up 6 months after bromocriptine. Pituitary appears quite normal. Residual "one-pixel" low-dense image (*arrow*)

Fig. 6.15a–c. Prolactinoma. Secondary failure of medical treatment. Amenorrhea and galactorrhea for 6 years in a 24-year-old woman. Prolactinemia is 200 ng/ml. **a** Contrast coronal CT scan. Bulging of the sellar diaphragm. Inhomogeneous sellar content without well-defined microadenoma. **b** CT follow-up 2 months after bromocriptine treatment (5 mg per day). No menses. Prolactinemia is still 120 ng/ml. Decrease of the pituitary height; concavity of the sellar diaphragm. **c** New CT follow-up. Bromocriptine is now 25 mg per day. No menses. Prolactinemia is 185 ng/ml. Sellar diaphragm is bulging again. Same pituitary pattern as before treatment. A prolactinoma was confirmed at surgery

Chapter 7
Pituitary, Prolactinomas, and Pregnancy

The Pituitary Gland During Pregnancy

Erdheim and Stumme have demonstrated an increase in weight as much as twofold of the pituitary gland during pregnancy. There is very little increase in the anteroposterior and transverse axes of the gland due to the limited space available for expansion. The increase is thus primarily in height, which has been variously reported to increase from a mean 5.9 mm in the nulliparous woman to 7.5 mm at term, with a maximum size of 11.5 mm. The increase in circulating estrogen levels during pregnancy probably stimulates both hyperplasia and hypertrophy of lactotropic cells. Furthermore, there is a parallel between the increase in estrogen and prolactin levels during pregnancy.

The growth of lactotropic cells and the increase in blood supply to the gland can thus account for the overall increase in pituitary volume shown by CT scans carried out during pregnancy. In coronal section, the superior pole of the gland is convex upward (Fig. 7.1); the maximal height of the pituitary on the midline can be twice that seen during the nongestational period; a height of 10–11 mm is not uncommon (Fig. 7.2a). In axial section, the superior pole of the pituitary occupies the bottom of the chiasmatic cistern (Fig. 7.2b). Pregnancy has little effect upon attenuation values of the gland, and the only change is a slight heterogeneity of enhancement after intravenous injection of iodinated contrast media.

Prolactinomas During Pregnancy

Development of pregnancy in patients with prolactin-secreting adenomas, generally after treatment with dopamine agonists or, where hyperprolactinemia is unrecognized, by agents inducing ovulation, brings about a massive rise in estrogen secretion with an increase in volume not only of the pituitary, but also of the prolactinoma. This tumoral growth during pregnancy can cause headache, and less frequently, diabetes insipidus or ophthalmoplegia due to compression of the intracavernous nerves. Less infrequently, bitemporal hemianopia or superior temporal quadranopsia may occur when the superior pole of the hypophysis impinges upon the optic chiasm. The risk of chiasmatic compression has been evaluated variously by different authors (Magyar and Marshall, Gemzell and Wang, Bergh and Nillius). Visual complications are both more frequent and more severe when the prolactinoma is not recognized or its volume not properly evaluated before the beginning of treatment of sterility and hyperprolactinemia. The risk of visual complications appears to be very limited with microadenomas, but is seen in as many as 25% of all macroadenomas.

A literature search shows that, in light of the above, a treatment protocol has evolved. For macroadenomas, transsphenoidal adenomectomy (rather than radiotherapy) is carried out before induction of pregnancy. For microadenomas, in the absence of surgery, pregnancy can be triggered with dopamine agonists provided that careful monitoring is maintained of prolactin levels and visual field.

Our attitude is somewhat different. We consider that systematic surgery for all macroadenomas is unjustified in view of the risk (however limited) of postoperatory hypopituitarism and thus FSH and LH deficiencies. The borderline between macroadenomas and microadenomas appears to be somewhat hazy and subjective.

Furthermore, there are differences in sensitivity of lactotropic cells to estrogen stimulation (Rjosk), and the progression of prolactinomas during pregnancy can be unpredictable. Monitoring of pregnant patients with prolactin adenomas thus cannot be based solely upon serial evaluation of prolactin levels or visual fields. We thus suggest that monitoring can be useful during pregnancy for rapid detection and control of complications (Bonneville, Portha, and Dietemann). Nevertheless, this does not provide absolute protection against pituitary apoplexy, which, promoted by hypervascularization, remains a risk throughout the duration of pregnancy (Le Pogamp, Linquette).

Such surveillance is nevertheless not required for small adenomas. We now consider that CT scanning during the 5th or 6th month of pregnancy may be justifiable where the size of the adenoma is greater than 6 mm, or where the maximal height of the pituitary gland exceeds 8 mm before pregnancy.

CT of Prolactinomas During Pregnancy

Pituitary scanning during pregnancy presents two types of theoretical risks: fetal irradiation and those associated with intravenous injection of an iodinated contrast medium.

The risk of irradiation appears negligible: ovarian doses during CT scanning of the head are less than 1 mrad per section (Dietemann), and abdominal protection with a lead apron further reduces these doses. By way of comparison, the International Commission on Radiological Protection recommends that fetal doses not exceed 1 rad during radiological investigations. Fetal risk is thus infinitesimal, and does not contraindicate computed tomography during pregnancy.

The risks posed by intravenous injection of water-soluble contrast media also appear miniscule. While iodine crosses the placental barrier, there have been no reports of hypothyroidism after intravenous urography in pregnant women, probably due to rapid excretion of contrast media. On the other hand, cases of hypothyroidism have been reported after amniofetography with fat-soluble media (Denavit).

The only true risk is anaphylactic shock during injection of contrast media, with placental hypovolemia and fetal distress. It has been our experience that direct coronal sections without injection of contrast media is sufficient to evaluate the topography and morphology of the superior pole of the pituitary gland in the chiasmatic cistern (Fig. 7.3). We currently propose this procedure, the native CT scan, during pregnancy. After targeting with a "scoutview," three or four frontal sections are generaly adequate to evaluate changes in the adenoma. Beyond the 7th month, a decubitus position with hyperextension of the head is preferable to procubitus. If only axial sections can be obtained, reconstructions are required for satisfactory visualization of the superior pituitary pole (Figs. 7.8 and 7.9).

For adenomas initially smaller than 6 mm, the control CT scan during the 6th month of pregnancy generally shows a doubling in tumor volume, while the size of the pituitary itself increases to a lesser extent (Figs. 7.4 and 7.5). This increase in volume of the adenoma is not absolutely constant. The absence of increase in pituitary volume during pregnancy could provide evidence for necrosis of the adenoma. Furthermore, there is frequently no increase in volume of the adenoma when dopaminergic agonist treatment continued for the full period of gestation (Fig. 7.6).

We now consider that CT monitoring is unnecessary where the adenoma is initially smaller than 6 mm, or where the total height of the gland is less than 8 mm; in these situations, a doubling of the volume of the lesion does not pose a threat of compression of the visual pathways. For adenomas larger than 6 mm, or where the height of the pituitary gland is greater than 8 mm, development is unpredictable and computed tomographic monitoring is useful. We have seen cases where CT scans during the 6th month of pregnancy show major tumoral growth with complete filling of the chiasmatic cistern and serious erosion of the sella turcica requiring reinstatement of dopamine agonists. Such treatment should lead to rapid shrinkage of the suprasellar extension (Figs. 7.7 and 7.8).

In conclusion, computed tomographic monitoring of prolactinomas greater than 6 mm is without danger during pregnancy. Several coronal sections taken without injection of contrast media are sufficient. This examination provides an objective index, and allows a more conservative and cautious approach, in particular where adenomas are sufficiently large for surgical exeresis to be considered before authorizing the patient to become pregnant.

Prolactinomas After Pregnancy

In the absence of nursing, prolactin levels normally return to baseline values 1 to 3 weeks after delivery. In nursing women, prolactin secretion remains elevated during this period. Changes in prolactinoma morphology after delivery parallel those in prolactin blood levels. Nursing prolongs the postpartum hyperprolactinemia phase: the adenoma size is then not modified after delivery in nursing women. Nevertheless, this drawback is not sufficiently great to preclude nursing in women presenting prolactinomas without major suprasellar expansion.

Aggravation of hyperprolactinemia during the postpartum period is very infrequent; we have never seen a durable increase in the volume of the tumor after pregnancy. Furthermore, in a substantial number of cases (variable between authors: 10% for Rjosk; 25% for Mornex), pregnancy leads to a virtually complete resolution, with normalization of blood prolactin levels and spontaneous development of menstrual cycles in the absence of any drug treatment. This normalization is not dependent upon the degree of initial hyperprolactinemia, the volume of the prolactinoma, or bromocriptine dosage before pregnancy. Resolution of hyperprolactinemia appears to be directly linked to pregnancy: the increase in pituitary vascularization is insufficient to fully compensate for the increased volume of the gland, leading to partial or total tumoral infarction. The possibility of such spontaneous resolution can be evoked where there is a sudden drop in prolactin levels during pregnancy. Even if relatively infrequent, the possibility of resolution of a prolactinoma requires that spontaneous normalization of blood prolactin levels be tested for during a temporary interruption of treatment.

A control CT scan is useful 3 months after delivery. In comparison with size of the adenoma before pregnancy, size can be unchanged (Figs. 7.4 and 7.9) or decreased (Fig. 7.7), sometimes even when bromocriptine treatment is not reinstated after delivery (Fig. 7.8). In patients where there is complete resolution of hyperprolactinemia in the absence of any drug treatment, the lesion can be completely invisible (Fig. 7.10) or can present a morphology very different from that seen before pregnancy (Fig. 7.6). Remineralization of the sellar floor can also occur (Figs. 7.6 and 7.7).

References

Andersen AN, Starup J, Tabor A, Jensen HK, Westergaard JG (1983) The possible prognostic value of serum prolactin increment during pregnancy in hyperprolactinaemic patients. Acta Endocrinol 102:1–5

Aubert B (1981) Mesures des doses en radiodiagnostic et tomodensitométrie. J Radiol 62:587–589

Batrinos L (1983) Prolactinomas and Pregnancy. In: Tolis G (ed) Prolactin and Prolactinomas. Raven Press, New York

Belforte L, Bruno M, Campagnoli C, Fessia L, Massara F, Molinatti GM (1980) Hormone pattern during bromocriptine. Induced pregnancy in hyperprolactinaemic patients. Eur J Obstet Gynec Reprod Biol 10:309–317

Bergh T, Nillius SJ, Wide L (1978) Clinical course and outcome of pregnancies in amenorrheic women with hyperprolactinemia and pituitary tumours. Br Med J 1:875–880

Bonneville JF, Dietemann JL, Cattin F, Poulignot D, Couturier M, Mollet E (1981) Microadenomes à prolactine traités par la bromocriptine. Surveillance tomodensitométrique. Nouv Presse Med 10:3725–3726

Bonneville JF, Poulignot D, Cattin F, Couturier M, Mollet E, Dietemann JL (1982) Apport des méthodes nouvelles dans l'exploration morphologique des tumeurs hypophysaires. Ann Endocrinol (Paris) 43:303–308

Bonneville JF, Cattin F, Poulignot D, Dietemann JL, Couturier M (1983) Value of high-resolution CT in the follow-up of prolactinomas treated with bromocriptine. In: Tolis G (ed) "Human Prolactin". Raven Press, New York

Burry KA, Schiller HS, Mills R, Harris B, Heinrich L (1978) Acute visual loss during pregnancy after bromocriptine-induced ovulation. Obstet Gynecol 52:19–22

Buvat J, Buvat-Herbaut M (1982) Prolactine, bromocriptine et fonction gonadique de la femme: données récentes. I. Physiologie de la prolactine, physiopathologie et diagnostic des hyperprolactinémies. J Gyn Obst Biol Repr 11:341–353

Buvat J, Buvat-Herbaut M (1982) Prolactine, bromocriptine et fonction gonadique de la femme: données récentes. II. Traitement des hyperprolactinémies féminines et autres indications de la bromocriptine. J Gyn Obst Biol Repr 11:509–521

Campagnoli C, Belforte L, Massara F, Peris C, Molnatti GM (1981) Partial remission of hyperprolactinaemic amenorrhea after bromocriptine-induced pregnancy. J Endocrinol Invest 4:85–91

Campagnoli C, Belforte L, Lotano M, Sandri A, Camanni F (1982) Prognosis of microprolactinoma and hyperprolactinemia in relation to pregnancy. In: Molinatti GM (ed) A clinical problem: Microprolactinoma. Diagnosis – treatment. Excerpta Medica, Amsterdam

Canales ES, Garcia IC, Ruig JE, Zarate A (1981) Bromocriptine as prophylactic therapy in prolactinoma during pregnancy. Fertil Steril 36:524

Chang JR, Keye WR, Young JR, Wilson CB, Jaffe RB (1977) Detection, evaluation and treatment of pituitary microadenoma in patients with galactorrhea and amenorrhea. Am J Obstet Gynecol 128:350

Corenblum B (1979) Successful outcome of ergocryptine-induced pregnancies in twenty-one women with prolactin-secreting pituitary adenomas. Fertil Steril 32:183–186

Corenblum B, Taylor PJ (1978) Bromocriptine in pituitary tumours. Lancet 2:786

Cowden EA, Thomson JA (1979) Resolution of hyperprolactinemia after bromocriptine-induced pregnancy. Lancet 1:613

Denavit MF, Lecointre C, Mallet E, De Menibus C, Rosier A (1977) Un accident de l'amniofoetographie: l'hypothyroidie. Arch Franç Ped 34:543–551

Dietemann JL, Bonneville JF (1985) Radiological Diagnosis of Pituitary Diseases. In: Imura H (ed) The pituitary gland. Raven Press, New York, pp 341–361

Dietemann JL, Portha C, Cattin F, Mollet E, Bonneville JF (1983) CT follow-up of microprolactinomas during bromocriptine-induced pregnancy. Neuroradiology 25:133–138

Emperaire JC, Riemens V, Dubecq JJ, Palmade J, Leuret JPH (1972) Hypophysectomie d'urgence à deux mois de grossesse après induction de l'ovulation. Bordeaux Med 15:1901–1904

Erdheim J, Stumme E (1909) Über die Schwangerschaftsveränderung der Hypophyse. Beitr Pathol Anat 46:1–122

Etling N, Gehin-Poupue F, Vielh JP, Gautray JP (1979) The iodine content of amniotic fluid and placental transfer of iodinated drugs. Obstet Gynecol 53:376–380

Gemzell C, Wang CF (1979) Outcome of pregnancy in women with pituitary adenoma. Fertil Steril 31:363–372

Griffith RW, Turkalj I, Braun P (1979) Pituitary tumours during pregnancy in mothers treated with bromocriptine. Br J Clin Pharmacol 7:393–396

Hancock KW, Scott JS, Gibson RM, Lamb JT (1978) Pituitary tumours and pregnancy. Br Med J 1:1487–1488

Hervet E, Barrat J, Pigne A, Darbois Y, Faguer C (1975) Adénome à prolactine. Hypophysectomie pendant la grossesse. Nouv Presse Méd 4:2392–2395

Husami N, Jewelewicz R, Vande Wiele RL (1977) Pregnancy in patients with pituitary tumors. Fertil Steril 28:920

Isaac AJ (1979) Resolution of hyperprolactinemia after bromocriptine-induced pregnancy. Lancet 1:784–785

Jewelewicz R, Vande Wiele RL (1980) Clinical course and outcome of pregnancy in 25 patients with pituitary microadenomas. Am J Obstet Gynaecol 136:339

Kletzky OA, Marrs RP, Howards WF, McCormick W, Mishell DR Jr (1980) Prolactin synthesis and release during pregnancy and puerperium. Am J Obstet Gynecol 136:545

Lamberts SW, Seldenrath HJ, Kwa HG, Birkenhäger JC (1977) Transient bitemporal hemianopsia during pregnancy after treatment of galactorrhea-amenorrhea syndrome with bromocriptine. J Clin Endocrinol Metab 44:180–184

Lamberts SW, Klijn JG, Lange SA, Singh R, Stefanko SZ, Birkenhäger JC (1979) The incidence of complications during pregnancy after treatment of hyperprolactinemia with bromocriptine in patients with radiologically evident pituitary tumors. Fertil Steril 31:614–619

Le Pogamp C, Grall JY, Massart C, Ramee A, Toulouse R (1979) Poussée évolutive d'adénome à prolactine au cours d'une grossesse après stérilité traitée. 2 observations. Nouv Presse Med 8:2009–2011

Linquette M, Buvat J, Gauthier A, Gasnault JP, Pagniez T, Decoulx M, Laine E (1977) Apoplexie hypophysaire révélatrice d'un adénome à cellules prolactiniques au cours d'une grossesse permise par la bromocriptine. Nouv Presse Med 6:3525–3531

Magyar DM, Marshall JR (1978) Pituitary tumors and pregnancy. Am J Obstet Gynecol 132:739

Mole RH (1974) Antenatal irradiation and childhood cancer. Br J Cancer 30:199–205

Mornex R, Orgiazzi J, Hugues B, Gagnaire JP, Claustrat B (1978) Normal pregnancies after treatment of hyperprolactinemia with bromoergocriptine, despite suspected pituitary tumors. J Clin Endocrinol Metab 47:290–295

Morrison JC, Boyd M, Ermans AM (1973) The effects of renografin 60 on the fetal thyroid. Obstet Gynecol 42:95

Nelson PB, Robinson AG, Archer DF, Maroon JC (1978) Symptomatic pituitary tumor enlargement after induced pregnancy: case report. J Neurosurg 49:283–287

Portha C (1982) Adénome à prolactine et grossesse. Intérêt du scanner. Thèse Med, Besançon

Ramussen AT (1934) The weight of the principal components of the normal hypophysis cerebri of the adult human femal. Amer J Anat 55:253

Reichlin S (1979) The prolactinoma problem. N Engl J Med 300:313–315

Rjosk HK, Fahlbusch R, Von Werder K (1982) Influence of pregnancies on prolactinomas. Acta Endocrinol (Copenh) 100:337–346

Rodesh F, Camus M, Ermans AM (1976) Adverse effect of amniofoetography on fetal thyroid function. Amer J Obstet Gynecol 126:723

Thorner MO, Edwards CRW, Charlesworth M, Dacie JE, Moult PJA, Rees LH, Jones AE, Besser GM (1979) Pregnancy in patients presenting with hyperprolactinemia. Br J Med J 2:771–774

Tubiana M (1979) Problèmes posés par l'irradiation des femmes enceintes. Effets des radiations ionisantes sur l'embryon et le foetus. Bull Cancer 66:155–164

Van Dalen JW, Greve EL (1977) Rapid deterioration of visual fields during bromocriptine-induced pregnancy in a patient with a pituitary adenoma. Br J Ophthalmol 61:728–733

Van Roon E, Van der Vijver JCM, Gerretsen G, Hekster REM, Wattendorff RA (1981) Rapid regression of a suprasellar extending prolactinoma after bromocriptine treatment during pregnancy. Fertil Steril 36:173–177

Ylikorkala O, Kivenen S, Ronnberg L (1980) Bromocriptine treatment during early human pregnancy: effect on the levels of prolactin, sex steroids and placental lactogen. Acta Endocrinol 95:412

Yuen BH (1978) Bromocriptine, pituitary tumours, and pregnancy. Lancet (Letter):1314

Zarate A, Canales ES, Alger M, Forsbach T (1979) The effect of pregnancy and lactation on pituitary secreting tumours. Acta Endocrinol (Copenh) 92:407–412

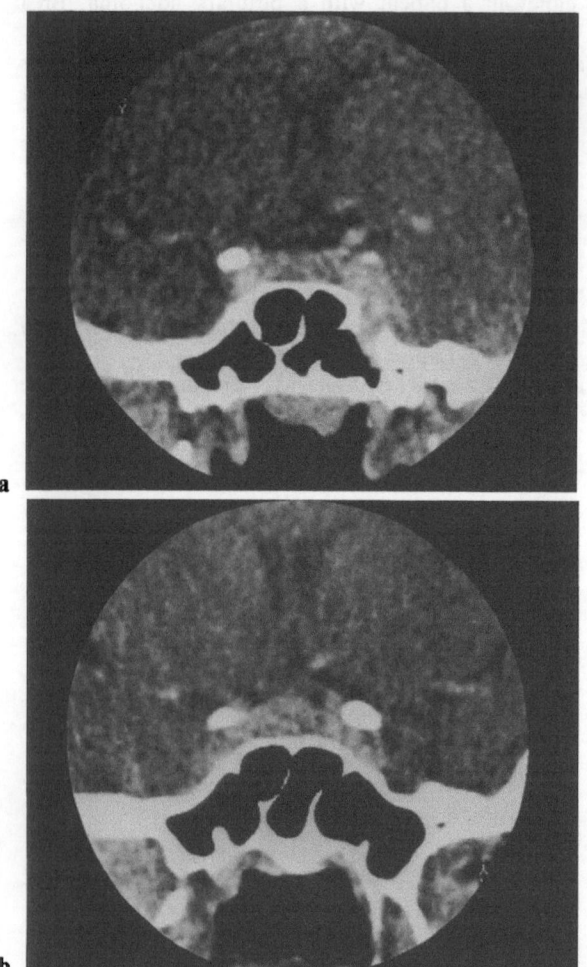

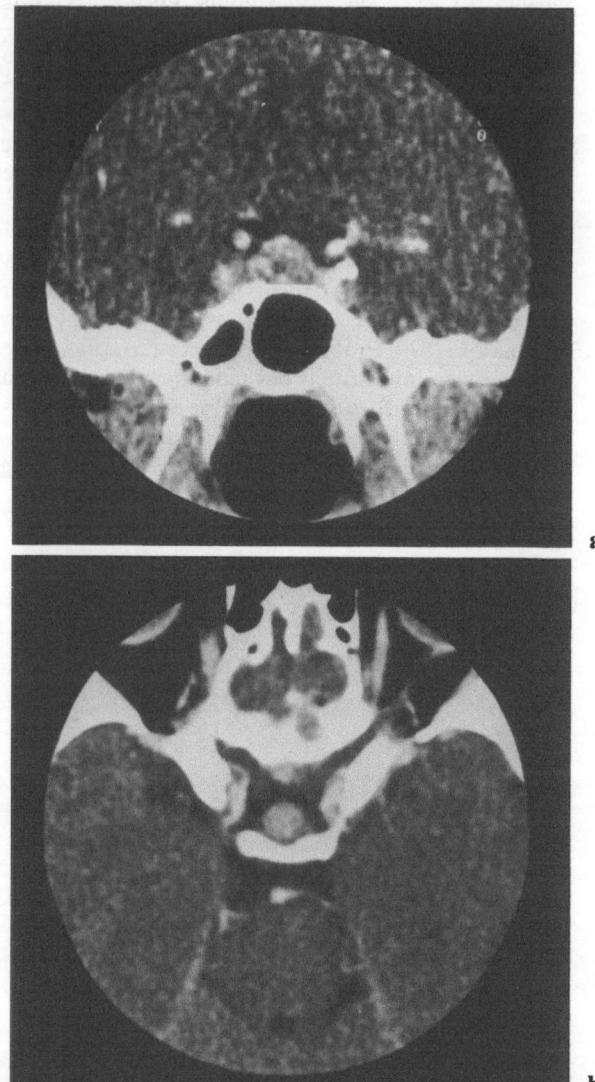

Fig. 7.1a, b. Normal pituitary gland **a** before pregnancy and **b** during pregnancy (6th month). Contrast-enhanced coronal CT. Bulging of the upper aspect of the gland is seen is **b**

Fig. 7.2a, b. Normal pituitary at the 6th month of pregnancy. **a** Contrast-enhanced CT scan. Bulging of the upper surface of the gland. The maximum height of the pituitary is 10 mm. **b** Axial CT scan; the enlarged pituitary occupies the bottom of the chiasmatic cistern

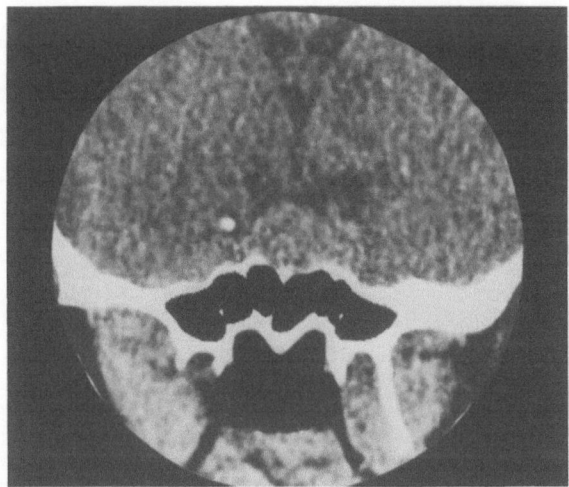

Fig. 7.3. Native CT scan during pregnancy (7th month) in a hyperprolactinemic woman treated with bromocriptine before pregnancy. The volume of the enlarged pituitary can be evaluated without contrast injection. Note the eroded sellar floor

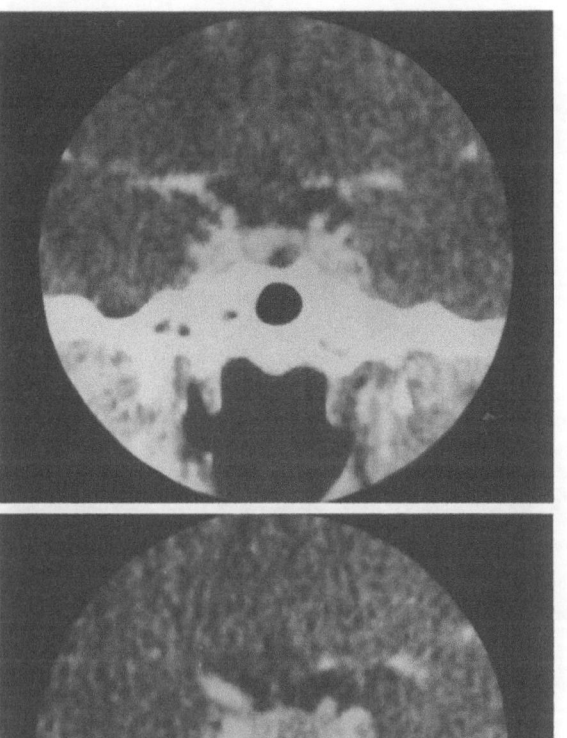

a

b

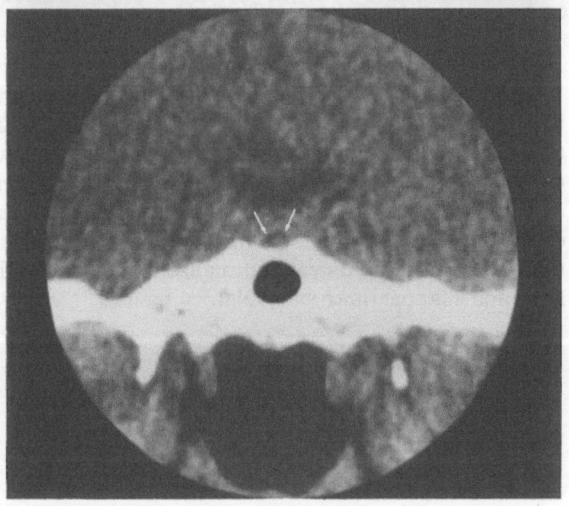

c

Fig. 7.4a–c. Microprolactinoma **a** before, **b** during, and ▶ **c** after pregnancy. Contrast-enhanced coronal CT (**a, b**), native CT (**c**). **a** Before pregnancy, left-sided hypodense lesion. **b** After medical treatment, a pregnancy occurs; bromocriptine is then withdrawn from the 1st month of pregnancy. CT follow-up at the 6th month shows a doubling of the prolactinoma size. Normal pituitary tissue on the right is also slightly enlarged. **c** Three months after delivery, a native CT scan is sufficient to find prolactinoma (*arrows*) and pituitary of the same size as before pregnancy

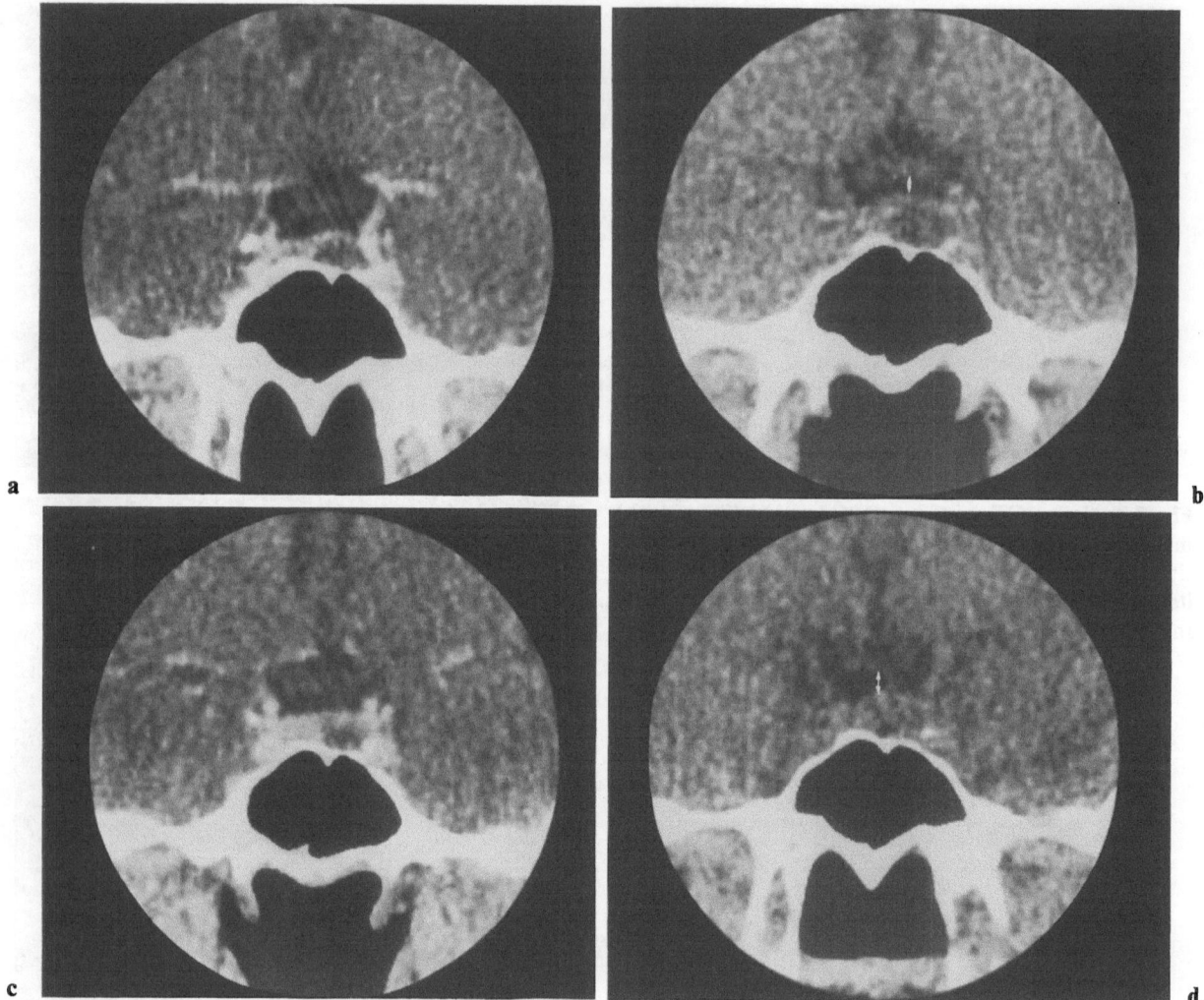

Fig. 7.5a–d. Microprolactinoma. Evolution with medical treatment and pregnancy. Amenorrhea-galactorrhea and hyperprolactinemia (40 ng/ml) in a 28-year-old woman. **a, b** Contrast-enhanced coronal CT scans; **c, d** native coronal CT scans. **a** Left-sided prolactinoma. Bulging of the diaphragma sellae on the left and corresponding erosion of the sellar floor. **b** Three months after institution of medical treatment, normalization of prolactin level and reappearance of ovulatory menses. Decrease of the adenoma size. **c** A pregnancy consequently occurs and bromocriptine is immediately withdrawn. Native CT control at the 6th month demonstrates increase of adenoma size and thinning of the sellar floor. Relationship of pituitary tumor and chiasm are nicely shown without contrast (*arrows*). Bromocriptine is then reintroduced to avoid optic compression. **d** One month later, a native CT scan is reassuring. Slight decrease of the pituitary height (*arrows*). Remineralization of the sellar floor

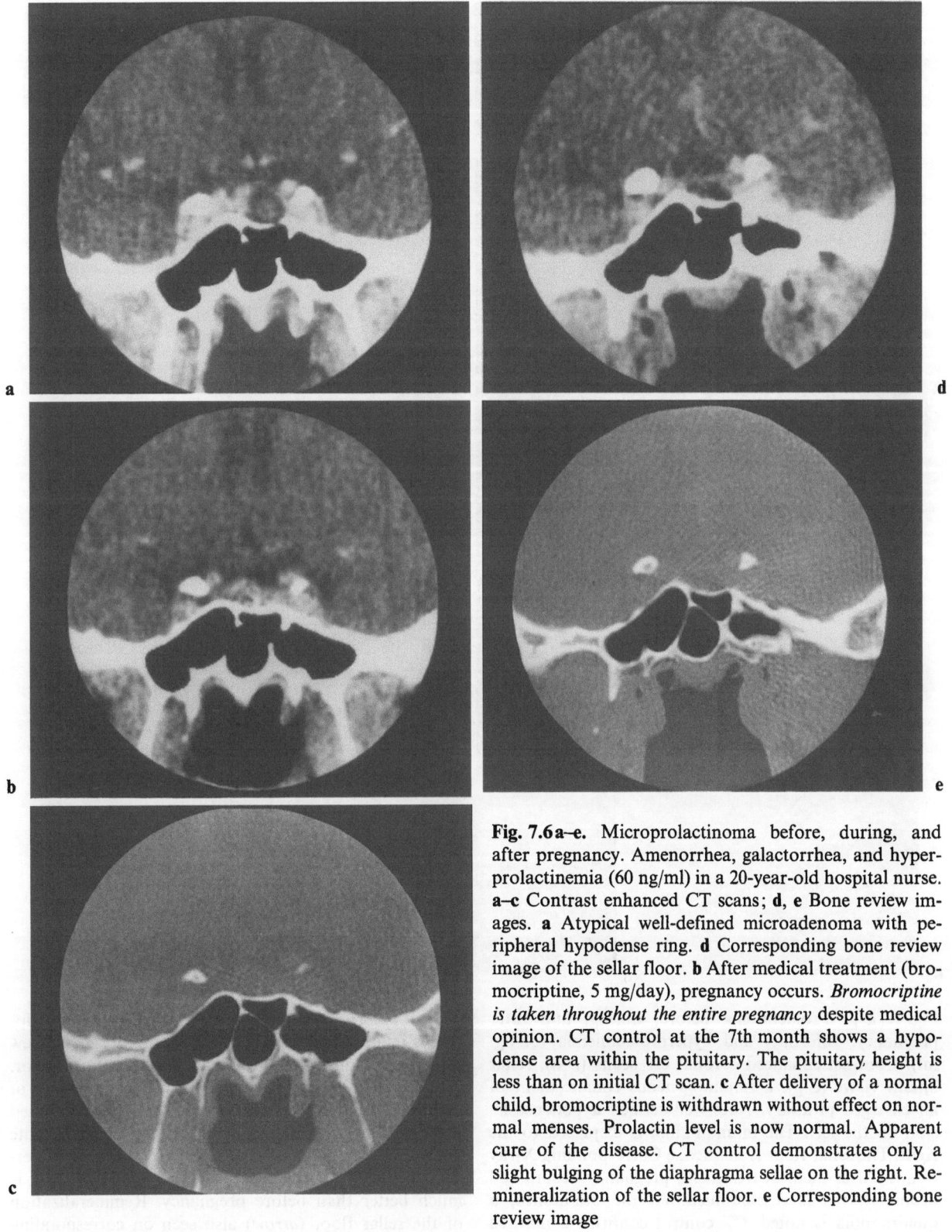

Fig. 7.6a–e. Microprolactinoma before, during, and after pregnancy. Amenorrhea, galactorrhea, and hyperprolactinemia (60 ng/ml) in a 20-year-old hospital nurse. **a–c** Contrast enhanced CT scans; **d, e** Bone review images. **a** Atypical well-defined microadenoma with peripheral hypodense ring. **d** Corresponding bone review image of the sellar floor. **b** After medical treatment (bromocriptine, 5 mg/day), pregnancy occurs. *Bromocriptine is taken throughout the entire pregnancy* despite medical opinion. CT control at the 7th month shows a hypodense area within the pituitary. The pituitary height is less than on initial CT scan. **c** After delivery of a normal child, bromocriptine is withdrawn without effect on normal menses. Prolactin level is now normal. Apparent cure of the disease. CT control demonstrates only a slight bulging of the diaphragma sellae on the right. Remineralization of the sellar floor. **e** Corresponding bone review image

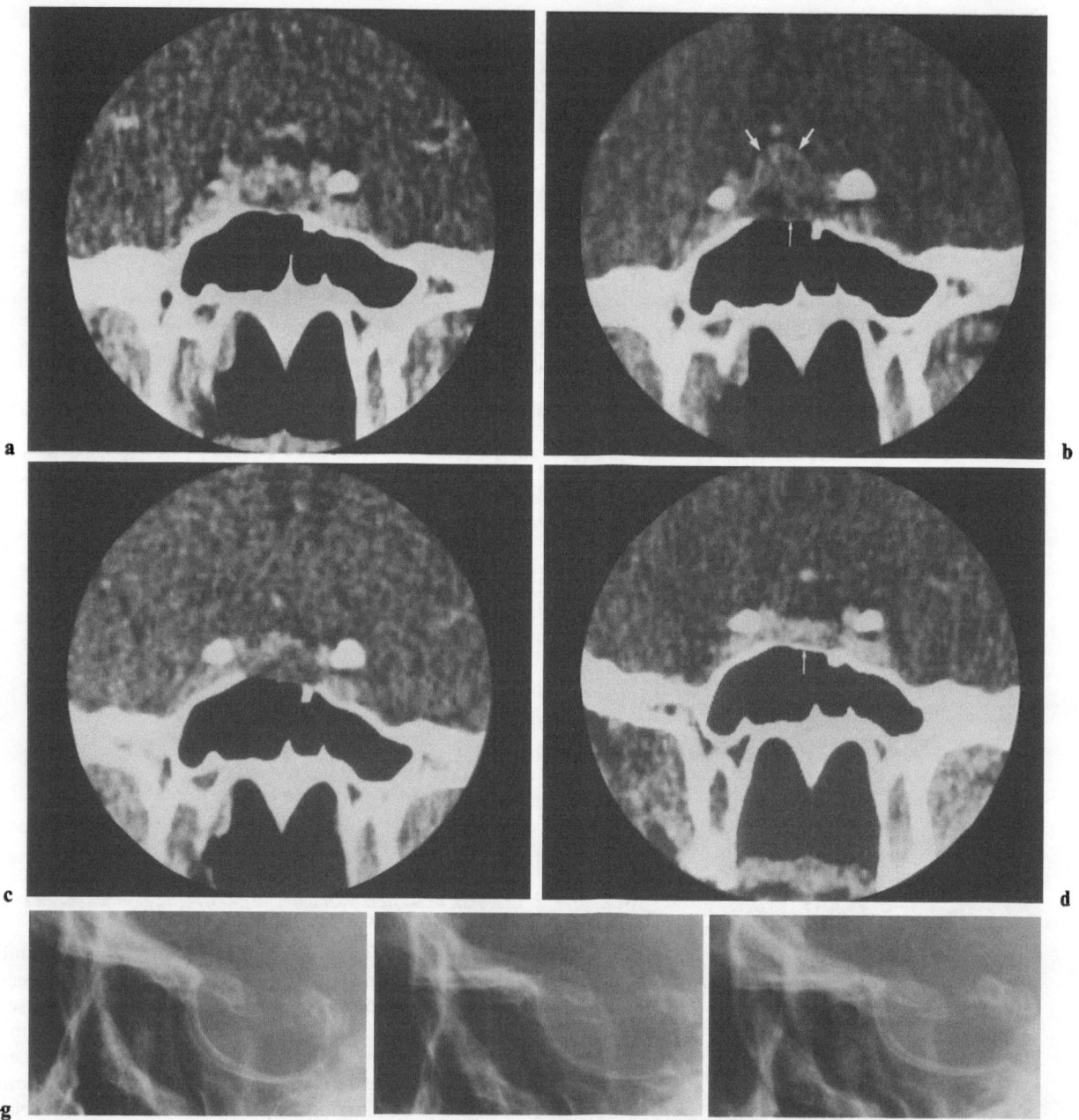

Fig. 7.7a–g. Prolactinoma before, during, and after pregnancy. Infertility and hyperprolactinemia (52 ng/ml). Contrast-enhanced coronal CT scans (**a–d**), sellar lateral views (**e–g**). **a** Inhomogeneous enhancement of an enlarged pituitary. Note well-defined hypodense lesion on the left. The infundibulum is displaced to the right. The sellar floor is thinned. **e** Corresponding X-ray film. **b** After medical treatment, pregnancy occurs and bromocriptine is discontinued. At the 5th month, a quadranopia is noted. CT control confirms a marked upward development of the tumor within the chiasmatic cistern (*arrows*). The sellar floor is still thinner (*small arrow*). **f** Corresponding sellar X-ray film. **c** Bromocriptine is reintroduced (5 mg/day). One month later a new CT scan demonstrates shrinkage of the pituitary tumor. The sellar floor is still very thin. **d** After delivery of a normal child, bromocriptine treatment is continued for 1 year. At this time, control CT scan shows a quite normal pituitary gland. The right part of the gland persists in being less enhanced, but the general pattern is much better than before pregnancy. Remineralization of the sellar floor (*arrow*) also seen on corresponding X-ray film (**g**)

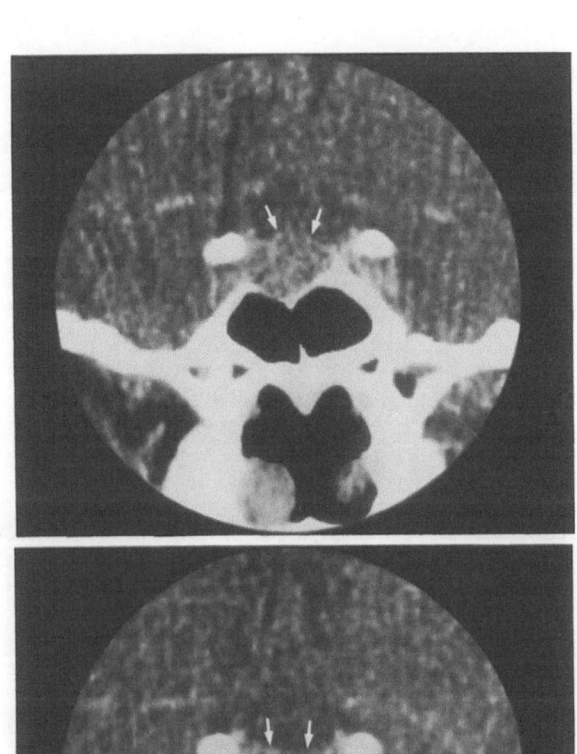

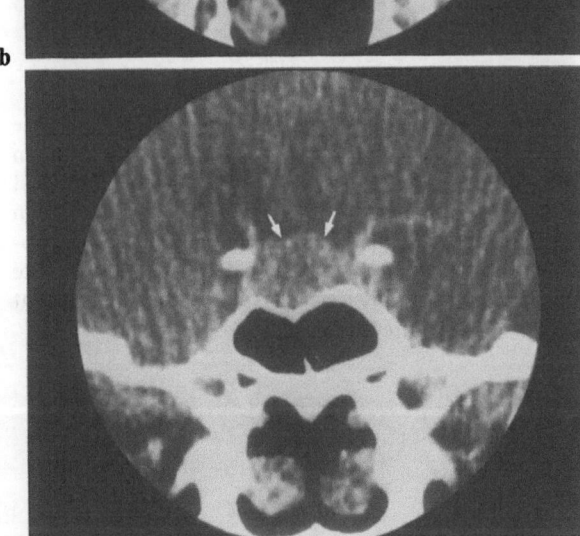

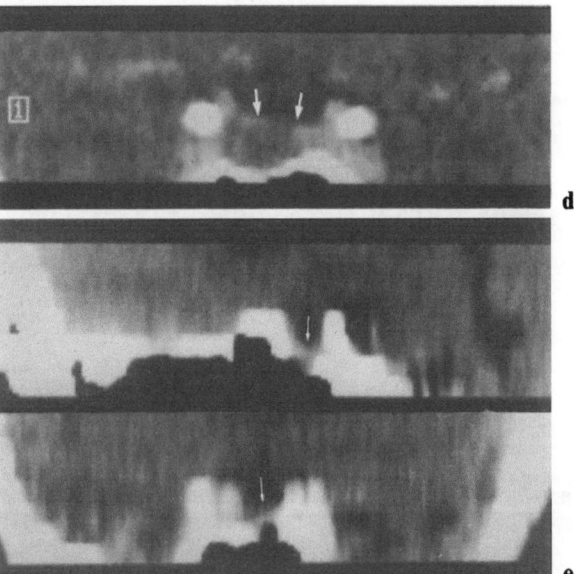

Fig. 7.8a–e. Prolactinoma before, during, and after pregnancy. Amenorrhea, infertility and hyperprolactinemia (45 ng/ml) in a 32-year-old woman. **a–c** Contrast-enhanced coronal CT scans; **d** coronal reformatted image; **e** coronal and sagittal reformatted image. **a** Enlarged pituitary which is less enhanced than usual. Angulated and thinned sellar floor. **b** Medical treatment is given. Four months later, the sellar diaphragm is depressed. Less enhancement of the right part of the intrasellar contents. **c** Pregnancy has begun and treatment is stopped. CT follow-up at the 5th month shows a marked enlargement of the pituitary with inhomogeneous density threatening the chiasm. **d** One month later, after reinstitution of bromocriptine, coronal reformatted image demonstrates shrinkage of the intrasellar contents. The prolactinoma itself is now better defined on the right (*arrows*). **e** Bromocriptine is given until delivery, then stopped. Improvement of medical status with normal menses and normoprolactinemia occurs. CT follow-up shows a partially empty sella. Necrosis of the adenoma favored by medical treatment and pregnancy is probable

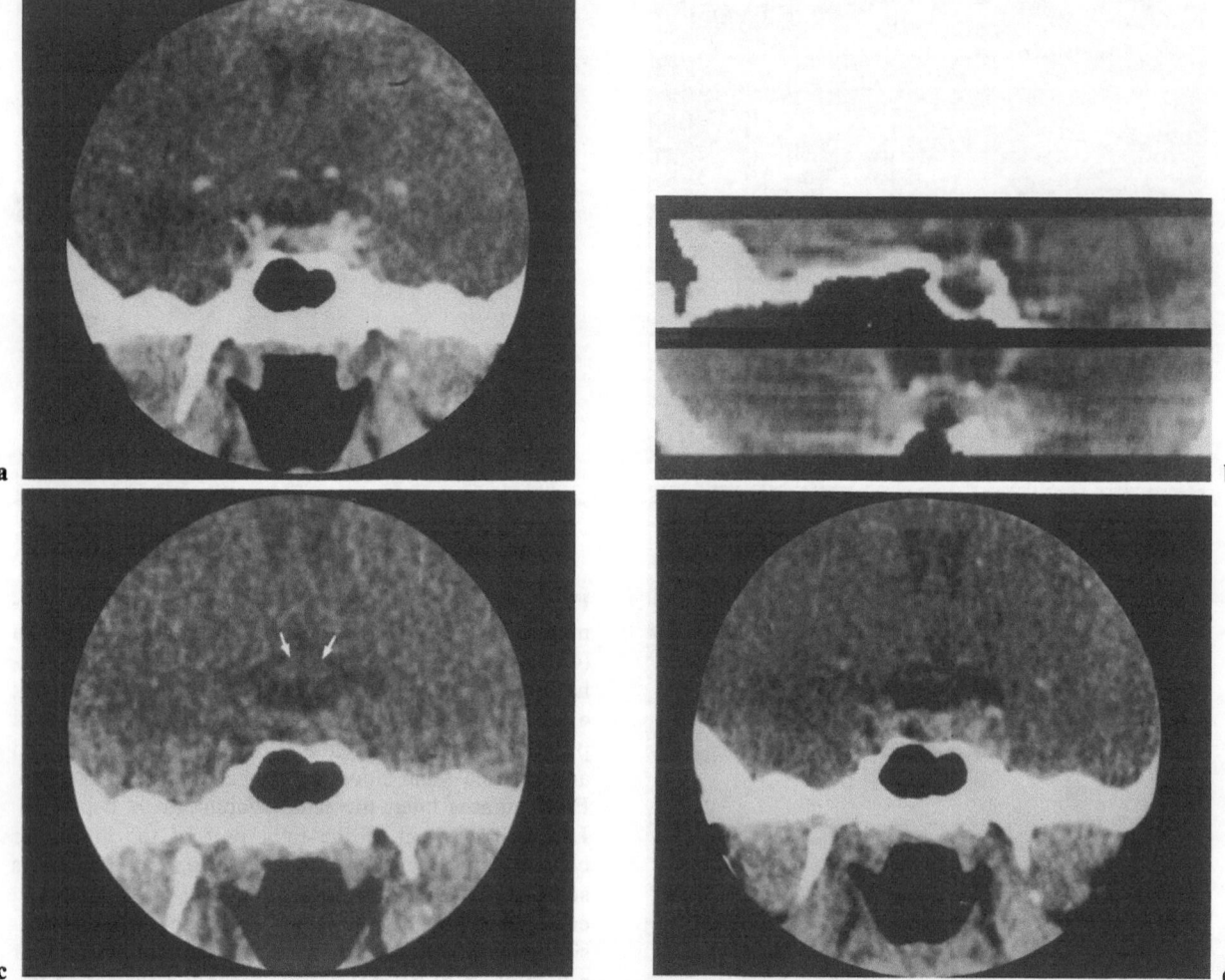

Fig. 7.9a–d. Microprolactinoma before, during and after pregnancy. Amenorrhea, galactorrhea, and hyperprolactinemia (30 ng/ml) in a 27-year-old woman. **a, c** Contrast-enhanced CT scans. **b** Sagittal and frontal reformatted CT scans. **d** Native coronal CT scan. **a** 3-mm right-sided microprolactinoma. **b** After medical treatment, pregnancy occurs. Bromocriptine is immediately withdrawn. CT scan at the 6th month shows an important increase of the tumor size with bulging of the diaphragma sellae. Bromocriptine is reinstituted. **c** One month later, the adenoma is slightly smaller than on initial CT scan. **d** Bromocriptine is given during the last 3 months of pregnancy and without discontinuation after delivery until the last native CT control. The prolactinoma is of the same size as before pregnancy. Note that as usual the optic chiasm is better defined without contrast injection (*arrows*)

Fig. 7.10a, b. Intrasellar prolactinoma before and after pregnancy in a 24-year-old woman. **a** Dynamic CT scan before pregnancy. Left-sided intrasellar prolactinoma pushing the capillary bed to the right ("tuft sign"). Depressed and eroded sellar floor (*arrows*). After medical treatment, pregnancy occurs. **b** After normal delivery, normoprolactinemia persists in the absence of any treatment. The patient is considered cured. The pituitary gland now appears normal and the sellar floor is remineralized

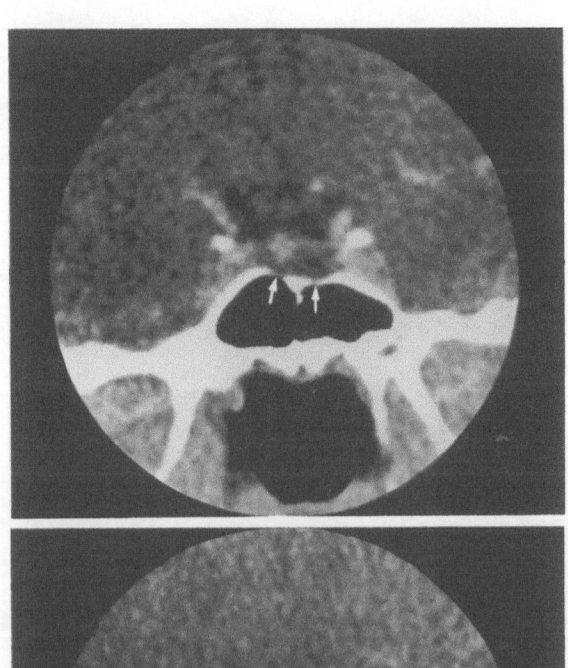

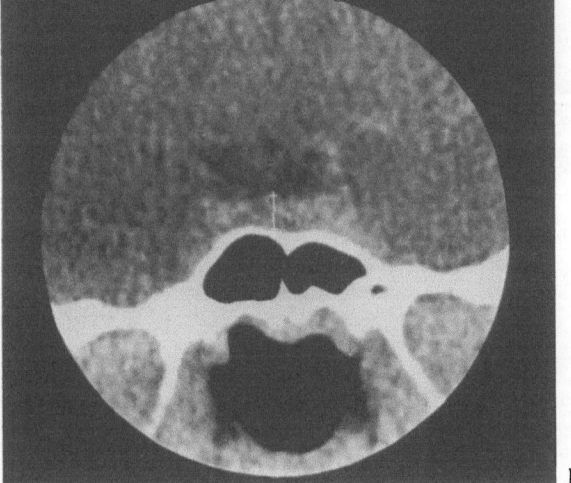

Chapter 8
Growth-Hormone Secreting Pituitary Adenomas

Acromegaly is mostly due to a GH-secreting pituitary adenoma. Such adenoma exceptionally develops during the prepubertal period giving rise to gigantism. The diagnosis of acromegaly is generally suggested after complete development of growth cartilage. Because of the very slow evolution of the dysmorphic syndrome, the diagnosis of acromegaly is often recognized only about the age of 40.

The clinical picture is dominated by a dysmorphic syndrome affecting mainly the extremities and the face: thickening and widening of the hands and feet, prognathism, and protrusion of the arcus superciliaris. A dorsal cyphosis, a sternal prominence, a hypertrophy of the cutaneous tissues, and a splanchnomegalia complete the clinical picture. Before outcome of the dysmorphic syndrome, diagnosis of acromegaly is rarely evoked, since clinical signs remain nonspecific: diffused headaches, nocturnal perspiration, and paresthesia of the hands. Raised blood pressure and diabetes mellitus are very common and account for the gravity of the disease.

The diagnosis of acromegaly is confirmed by radio-immunoassay of plasma growth hormone; normally, the GH level on fasting and at rest is less than 5 ng per ml.

CT scan demonstrates the pituitary adenoma responsible of acromegaly and of bony and soft-tissue changes secondary to the hypersecretion of GH.

CT Signs of GH-Secreting Pituitary Adenomas

Methods

CT scan with coronal sections of an acromegalic patient is technically difficult especially if spinal deformations are important. The projection obtained is rarely OM −90°, but more frequently OM −30° or OM −45°, such that the maxillary sinuses may be visible on sections through the sella turcica (Fig. 8.1). The prognathism and the opening of the mandibular angle (Fig. 8.2a) displace the pituitary region upwards, relative to the field of view. This results in a poorly enhanced CT image (Fig. 8.2b).

This poorly enhanced CT image is also due to the thickening of the vault and mandibular symphysis. That is the reason why reformatted coronal images, realized from thin axial sections, may in this case be better than coronal sections.

Results

A GH-secreting pituitary adenoma may be strictly intrasellar: in our experience, the smallest GH-secreting adenoma measured 3 mm. But generally a GH-secreting adenoma is much more voluminous by the time of its discovery. It extends preferentially downwards, into the sphenoid sinus and much less frequently upwards, towards the chiasmatic cistern. Such inferior extension differentiates GH-secreting adenomas from nonsecreting adenomas and from prolactinomas with suprasellar extension. Suprasellar extension is rarely sufficient to produce a chiasmatic compression.

Intrasellar Adenomas
GH-secreting pituitary adenoma is more frequently intrasellar. The enlarged pituitary fossa is occupied by an abnormal enhanced tissue after intravenous injection of contrast medium, less intensely enhanced than the normal pituitary gland, and generally homogeneous. The

limit between adenomatous tissue and normal pituitary tissue is often imprecise (Fig. 8.3). Low-density areas, more or less well limited and corresponding to areas of necrosis may be observed in the inferior part of the adenoma (Fig. 8.5a). The superior pole of the adenoma may bulge slightly into the chiasmatic cistern, without producing a true suprasellar extension (Figs. 8.3 and 8.4). The sellar floor is more often depressed, but the cortical is either slightly thinned or normal (Figs. 8.3 and 8.4). In axial sections, the anterior wall is rectilinear or concave towards the back, slightly thinned or normal; the dorsum sellae is stiff (Fig. 8.5a); the anterior clinoid processes are thick and stout. Hypertrophy of the tuberculum sellae, or "bec acromégalique" may be visualized as a structure of bone density, situated on the midline, between both anterior clinoid processes (Fig. 8.5b).

The GH-secreting adenoma may also occupy only part of the sella turcica. Its image is then comparable to those of intrasellar prolactinomas. It is thus seen as a very delineated low-attenuation image within the pituitary gland (Figs. 8.6 and 8.7). In rare cases, the GH-secreting adenoma is nearly invisible and the diagnosis of acromegaly is retained, in the presence of suggestive biological and clinical features, upon an aspect of moderate increase in volume of the gland, with bulging of its upper pole and inhomogeneous gland density (Fig. 8.8). Dynamic scan may make an important contribution to the diagnosis (Figs. 8.6, 8.21).

The general tendancy of intrasellar GH-secreting adenomas is to grow downwards. Adenomas are often seen, distorting and eroding locally the sellar floor, without modifying the superior pole of the pituitary which remains flat (Fig. 8.9). Calcifications seem to be still less common than in other adenomas.

Adenomas with Intrasphenoidal Extension

Intrasphenoidal extension proceeds preferentially downwards and sometimes forwards. Distortion of the sellar floor is well visualized in coronal sections. When such sections are impossible, the presence of an adenoma with slight intrasphenoidal extension may be expressed in axial sections by partial obliteration of the sphe-

noid sinus (Fig. 8.10). When the adenoma displays an anterior extension, an important bulging of the anterior wall may be observed; the absence or the modesty of the cortical bone erosion contrasts with the importance of the distortion (Fig. 8.11). GH-secreting adenomas may show an invasive pattern: destruction of the cortical bone of the sella turcica and high attenuation value within the sphenoid sinus are then observed (Fig. 8.12). Invading adenomas may also extend laterally towards the cavernous sinuses. But diagnosis between laterally extending adenoma, adenoma invading the cavernous sinuses, and mere enlargement of the cavernous sinus due to enlarged carotid siphons, may be difficult. Dynamic scan is here fundamental.

Adenomas with Suprasellar Extension

Adenomas showing suprasellar extension are rare during acromegaly; such adenomas are distinguished from others by the poor degree of suprasellar extension and the only slightly eroded or even normal sellar floor. Enhancement is generally homogeneous. Areas of intratumoral necrosis may also be observed (Fig. 8.13). Ring enhancements are rarer than in prolactinomas (Fig. 8.14).

Other Types of GH-Secreting Adenomas

It is not uncommon to observe a partially empty sella turcica while investigating an acromegaly. Such empty sellae turcicae are more frequently secondary to a partial or total necrosis of a GH-secreting adenoma, occurring either spontaneously or after X-ray therapy.

The seemingly paradoxal association of an intrasellar extension of subarachnoidal spaces and bone signs evoking a GH-secreting pituitary adenoma, must suggest the diagnosis of a necrosed adenoma (Fig. 8.15). In other cases, the diagnosis of partial necrosis will be based upon discordance between major bone changes and the small size of the adenoma (Fig. 8.16).

CT Follow-Up of GH-Secreting Pituitary Adenoma After Treatment

Treatment of GH-secreting adenomas should be radical. Depending upon the treating teams and the clinical symptoms such treatment is surgical and/or radiotherapeutic. However, the hyposomatotropic effect of bromocriptine may be put into contribution, either before radical treatment in an attempt to reduce tumoral mass and to render surgical treatment easier, or after radical treatment when a high plasmatic level of GH still persists. By using very high doses of bromocriptine (up to 60 mg daily), a reduction in tumoral mass may be observed (Fig. 8.17), but GH plasmatic level is rarely normalized. Interruption of the treatment will provoke, in most cases, a new rise in GH level. Treatment by dopaminergic drugs may be moreover justified in mixed adenomas, secreting GH and prolactin.

After surgical cure of the adenoma, most patients are improved, but biological failures are not uncommon. Such failures are due to local invasion of the sphenoid sinus and/or the cavernous sinus (see Chap. 14).

After surgery, a reconstruction of the sellar walls may be observed. The cortical bone of the anterior wall and sellar floor may appear thickened (Fig. 8.18).

Ancillary Signs of Acromegaly

Most of the craniofacial osseous and tissulary modifications due to GH-hypersecretion may be visualized on CT scan (Fig. 8.19).

The vault is thickened; internal frontal hyperostosis and hypertrophy of the external occipital protuberance are frequently observed. CT scan also shows clearly extensive pneumatization of the facial sinuses and mastoids as well as soft tissue thickening of the scalp and muscles.

Arterial lengthenings and distensions are frequent during acromegaly. Distension of the carotid siphons brings an enlargement of the cavernous sinuses. Due to their lengthening, the carotid siphons may be displaced inside the sella turcica itself (Fig. 8.20). In these cases, the upper surface of the pituitary may be convex upwards. In the absence of dynamic CT, intrasellar carotid arteries may then simulate an abnormal pituitary gland.

Bilateral and symmetrical extraocular muscle enlargement is frequently observed during acromegaly (Rothfus, Dal Pozzo) (Fig. 8.22). This enlargement is fusiform and respects the tendinous portion of the muscle; such enlargement is more moderate than in Graves's disease and it gives rise neither to exophthalmia nor to optic nerve or superior ophthalmic vein compression. Enhancement is homogeneous within the muscles, but these muscles appear less dense than the brain, whereas they are denser in Graves's disease and in myositis. Muscular enlargement generally does not regress even when biological recovery is obtained. Lastly, in our experience, optic nerves seem enlarged, but remain rectilinear (Fig. 8.22).

References

Bamberger C, Rosier J (1980) Examen tomodensitométrique des lésions sellaires et parasellaires. Feuillets de radiologie 20:197–208

Banna M (1980) CT arteriography of microadenomas. J Comp Ass Tomog 4:690–692

Banna M, Baker HL Jr, Houser OW (1980) Pituitary and parapituitary tumours on CT. Br J Radiol 53:1123–1143

Barrow DL, Tindall GT, Kovacs K, Thorner MO, Horvath E, Hoffman JC (1984) Clinical and pathological effects of bromocriptine on prolactin-secreting and other pituitary tumors. J Neurosurg 60:1–7

Baskin DS, Boggan JE, Wilson CB (1982) Transsphenoidal microsurgical removal of growth-hormone-secreting pituitary adenomas. A review of 137 cases. J Neurosurg 56:634–641

Bonneville JF, Dietemann JL (1981) Radiology of the sella turcica. Springer Verlag, Berlin Heidelberg New York

Bonneville JF, Poulignot D, Cattin F, Couturier M, Mollet E, Dietemann JL (1982) Apport des méthodes nouvelles dans l'exploration morphologique des tumeurs hypophysaires. Ann Endocrinol (Paris) 43:303–308

Bonneville JF, Cattin F, Moussa-Bacha K, Portha C (1983) Dynamic computed tomography of the pituitary gland: the "tuft sign". Radiology 149:145–148

Burrow GN, Wortzman B, Rewcastle NB, Holgate RC, Kovacs K (1981) Microadenomas of the pituitary and abnormal sellar tomograms in an unselected autopsy series. N Engl J Med 304:156–158

Cabanis EA, Van Effenterre R, Iba-Zizen MT (1979) CT in parasellar space-occupying lesions and therapeutic decision. Acta Neurochirurgica (suppl 28) 28:329–333

Citrin CM, Davis DO (1977) CT in the evaluation of pituitary adenomas. Invest Radiol 12:27–35

Daniels DL, Williams AL, Thornton RS, Meyer GA, Cusick JF, Haughton VM (1981) Differential diagnosis of intrasellar tumors by CT. Radiology 141:697–701

Derome PJ (1982) Les adénomes hypophysaires. Encycl Med Chir 17340 A 10

Derome PJ, Jedynak CP, Peillon F (1980) Pituitary adenomas. Biology, physiopathology and treatment. Asclepios Publishers, Paris

Dietemann JL, Bonneville JF (1985) Radiological Diagnosis of Pituitary Diseases. In: Imura H (ed) The pituitary gland. Raven Press, New York, pp 341–361

Drayer BP, Kattah J, Rosenbaum A, Kennerdell J, Maroon J (1979) Diagnostic approaches to pituitary adenomas. Neurology 29:161–169

Eresue J, Drouillard J, Philippe JC, Guibert JL, Poux P, Tavernier J (1982) L'exploration des adénomes hypophysaires par scanographie à haute résolution et angioscanographie. Ann Radiol 25:509–517

Faglia G, Giovanelli MA, MacLeod RM (1980) Pituitary microadenomas. Academic Press, London

Fargason RD, Jacques S, Rand RW, Shelden CH, McCann GD, Linn P (1981) Visualization and three-dimensional reconstruction of pituitary microadenomas from CT data: a technical report. Surg Neurology 15:450–454

Gardeur D, Nachanakian A, Kulesza E, Metzger J (1979) La tomodensitométrie dans les adénomes hypophysaires. Ann Radiol 22:489–499

Gardeur D, Naidich TP, Metzger J (1981) CT analysis of intrasellar pituitary adenomas with emphasis on patterns of contrast enhancement. Neuroradiology 20:241–242

Ghoshhajra K (1981) High-resolution metrizamide CT cisternography in sellar and suprasellar abnormalities. J Neurosurg 54:232–239

Hardy J, Robert F, Somma M, Vezina JL (1973) Acromégalie-gigantisme. Traitement chirurgical par exérèse transsphénoïdale de l'adénome hypophysaire. Neurochirurgie 19 (supplément 2)

Hemminghytt S, Kalkhoff RK, Daniels DL, Williams AL, Grogan JP, Haughton VM (1983) Computed tomographic study of hormone-secreting microadenomas. Radiology 146:65–69

Hori T, Muraoka K, Hokama Y, Takami M, Saito Y (1982) A growth-hormone-producing pituitary adenoma and an internal carotid artery anevrysm. Surg Neurol 18:108–111

Jacobi JD, Fishman LM, Daroff RB (1974) Pituitary apoplexy in acromegaly followed by partial pituitary insufficiency. Arch Intern Med 134:559–561

Kleinberg DL, Boyd AE, Wardlaw S, Frantz AG, George A, Bryan N, Hilal S, Greising J, Hamilton D, Seltzer T, Sommers CJ (1983) Pergolide for the treatment of pituitary tumors secreting prolactin or growth hormone. N Engl J Med 22:704–709

Kovacs K, Ryan N, Howath E, Singer W, Exrin C (1980) Pituitary adenomas in old age. J Gerontol 35:16–22

Kricheff II (1979) The radiologic diagnosis of pituitary adenoma. Radiology 131:263–265

Kuuliala I (1981) CT of pituitary adenomas. Clinical Radiology 32:259–264

Lamberg BA, Pelkonen R, Gordin A, Haltia M, Wahlström T, Paetau A, Leppäluoto J (1983) Hyperthyroidism and acromegaly caused by a pituitary TSH- and GH-secreting tumour. Acta Endocrinol 103:7–14

Lemaître G, Linquette M, Fossati P, Cappoen JR (1982) Détection des adénomes hypophysaires sécrétants par tomodensitométrie. Ann Med Int 133:33–34

Linfoot JA (1979) Recent advances in the diagnosis and treatment of pituitary tumors, Raven Press, New York

Linquette M, Fossati P (1980) Les adénomes hypophysaires sécrétants. Rev Prat, Paris 30:3017–3040

McLachlan MSF, Wright AD, Doyle FH (1970) Plain film and tomographic assessment of the pituitary fossa in 140 acromegalic patients. Br J Radiol 43:360–369

Metzger J, Gardeur D, Houlbert D, Thibierge M (1982) Surveillance neuro-radiologique des adénomes hypophysaires sécrétants. Ann Med Int 133:29–32

Mohr G, Hardy J (1982) Hemorrhage, necrosis, and apoplexy in pituitary adenomas. Surg Neurol 18:181–189

Naidich TP, Pinto RS, Kushner MJ, Lin JP, Kricheff II, Leeds NE, Chase NE (1976) Evaluation of sellar and parasellar masses by computed tomography. Radiology 120:91–99

Racadot J (1980) Adénomes de l'antéhypophyse: histoire naturelle, classification et histopathologie. Rev Prat (Paris) 30:2981–3002

Raji MR, Kishore PRS, Becker AP (1981) Pituitary microadenoma: a radiological surgical correlative study. Radiology 139:95–99

Richmond IL, Wilson CB (1978) Pituitary adenomas in childhood and adolescence. J Neurosurg 49:163–168

Rilliet B, Mohr G, Robert F, Hardy J (1981) Calcification in pituitary adenomas. Surg Neurol 15:249–255

Robertson WD, Newton TH (1978) Radiologic assessment of pituitary microadenomas. AJR 131:489–492

Rothfus WE, Curtin HD (1984) Extraocular muscle enlargement: a CT review. Radiology 151:677–681

Syvertsen A, Haughton VM, Williams AL, Cusick JF (1979) The computed tomographic appearance of the normal pituitary gland and pituitary microadenomas. Radiology 133:385–391

Taylor S (1982) High resolution computed tomography of the sella. Radiologic clinics of North America 20:207–236

Thibaut A, Rausin L, Stenevaert A (1981) Correlative study of the sella turcica with the site and size of 82 secreting microadenomas. Neuroradiology 20:281–285

Tindall GT, McLanahan CS (1980) Hyperfunctional pituitary tumors: pre- and postoperative management considerations. Clin Neurosurg 27:48–82

Tolis G, Koutsilieris M, Bertrand G (1984) Endocrine

diagnosis of growth hormone-secreting pituitary tumors. In: Peter MCL, Black et al. (eds) Secretory tumors of the pituitary gland. (Progress in Endocrine Research and Therapy, vol 1.) Raven Press, New York

Valenta LJ, Sostrin RD, Eisenberg H, Tamkin JA, Elias AN (1982) Diagnosis of pituitary tumors by hormone assays and computerized tomography. Am J Med 72:861–873
Warner BA, Santen RJ, Page RB (1982) Growth hormone and prolactin secretion by a tumor of the pharyngeal pituitary. Ann Intern Med 96:65–66

Wass JA, Moult PJ, Thorner MO (1979) Reduction of pituitary-tumour size in patients with prolactinomas and acromegaly treated with bromocriptine with or without radiotherapy. Lancet 2:66–69
Weiss MH (1981) Medical and surgical management of functional pituitary tumors. Clin Neurosurg 28:374–383
Wilson CB, Dempsey LC (1978) Transsphenoidal microsurgical removal of 250 pituitary adenomas. J Neurosurg 48:13–22

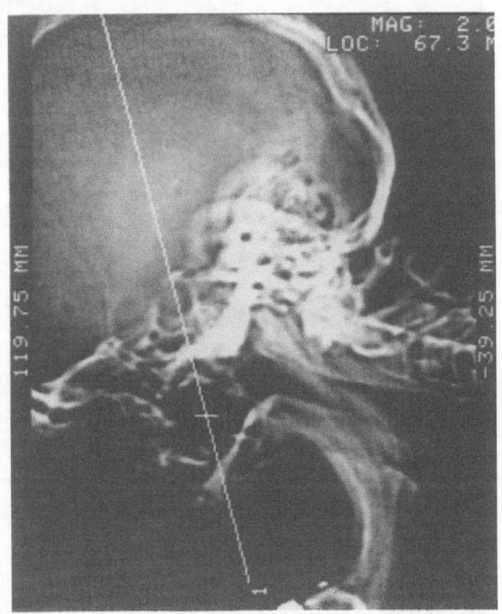
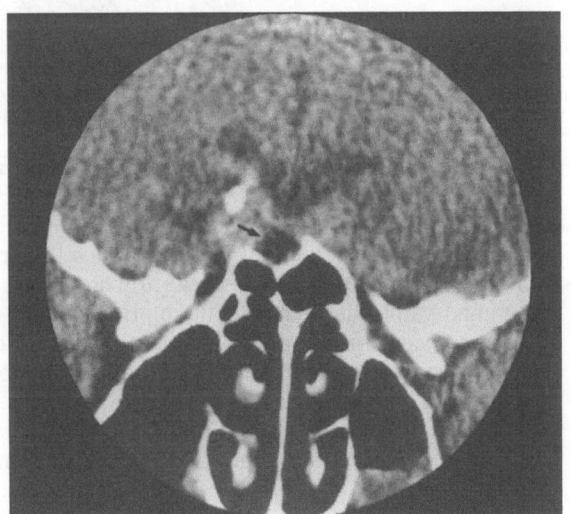

Fig. 8.1 a, b. Adequate head positioning hardness in an acromegalic patient. **a** Lateral digital localizer image. Prognathism and severe spondylotic changes impede correct head positioning for coronal scanning. **b** The result is not a true coronal view. The section is through the maxillary sinuses. Hypodense area within the sella represents the GH-secreting adenoma (*arrow*)

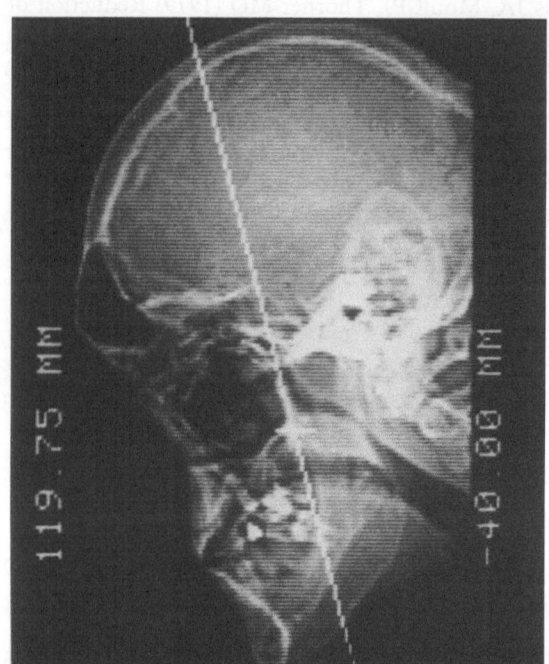

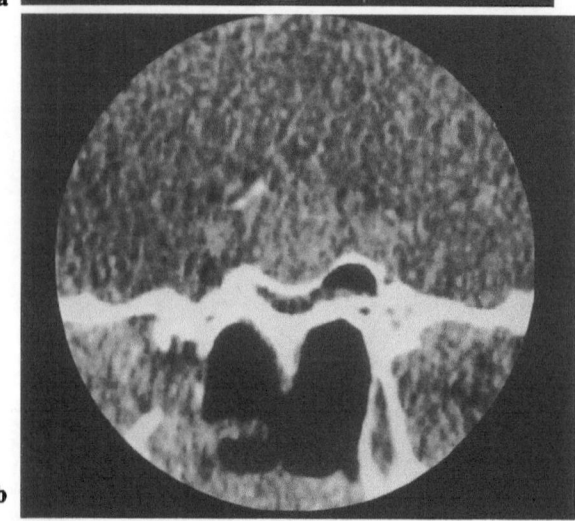

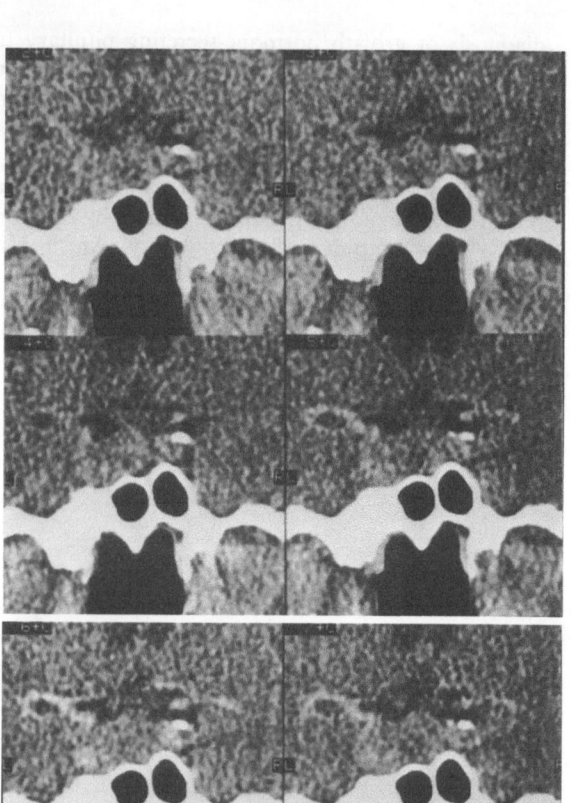

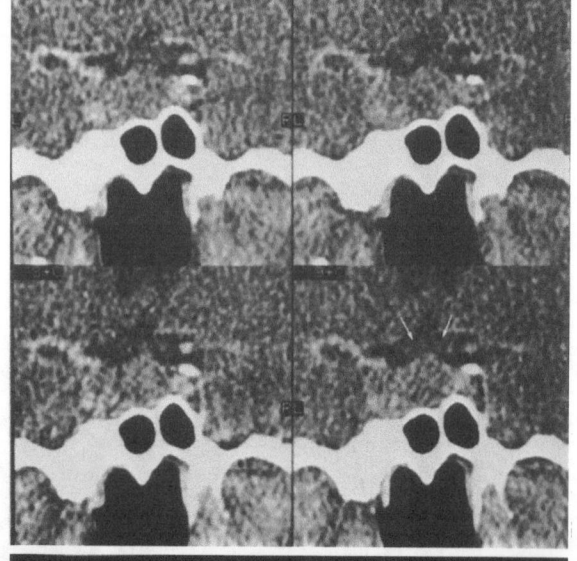

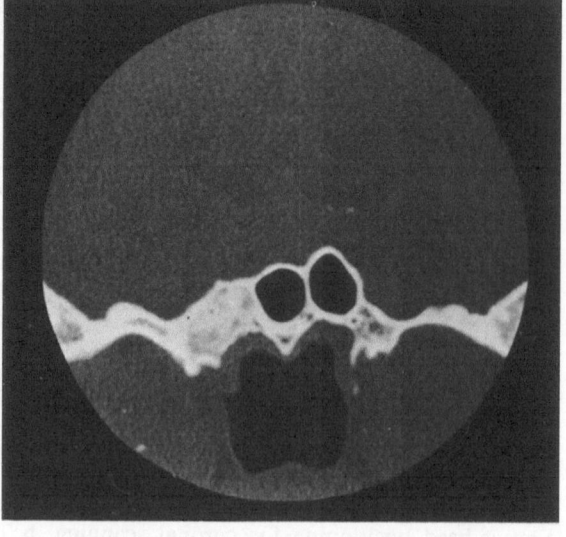

Fig. 8.2a, b. Blurred coronal image of the pituitary region in an acromegalic patient. **a** The lateral localizer image demonstrates that opening of mandibular angle in acromegalic patient impedes to putting the sellar area into the center of the field of view. **b** Thickening of the cranial vault contributes to the blurred pattern of the coronal view and the poor attenuation value differences between brain and pituitary tumor in spite of contrast infusion

Fig. 8.3a–c. GH-secreting pituitary adenoma. Acromegalic symptoms for 8 years in a 43-year-old man. **a, b** Coronal dynamic CT scan. Enlargement of the sellar content with bulging of the upper surface of the gland. The optic chiasm is not raised (*arrows*). Faint enhancement of the mass after contrast. **c** Bone review image. Lateralized depression of the sella floor without thinning of the cortical bone

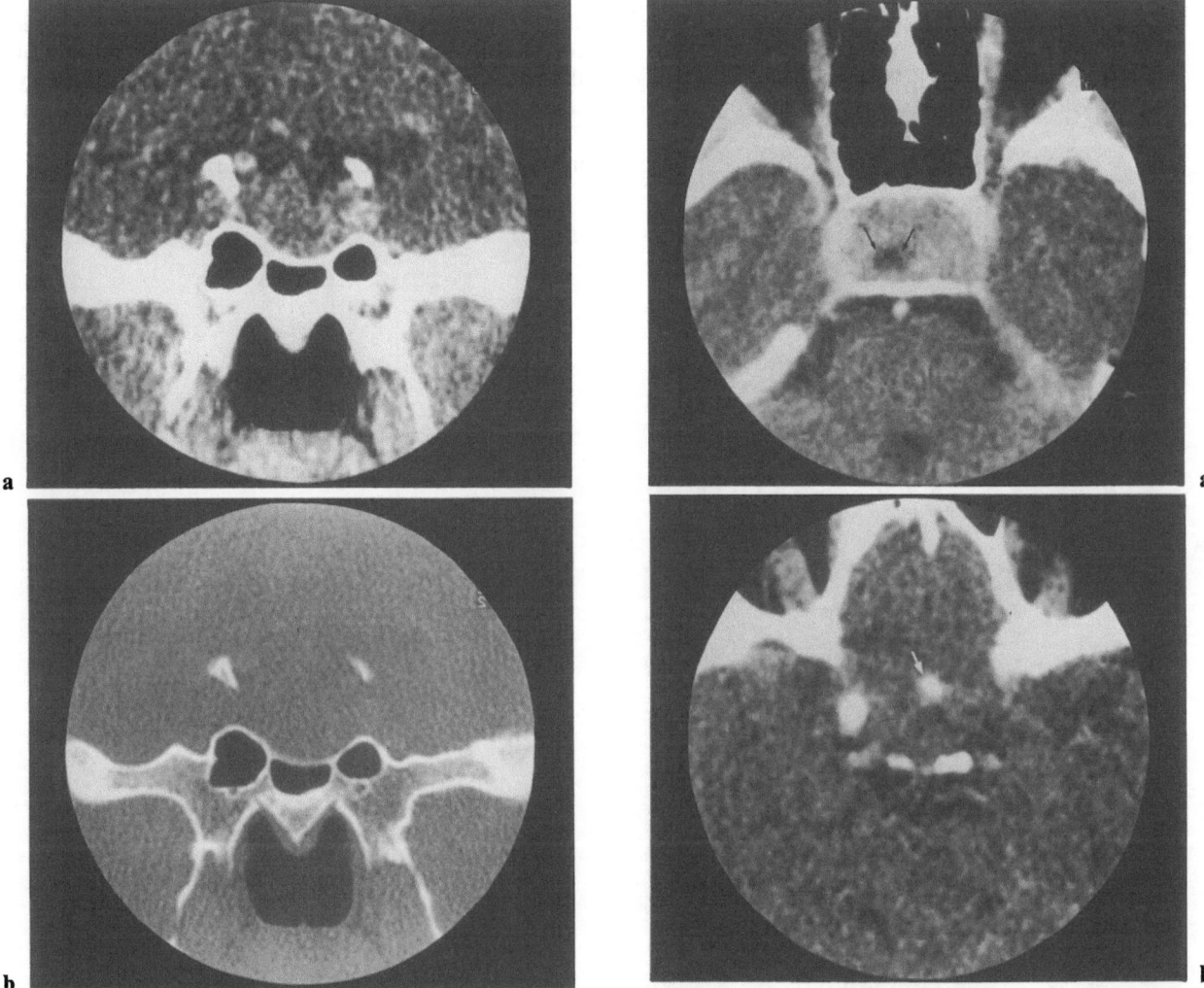

Fig. 8.4a, b. Acromegaly in a 46-year-old man. GH level is 13 ng/ml. **a** Coronal contrast CT scan. Enlargement of the pituitary fossa. Homogeneous enhancement of a diffuse pituitary adenoma. The upper pole of the tumor is pointing into the chiasmatic cistern. **b** Bone review image. Thinning of the midpart of the sellar floor. Erosive changes of the clinoid processes indicate that the tumor presented previously a suprasellar extension (spontaneous tumor shrinkage)

Fig. 8.5a, b. Acromegaly in a 28-year-old woman. **a** Axial contrast CT scan. Enlargement of the sellar content pushing the carotid arteries laterally. Small necrotic area on the right (*arrows*). Note the straight dorsum sellae. **b** Axial section 12 mm above **a.** Note the prominent acromegalic "bec" (*arrow*)

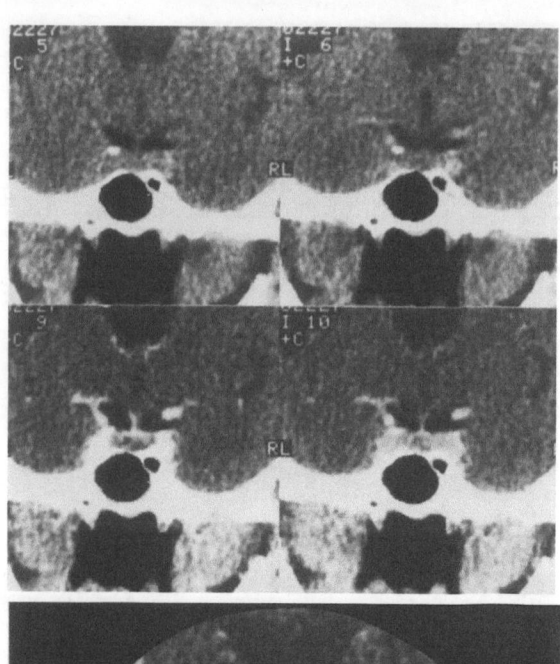

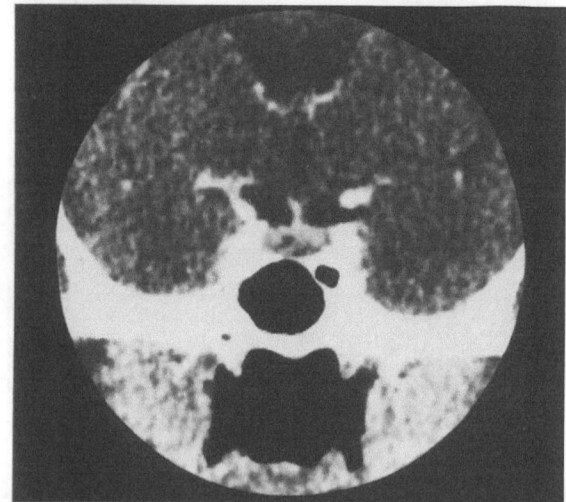

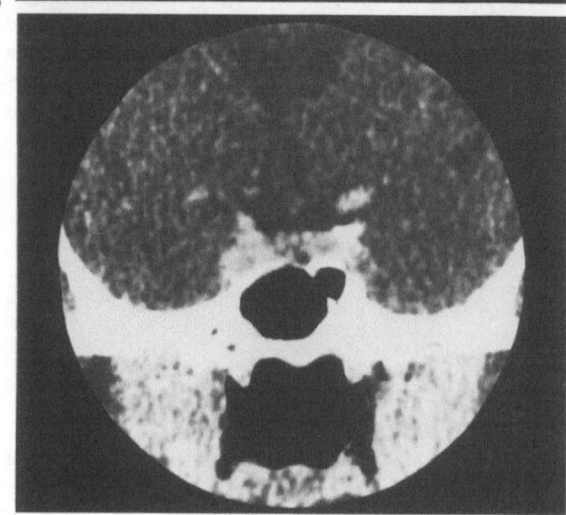

Fig. 8.6a–c. Acromegalic symptoms for 3 years in a 10-year-old boy. **a** Dynamic coronal CT scan. **b** Magnified view of part of **a**: compression of the bottom of the pituitary tuft without displacement of the pituitary stalk. **c** Contrast CT scan 1 min after bolus injection. 2-mm hypodense lesion on the midline: a GH-secreting microadenoma was confirmed at surgery

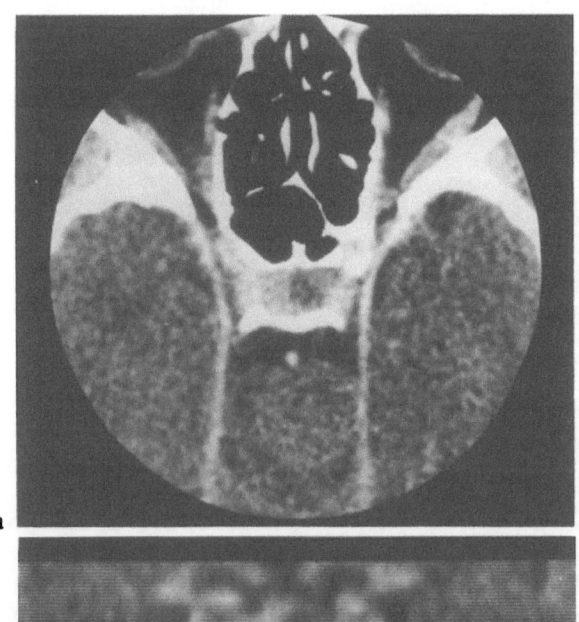

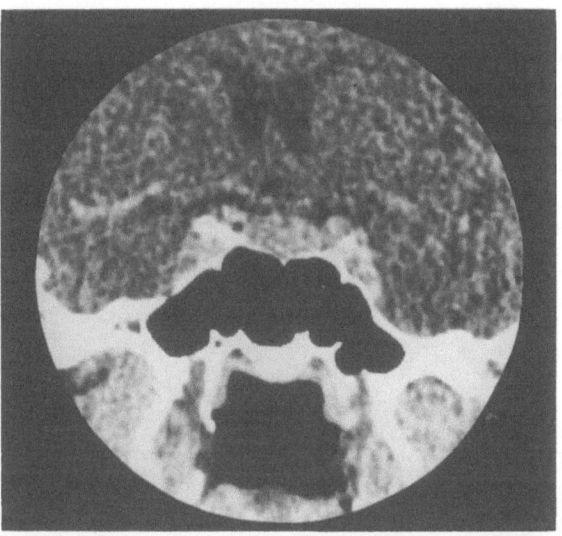

Fig. 8.8. GH-secreting pituitary microadenoma. Acromegaly for 15 years in a 70-year-old woman. Coronal contrast CT scan. Inhomogeneous enhancement of an enlarged pituitary gland; erosion of the sellar floor. A 5-mm right-sided microadenoma was found at surgery

Fig. 8.7a, b. Acromegaly in a 53-year-old man. **a** Axial contrast CT scan (coronal CT scan was not diagnostic). Rounded defect in enhancement on the left. **b** Coronal reformatted image. Slight bulging of the upper pole of the gland above a left-sided microadenoma proven at surgery

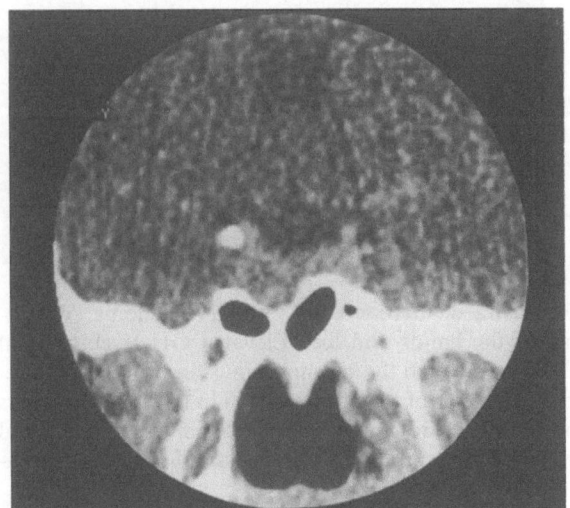

Fig. 8.9. GH-secreting pituitary microadenoma with intrasphenoidal extension in a 34-year-old man. Coronal contrast CT scan. Marked depression of the sellar floor on the right. The pituitary adenoma in front of the bony changes is less enhanced than the normal pituitary gland. No bulging of the upper pole of the sellar content

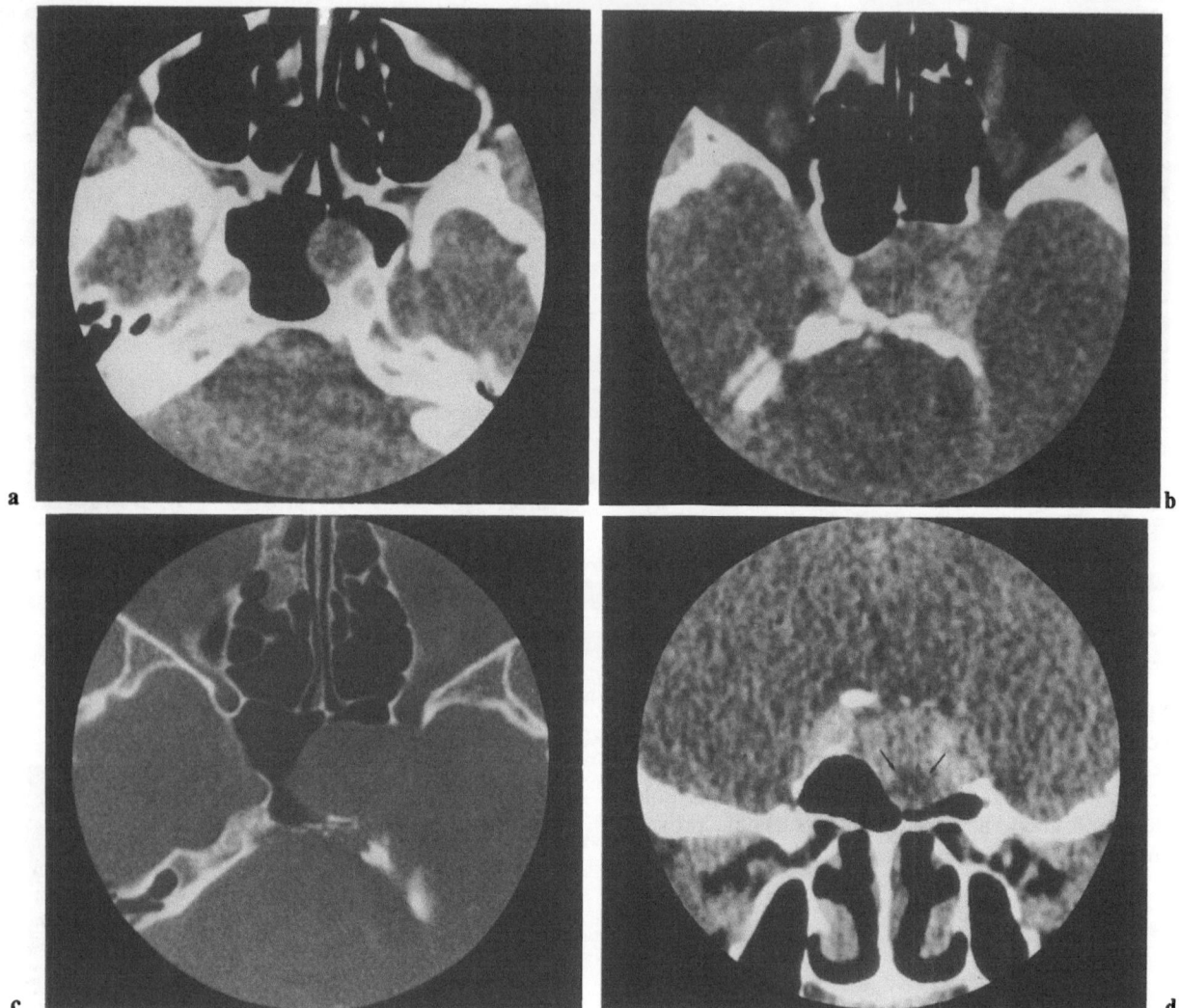

Fig. 8.10a–d. GH-secreting pituitary adenoma. Widening of the hands and feet. Elevated GH level. **a** Axial CT scan through the sphenoid sinus. Imprint of the inferior extension of the adenoma within the sphenoid sinus on the left. **b** Bone review image 5 mm above **a**. **c** Axial contrast CT scan through the sella. Inhomogeneous enhancement of the enlarged pituitary fossa. **d** Coronal contrast CT scan. Marked depression and thinning of the sellar floor on the left. Increase of the pituitary height with bulging of the diaphragma sellae. A low density within the adenoma (*arrows*) corresponds to a necrotic area

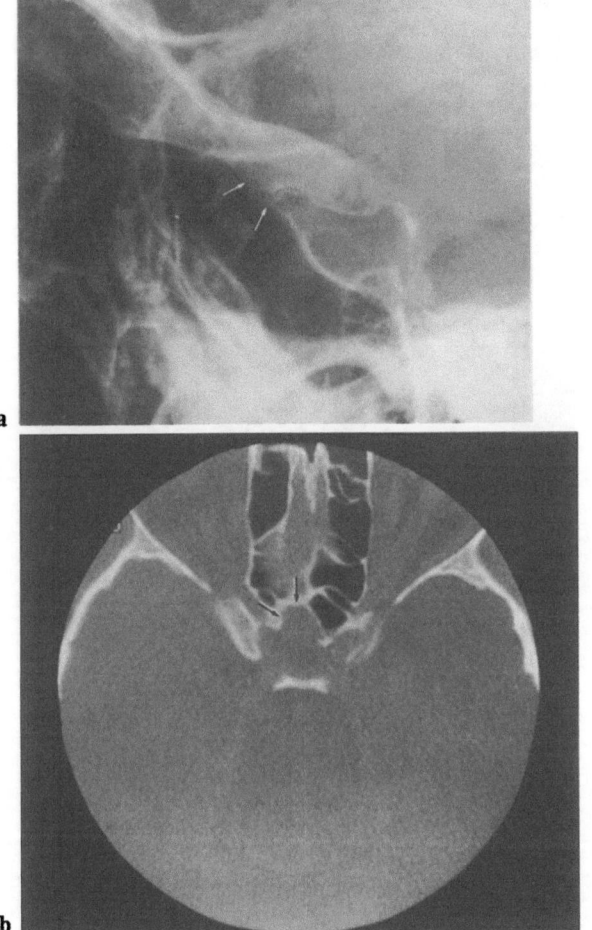

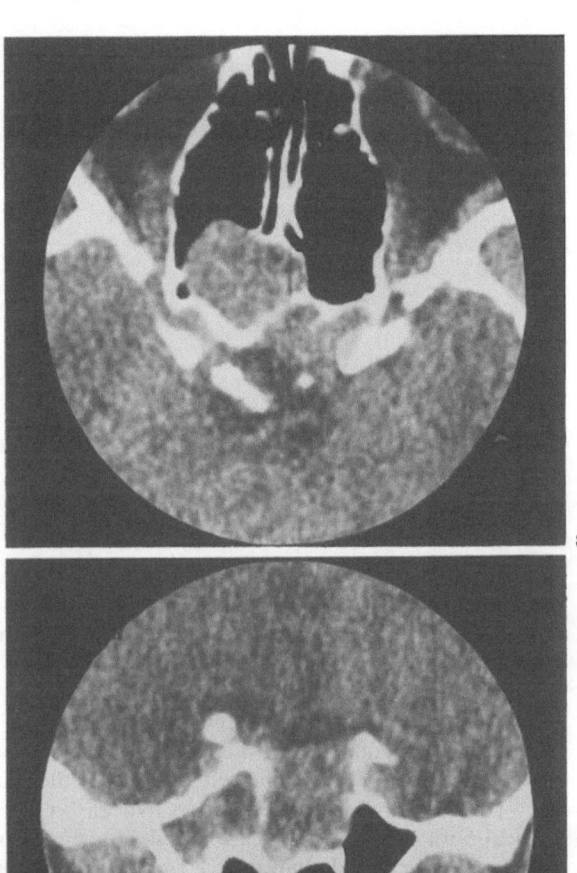

Fig. 8.11a, b. Anteriorly located GH-secreting pituitary adenoma. **a** Sella turcica lateral view. **b** Axial bone review image. Anterior ballooning and thinning of the upper part of the anterior wall of the sella (*arrows*) more pronounced on the right

Fig. 8.12a, b. GH-secreting invasive pituitary adenoma in a 38-year-old acromegalic man. Very elevated GH levels. **a** Axial contrast CT scan. Soft tissue density within the sphenoid sinus on the right. **b** Coronal contrast enhanced CT scan. Homogeneous soft tissue density occupying the pituitary fossa and the sphenoid sinus. The sellar floor is totally destroyed

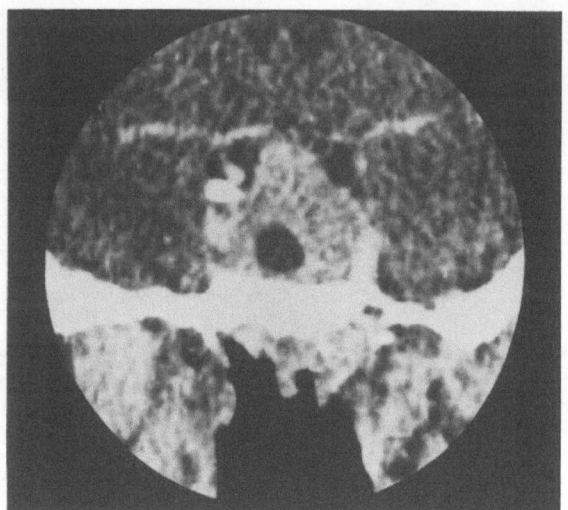

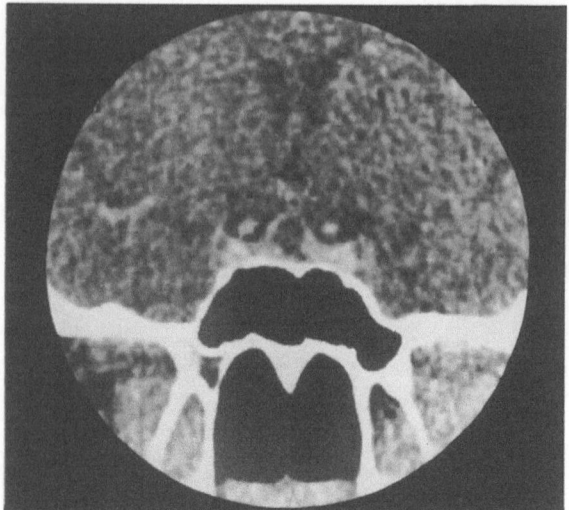

Fig. 8.13. GH-secreting pituitary adenoma in a 28-year-old woman. Coronal contrast CT scan. Enlarged pituitary fossa. Marked suprasellar extent. Necrotic area within the inferior part of the tumor

Fig. 8.14a–c. GH-secreting pituitary adenoma in a ▶ 57-year-old woman. **a** Contrast coronal CT scan. Hypodense focal area on the midline with bulging of the pituitary contents. **b** Bone review image. No pathological cortical bone erosion. Deep carotid sulci (normal variant) (*arrows*). **c** Axial contrast CT scan. The pituitary adenoma occupies the bottom of the suprasellar cistern

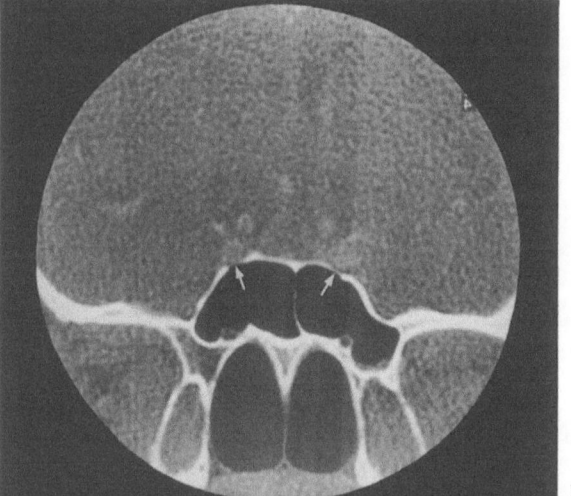

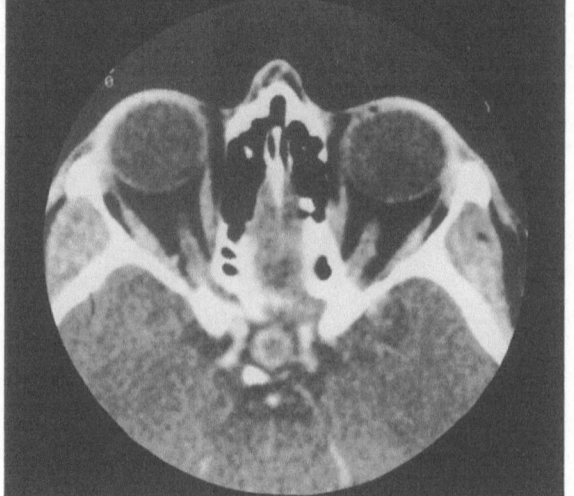

Fig. 8.15 a–c. Acromegaly in a 57-year-old woman. Bromocriptine was given for 2 years (7.5 mg/day). **a** Sella turcica lateral magnified view. Asymmetrical depression of the sellar floor (*arrows*). **b** Bone review image. The sellar floor is depressed on the left, but is not eroded (*arrow*). **c** Coronal contrast CT scan. Partially empty sella. The diaphragma sellae is depressed on the left. No visible tumor. Probable shrinkage of a GH-secreting adenoma after bromocriptine treatment

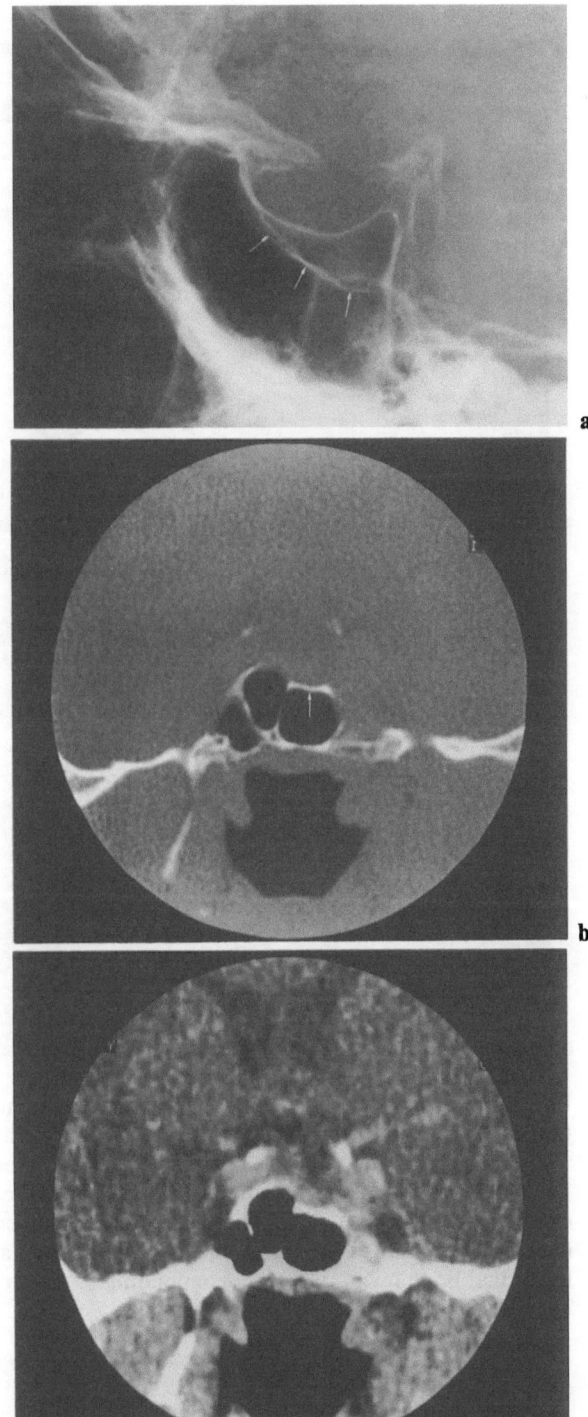

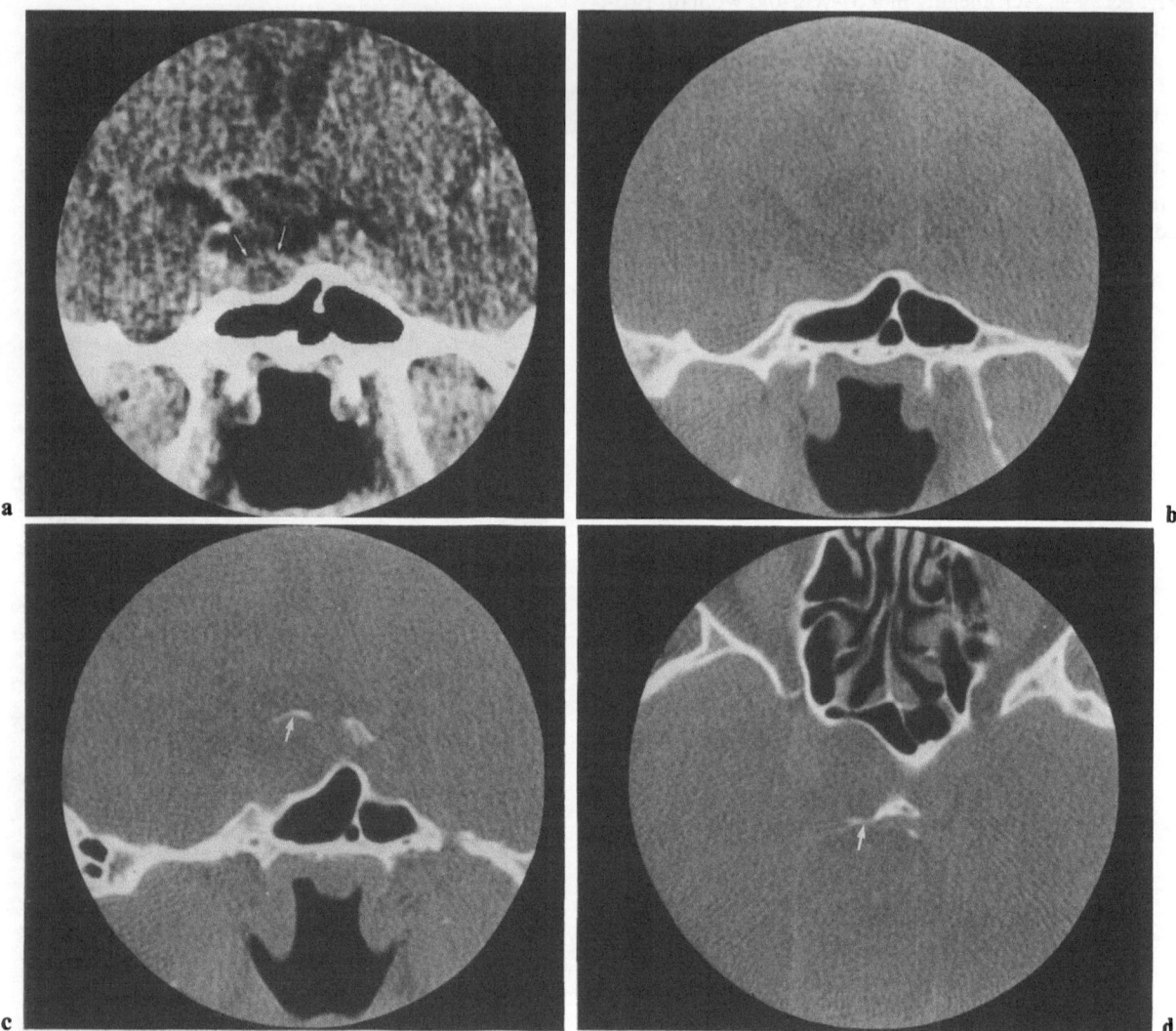

Fig. 8.16a–d. GH-secreting pituitary adenoma. Evolution with X-ray therapy in a 52-year-old acromegalic woman. **a** Coronal contrast CT scan. Hypodense area on the right with slight bulging of the upper surface of the gland (*arrows*). Note the discordance between marked bony changes and the small size of the adenoma. **b** Coronal bone review image (corresponding to **a**). Depression without thinning of the sellar floor on the right. **c** Coronal bone review image (5 mm behind **b**). Depression of the posterior part of the sellar floor on the right. Raising of the modified right posterior clinoid process (*arrow*). **d** Axial bone review image. Right ballooning of the anterior wall of the sella and erosive changes of the dorsum sellae (*arrow*)

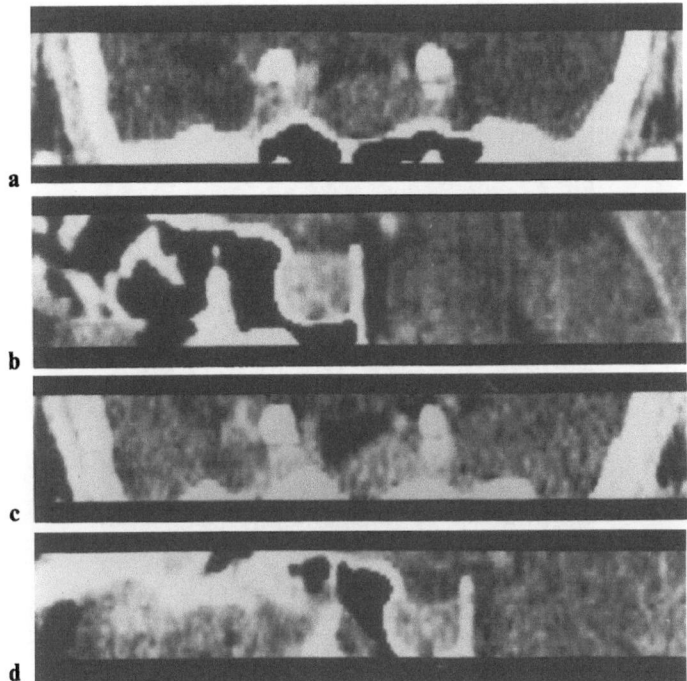

Fig. 8.17a–d. GH-secreting pituitary adenoma in a 44-year-old man, before and after bromocriptine. **a** Coronal and **b** sagittal reformatted image. Before treatment, enlargement of the sellar content with marked bulging of the upper surface of the pituitary gland. Note in **a** the depression of the sellar floor and in **b** the rectilinear aspect of the dorsum sellae. **c** Coronal and **d** sagittal reformatted image 6 months after institution of bromocriptine treatment (10 mg per day): decrease of the sellar content height

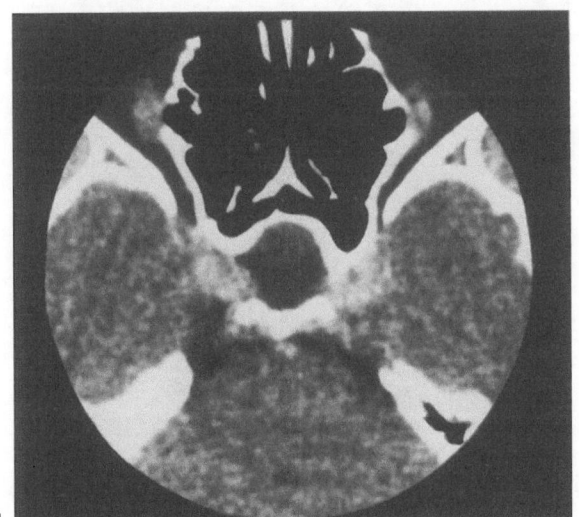
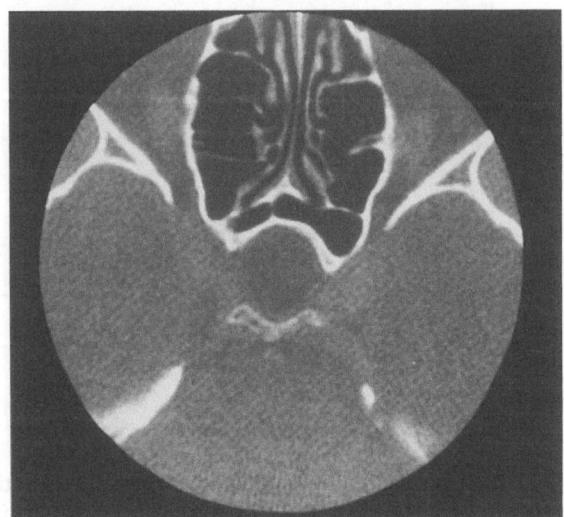

Fig. 8.18a, b. Acromegaly. Postoperative changes. **a** Axial contrast CT scan. Hypodense sellar content corresponding to a secondary empty sella. The nonvisible pituitary stalk was probably cut during surgery. **b** Axial bone review image. Bulging and thickening of the anterior wall of the sella

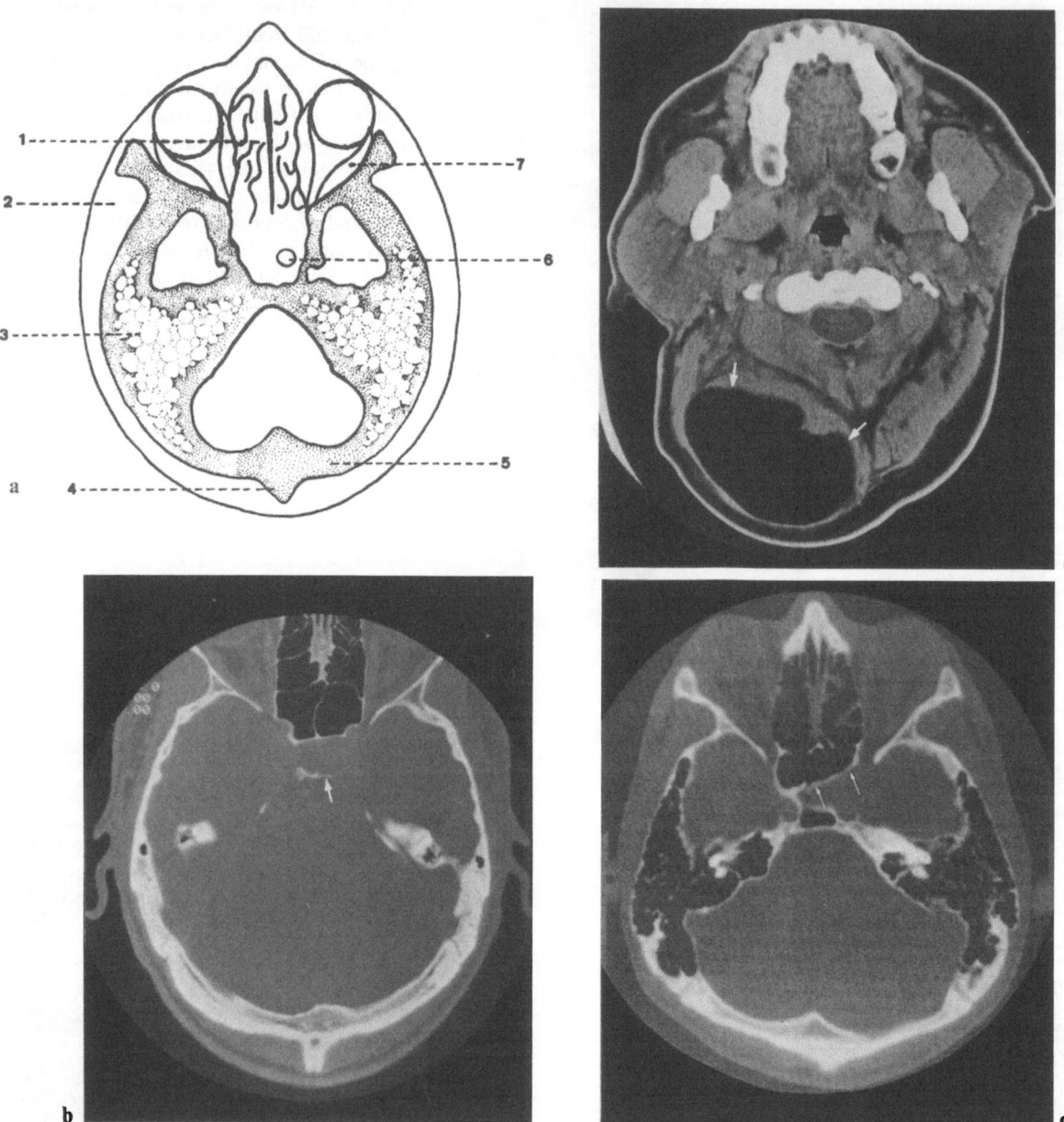

Fig. 8.19a–d. Ancillary signs of acromegaly. **a** Schematic drawing of ancillary signs of acromegaly: 1, extensive pneumatization of the ethmoidal cells; 2, soft tissue thickening; 3, extensive pneumatization of the mastoids; 4, hypertrophy of the external occipital protuberance; 5, thickening of the vault; 6, imprint of the inferior extension of the adenoma within the sphenoid sinus; 7, enlargement of the extraocular muscles. **b** Axial CT scan. Marked soft-tissue thickening; enlarged temporal, occi- pital, and pharyngeal muscles. Incidental demonstration of a large subcutaneous occipital lipoma (*arrows*). **c** Axial bone review image. Extensive pneumatization of the ethmoidal cells and mastoids. The left part of the anteri- or wall of the sella is pushed forwards (*arrows*). **d** Axial bone review image. Thickening of the vault and hyper- trophy of the external occipital protuberance. Erosion of the left part of the dorsum sellae (*arrow*)

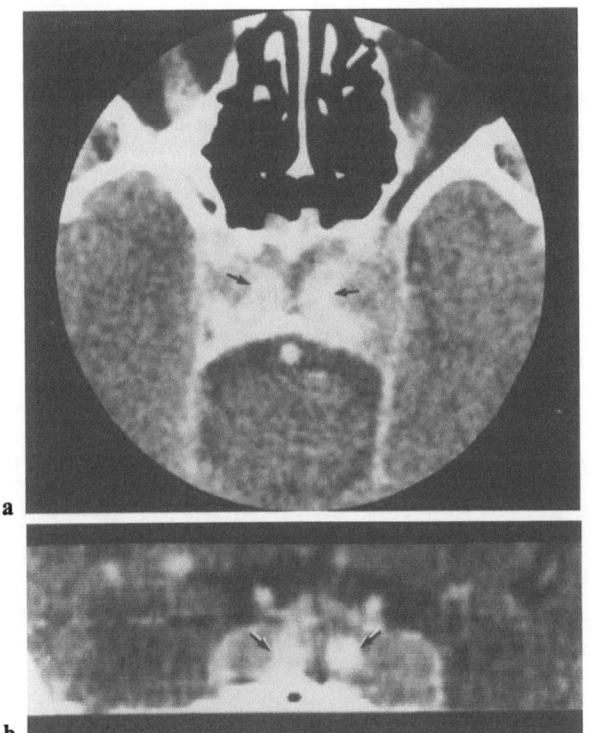

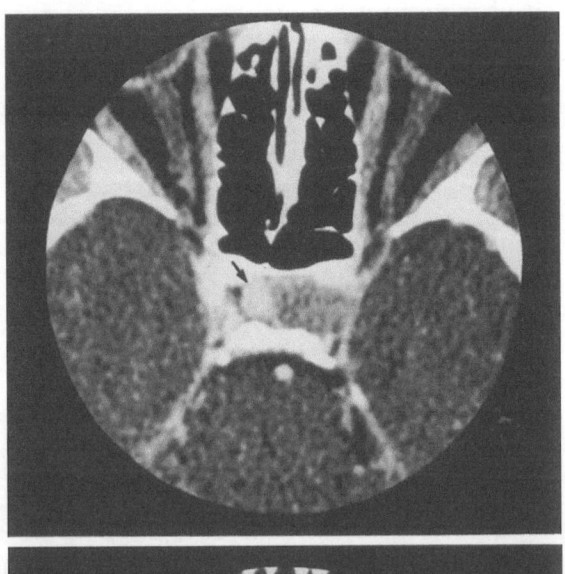

Fig. 8.20 a, b. Carotid siphons in acromegaly. **a** Axial contrast CT scan and **b** coronal reformatted image. Medial displacement of the elongated and enlarged carotid siphons towards the pituitary fossa (*arrows*)

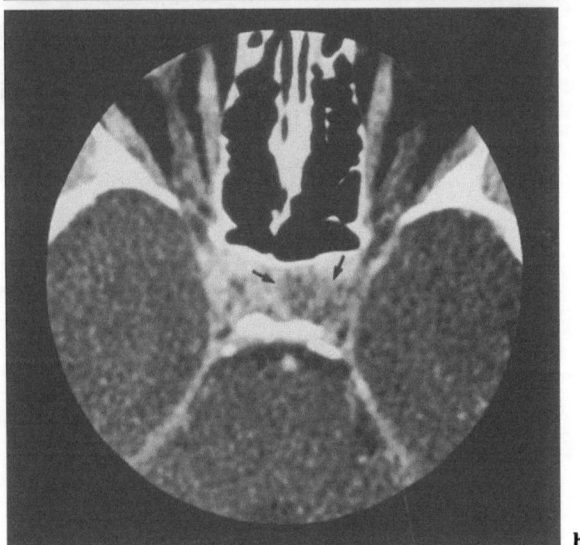

Fig. 8.21 a, b. Acromegaly in a 37-year-old man. Pituitary microadenoma. **a** Axial dynamic CT scan. Displacement to the right of normal pituitary tissue here mimicking intrasellar carotid siphon (*arrow*). **b** Axial CT scan. The pituitary adenoma on the left is well shown (*arrows*)

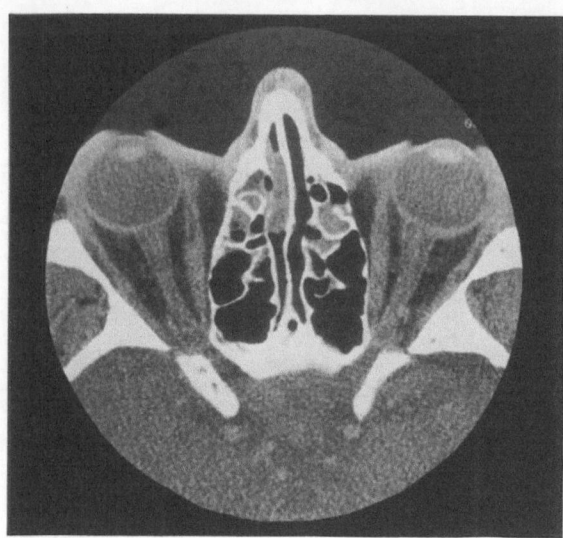

Fig. 8.22. Extraocular muscle changes in acromegaly. Axial CT scan. Slight bilateral and symmetrical enlargement of the extraocular muscles, especially the medial recti, without proptosis. The optic nerves are also moderately enlarged

Chapter 9
ACTH-Secreting Pituitary Adenomas

Cushing's disease is most of the time due to a pituitary corticotropic-cell adenoma. The treatment of this disease is surgical: it consists in a selective adenoma removal. Detection by computed tomography is thus essential, but such detection is not easy because on the one hand the lesion is often very small and on the other the obesity, the buffalo neck, and the backache characteristic of Cushing's disease, render direct coronal sections and dynamic scan difficult. Moreover, the tuft sign may be uncertain or even absent for certain corticotropic-cell microadenomas localized on the midline.

Clinical and Biological Reminder of Cushing's Disease

Cushing's disease represents 70% of all endogenous hypercortisolisms in adults. The hypercortisolism is secondary to an inappropriate secretion of ACTH, responsible for a bilateral adrenocortical hyperplasia. The hypercortisolism accounts for numerous clinical manifestations. The anomalous distribution of fat gives an obesity confined to the face, neck, and trunk, outstanding by an untripeted distribution of fat. The amyotrophy concerns the origin of the limbs, backside, and calves. Skin lesions are often characteristic: reddish face, acne, seborrhea, hypertrichosis. The purplish striae, localized on the abdomen and on the origin of the upper limbs, are very suggestive of Cushing's disease. Asthenia, arterial hypertension, and osteoporosis responsible for bone aches, account for the seriousness of the disease. Biological investigation confirms the existence of a hypercortisolism. The most convincing arguments in favor of the central origin of the disease are:

1. A response exceeding 40% to a high-dose dexamethasone suppression test; an explosive response to a metopyrone test
2. A normal or increased plasma ACTH level, but not exceeding 20 pg per liter
3. Lastly, a normal or increased beta-LPH level.

CT Signs of ACTH-Secreting Pituitary Adenomas

The importance of the detection of a corticotropic-cell adenoma in Cushing's disease calls for a rigorous and complete investigation in order to confirm the localization of the adenoma on several views. Such investigation thus ideally requires coronal sections with dynamic scan and axial sections with coronal and sagittal reconstructions. If direct coronal sections are not possible, thin axial cuts of 1.5 mm every millimeter, with overlapping, will be realized in order to obtain the most reliable reconstructions possible (Fig. 9.1).

Our experience is based on CT exploration of 48 patients displaying Cushing's disease of central origin. A clearly defined adenoma was found in 55% of our patients. An abnormal pituitary gland, but showing no clearly defined microadenoma, was seen in 15% of our cases. Thus, a total of 70% of abnormal pituitary glands are noted. In all cases where the CT was abnormal and where the patients had been operated, a corticotropic-cell adenoma was seen either directly during the operation or secondarily during histologic examination. In 8% of cases, the investigation was judged insufficient. In 22% of cases, that is, in 11 patients, the pituitary gland was considered normal. A surgical exploration was carried out in six of these patients. One adenoma of 3 mm was found during operation. Three very small adenomas were found only upon pathological examination, two explorations proved negative. On the whole, CT investigation of pituitary gland during Cushing's disease of central origin permits a diagnosis of hypophyseal anomaly in 70% of cases, thus guiding and facilitating surgery.

The pituitary corticotropic-cell adenomas seen on CT scan are of small size measuring 2–10 mm in diameter; most of the time they are strictly intrasellar causing at most a bulging of the upper limit of the gland (Fig. 9.2). We have not observed any intraadenomatous calcification or necrotic modifications. The repercussion on the sellar floor is absent or modest (Fig. 9.3). The thinning of the floor, when it exists, is always difficult to appreciate due to associated osteoporosis. In our series, changes of the sellar floor secondary to the presence of an adenoma have been observed in 17% of all cases.

Most of the time a corticotropic-cell microadenoma appears as a round or oval intrahypophyseal low-density area, causing sometimes a bulging of the upper part of the gland facing the lesion (Figs. 9.4, 9.5a, and 9.6). The difference in density between the normal pituitary tissue and the adenoma can be very small: dynamic scan here becomes fundamental, allowing visualization of a deviation of the pituitary capillary bed or a delayed contrast enhancement in an area of the gland (Figs. 3.17, 9.5b, and 9.7); 50% of microadenomas appear lateralized within the gland. Another 50% are on the midline or paramedian. Adenomas located on the midline may cause no deviation of the pituitary capillary bed, thus rendering their diagnosis more difficult. The adenoma reaches 10 mm in diameter only in about 10% of cases (Fig. 9.8). The upper limit of the gland is then really convex, but in our experience it never constitutes a threat for the chiasm. It is generally in these relatively large adenomas that changes of the sellar floor may be clearly recognized.

Exceptionally the adenoma, whatever its size, may be of the same density as the normal pituitary gland after injection of iodinated contrast medium (Figs. 3.18 and 9.3). If the volume of the lesion is insufficient to provoke a bulging of the upper limit of the gland or a modification of the sellar floor, the only visible sign can then be a deviation of the pituitary capillary bed on dynamic scan (Fig. 3.17). However, the microadenoma may completely escape CT investigation if its diameter does not exceed 1 to 2 mm and specially if the examination is not perfectly reliable and if dynamic scan has not been obtained. Thus one must bear in mind that a normal CT of the pituitary gland must not contraindicate a surgical exploration of the gland if there is a formal diagnosis of Cushing's disease of central origin.

In our experience, CT exploration of Cushing's disease, after treatment with OP'DDD and/or X-ray therapy, often reveals a low-heighted pituitary gland, about 1 to 2 mm on the midline, with a concave upper limit and no disclosed adenoma (Figs. 9.9).

Other CT Findings in Cushing's Disease

The essential aim of CT investigation is to demonstrate the corticotropic-cell adenoma. CT also permits visualization of certain signs relative to hypercortisolism. In severe cases of Cushing's disease, CT clearly demonstrates osteoporosis; it consists of demineralization of the sellar floor as well as the whole cranial base and vault (Figs. 9.7 and 9.9). The CT scan also shows the anomalous distribution of fat in Cushing's disease. In agreement with Bachow's description, we have observed in 30% of cases of Cushing's disease, an abnormal fat deposit within the cavernous sinuses. Fat deposits, nodular or roughly triangular in shape, with a density between -30 and -60 Hounsfield units, can be physiologically observed near the superior orbital fissure. These fat deposits, when located in the medium or especially posterior part of the cavernous sinuses, seem more specific of hypercortisolism (Figs. 9.9 and 9.10). These fat deposits should not be taken for an intracavernous cranial nerve whose volume is less important and whose density is higher. We have lastly observed, in one case of Cushing's disease, a voluminous fat deposit located not within the cavernous sinus, but inside the pituitary fossa between the cavernous sinus and the pituitary gland (Fig. 9.9).

References

Bachow TB, Hesselink JR, Aaron JO, Davis KR, Taveras JM (1984) Fat deposition in the cavernous sinus in Cushing's disease. Radiology 153:135–136

Bamberger C, Rosier J (1980) Examen tomodensitométrique des lésions sellaires et parasellaires. Feuillets de radiologie 20:197–208

Barrow DL, Tindall GT, Kovacs K, Thorner MO, Horvath E, Hoffman JC (1984) Clinical and pathological effects of bromocriptine on prolactin-secreting and other pituitary tumors. J Neurosurg 60:1–7

Bigos ST, Robert F, Pelletier G, Hardy J (1977) Cure of Cushing's disease by transsphenoidal removal of a microadenoma from a pituitary gland despite a radiologically normal sella turcica. J Clin Endocrinol 45:1251–1260

Bonneville JF, Dietemann JL (1981) Radiology of the sella turcica. Springer Verlag, Berlin Heidelberg New York

Bonneville JF, Poulignot D, Cattin F, Couturier M, Mollet E, Dietemann JL (1982) Apport des méthodes nouvelles dans l'exploration morphologique des tumeurs hypophysaires. Ann Endocrinol (Paris) 43:303–308

Bonneville JF, Cattin F, Moussa-Bacha K, Portha C (1983) Un plus dans l'exploration de l'hypophyse: l'angioscan. Presse Méd 12:1669

Bonneville JF, Cattin F, Moussa-Bacha K, Portha C (1983) Dynamic computed tomography of the pituitary gland: the "tuft sign". Radiology 149:145–148

Burrow GN, Wortzman B, Rewcastle NB, Holgate RC, Kovacs K (1981) Microadenomas of the pituitary and abnormal sellar tomograms in an unselected autopsy series. N Engl J Med 304:156–158

Citrin CM, Davis DO (1977) CT in the evaluation of pituitary adenomas. Invest Radiol 12:27–35

Derome PJ (1982) Les adénomes hypophysaires. Encycl Med Chir 17340 A 10

Derome PJ, Jedynak CP, Peillon F (1980) Pituitary adenomas. Biology, physiopathology and treatment. Asclepios Publishers, Paris

Doppmann JL, Oldfield E, Krudy AG, Chrousos GP, Schulte HM, Schaaf M, Loriaux LD (1984) Petrosal sinus sampling for Cushing syndrome: anatomical and technical considerations. Radiology 150:99–103

Eresue J, Drouillard J, Philippe JC, Guibert JL, Poux P, Tavernier J (1982) L'exploration des adénomes hypophysaires par scanographie à haute résolution et angioscanographie. Ann Radiol 25:509–517

Faglia G, Giovanelli MA, MacLeod RM (1980) Pituitary microadenomas. Academic Press, London

Ghoshhajra K (1981) High-resolution metrizamide CT cisternography in sellar and suprasellar abnormalities. J Neurosurg 54:232–239

Guthrie FW Jr, Ciric I, Hayashida S, Kerr WD Jr, Murphy ED (1981) Pituitary Cushing's syndrome and Nelson's syndrome: diagnostic criteria, surgical therapy and results. Surg Neurol 16:316–323

Hemminghytt S, Kalkhoff RK, Daniels DL, Williams AL, Grogan JP, Haughton VM (1983) Computed tomographic study of hormone-secreting microadenomas. Radiology 146:65–69

Kricheff II (1979) The radiologic diagnosis of pituitary adenoma. Radiology 131:263–265

Kuuliala I (1981) CT of pituitary adenomas. Clinical Radiology 32:259–264

Lemaître G, Linquette M, Fossati P, Cappoen JR (1982) Détection des adénomes hypophysaires sécrétants par tomodensitométrie. Ann Med Int 133:33–34

Levy SR, Wynne CV, Lorentz WB (1960) Cushing's syndrome in infancy secondary to pituitary adenoma. Am J Dis Child 136:1605–1607

Linquette M, Fossati P (1980) Les adénomes hypophysaires sécrétants. Rev Prat, Paris 30:3017–3040

Manni A, Latshaw RF, Page R, Santen RJ (1983) Simultaneous bilateral venous sampling for adrenocorticotropin in pituitary-dependent Cushing's disease: evidence for lateralization of pituitary venous drainage. J Clin Endocrinol Metab 57:1070–1073

Metzger J, Gardeur D, Houlbert D, Thibierge M (1982) Surveillance neuro-radiologique des adénomes hypophysaires sécrétants. Ann Med Int 133:29–32

Racadot J (1980) Adénomes de l'antéhypophyse: histoire naturelle, classification et histopathologie. Rev Prat (Paris) 30:2981–3002

Raji MR, Kishore PRS, Becker AP (1981) Pituitary microadenoma: a radiological surgical correlative study. Radiology 139:95–99

Robert F, Pelletier G, Hardy J (1978) Pituitary adenomas in Cushing's disease. Arch Pathol Lab Med 102:448–455

Robertson WD, Newton TH (1978) Radiologic assessment of pituitary microadenomas. AJR 131:489–492

Rovit RL, Duane TD (1969) Cushing's syndrome and pituitary tumors. Am J Med 46:416–427

Syvertsen A, Haughton VM, Williams AL, Cusick JF (1979) The computed tomographic appearance of the normal pituitary gland and pituitary microadenomas. Radiology 133:385–391

Taylor S (1982) High resolution computed tomography of the sella. Radiologic clinics of North America 20:207–236

Tindall GT, McLanahan CS (1980) Hyperfunctional pituitary tumors: pre- and postoperative management considerations. Clin Neurosurg 27:48–82

Von Werder K, Brendel C, Eversmann T (1980) Medical therapy of hyperprolactinemia and Cushing's disease associated with pituitary adenomas. In: Faglia G, Giovanelli MA, MacLeod RM (eds) Pituitary microadenomas. Academic Press, New York, p 383–397

Weiss MH (1981) Medical and surgical management of functional pituitary tumors. Clin Neurosurg 28:374–383

Wilson CB, Dempsey LC (1978) Transsphenoidal microsurgical removal of 250 pituitary adenomas. J Neurosurg 48:13–22

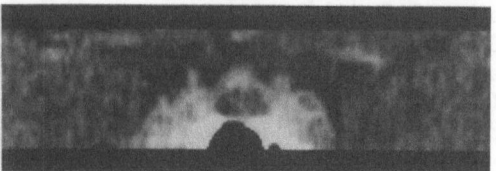

Fig. 9.1. Cushing's disease in a 27-year-old man. Elevated ACTH and beta-LPH levels. Coronal reformatted CT scan after contrast. Increased pituitary height. Bulging of the sellar diaphragm above a hypodense intrasellar ACTH-secreting adenoma confirmed at surgery

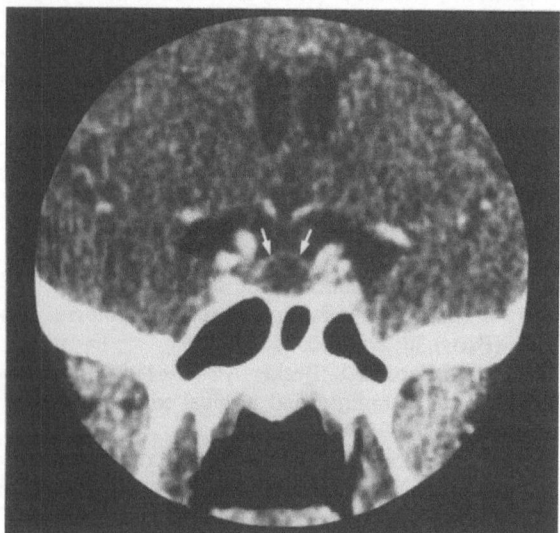

Fig. 9.2. ACTH-secreting pituitary adenoma in a 36-year-old woman. Elevated ACTH and beta-LPH levels. Coronal contrast CT scan. Hypodense 8-mm holosellar adenoma with ring enhancement (*arrows*) confirmed at surgery. Depressed sellar floor

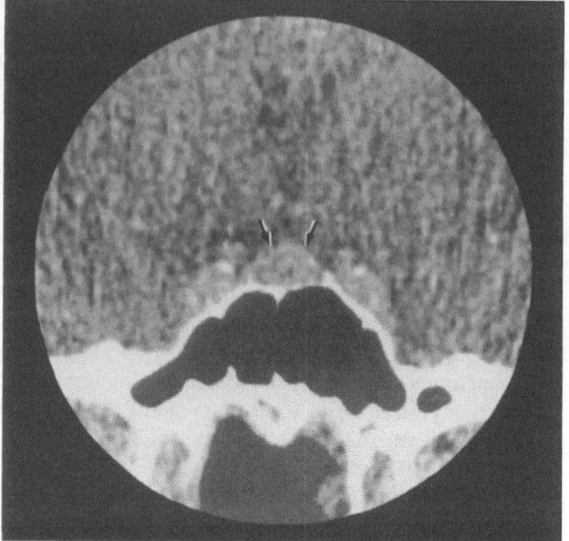

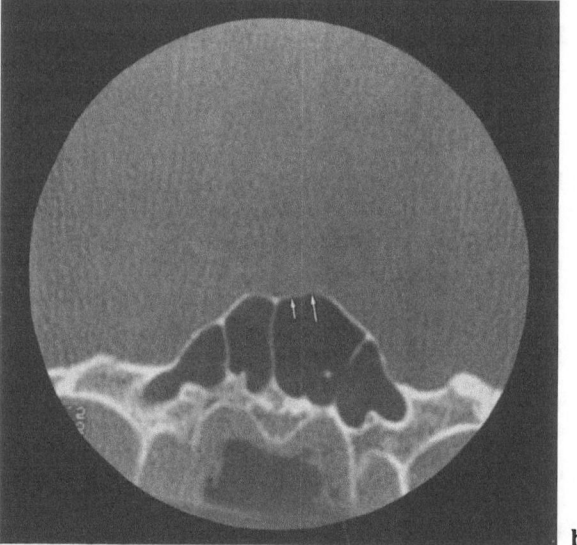

Fig. 9.3a, b. Cushing's disease in a 48-year-old man. **a** Contrast coronal CT scan. Enlarged pituitary content without well-defined pituitary tumor (*arrows*). **b** Bone review algorithm image: Angulated sellar floor, the cortical bone is thinned on the left (*arrows*). A 5-mm ACTH-secreting pituitary adenoma was found at surgery

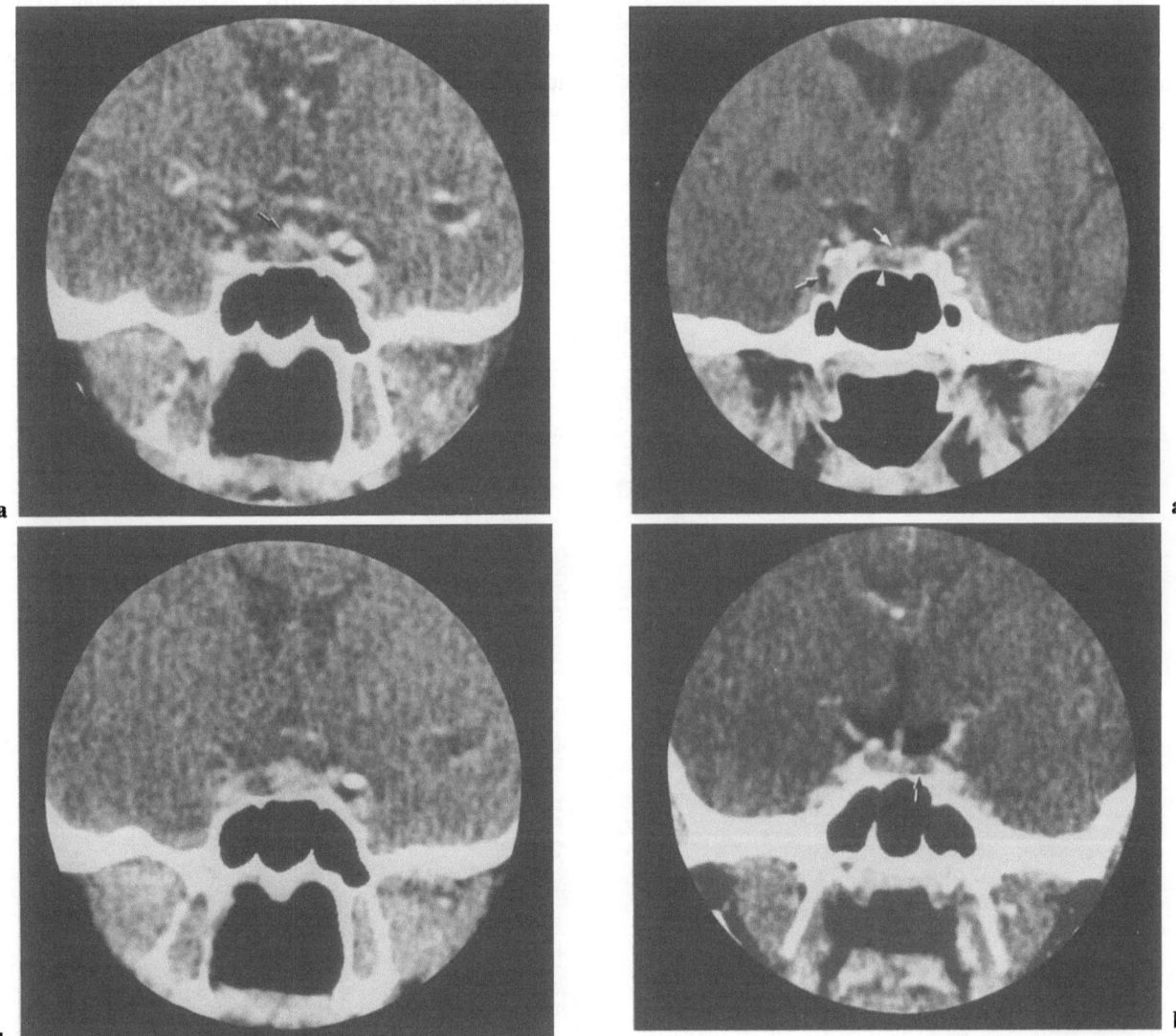

Fig. 9.4a, b. ACTH-secreting microadenoma in a 34-year-old woman. **a** Dynamic coronal CT scan. Delayed enhancement of the right part of the pituitary. Displacement of the capillary bed to the left (*arrow*). **b** Coronal CT scan, 30 s after bolus injection. Right-sided 4-mm microadenoma proven at surgery

Fig. 9.5a, b. ACTH-secreting pituitary microadenomas (proven at surgery). **a** Dynamic coronal CT scan, 30 s after bolus. Flat hypodense tumor on the right (*arrowhead*). The capillary bed is here not displaced (*white arrow*). Focal hypodensities within right cavernous sinus may represent fat deposits (*black* and *white arrow*). **b** Dynamic CT scan, 30 s after bolus (other patient). Displacement to the right of the pituitary stalk and secondary capillary bed by a hypodense 2-mm left-sided microadenoma (*arrow*)

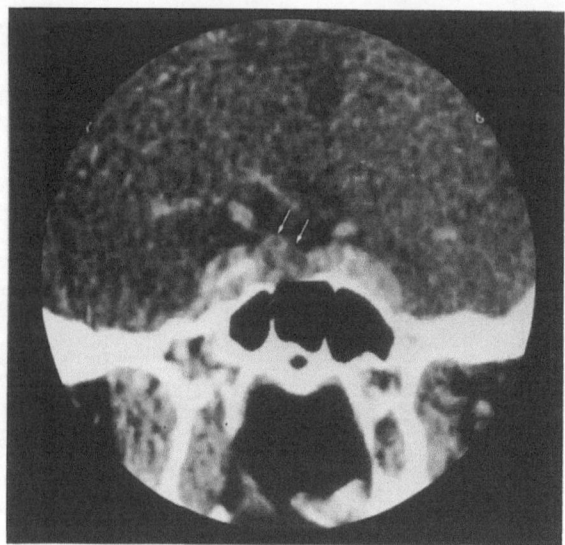

Fig. 9.6. Cushing's disease in a 27-year-old woman. Contrast coronal CT scan. Focal bulging of the sella contents on the right (*arrows*). A 8-mm less-enhanced pituitary adenoma is demonstrated on the right above an eroded sellar floor (confirmed at surgery)

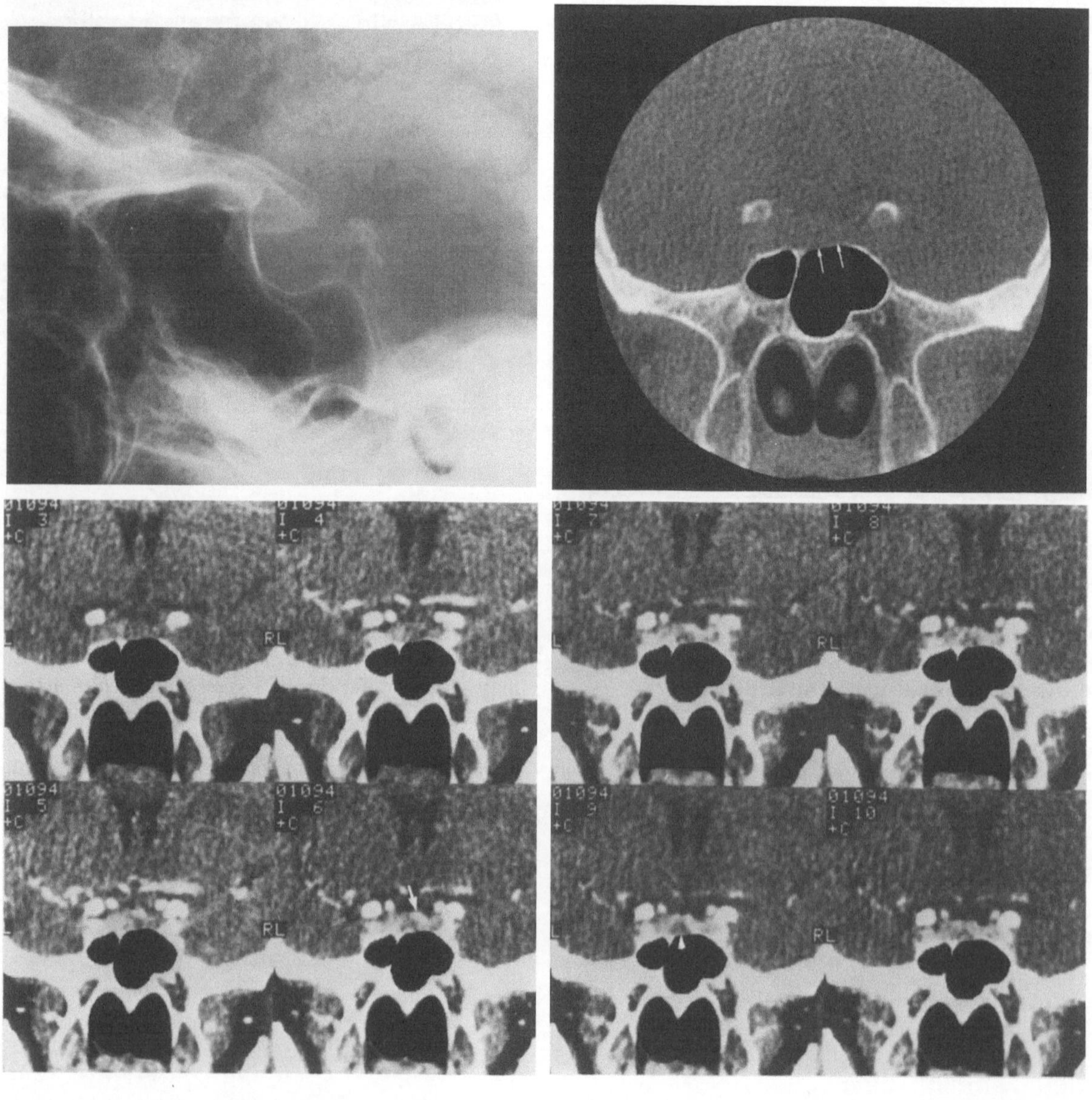

a

b

c

d

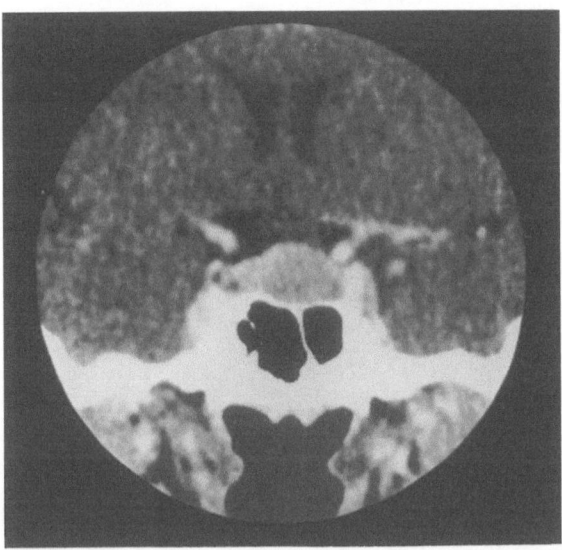

Fig. 9.8. ACTH-secreting pituitary adenoma in a 21-year-old woman. Contrast coronal CT scan; 9-mm less-enhanced pituitary adenoma on the left. Depressed sellar floor

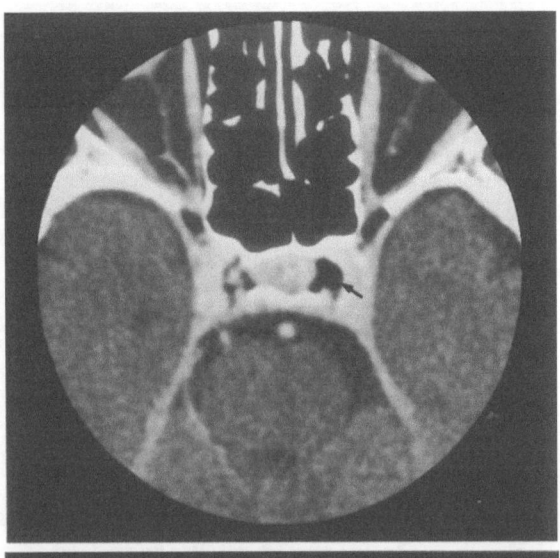

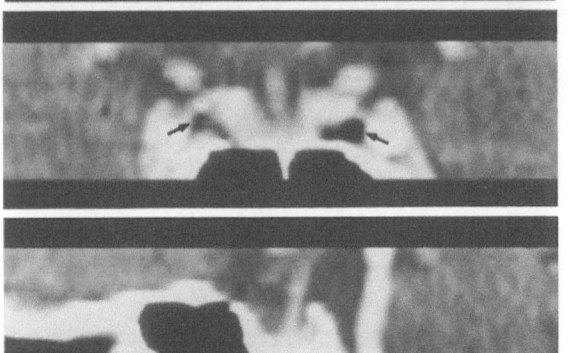

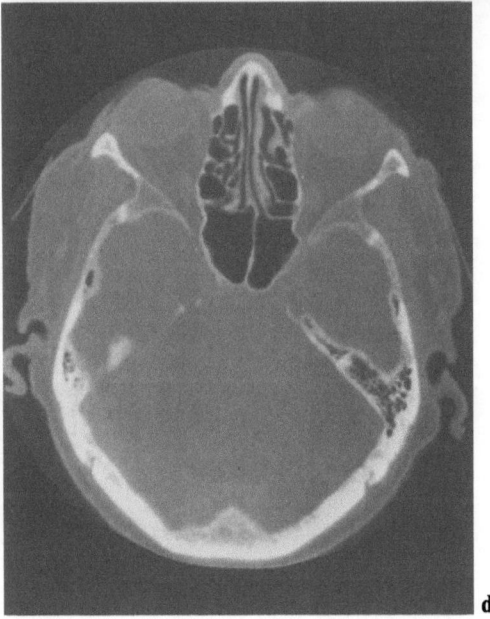

Fig. 9.9a–d. Cushing's disease in a 57-year-old woman▶ previously treated with radiotherapy. **a** Axial contrast CT scan. Fat deposits between pituitary gland and cavernous sinus on the left (*arrow*). **b, c** Coronal and sagittal reformatted images. Intrasellar extension of the subarachnoid spaces on the midline. The pituitary gland is small without focal lesion. Bilateral fat deposits are well shown (*arrows*). **d** Bone axial image: Marked demineralization of the skull

◀ **Fig. 9.7a–d.** Cushing's disease in a 47-year-old woman. Elevated ACTH and beta-LPH levels. **a** Sella turcica magnified lateral view. Generalized osteoporosis without definite focal erosive change. **b** Coronal review image: Slanted thinned sellar floor (*arrows*). **c, d** Dynamic coronal CT scan. Focal midline lesion of delayed enhancement. Displacement of the pituitary capillary bed (*arrow*). The ACTH-secreting microadenoma is well shown as a focal hypodense lesion relative to the surrounding pituitary (*arrowhead*) (proven at surgery)

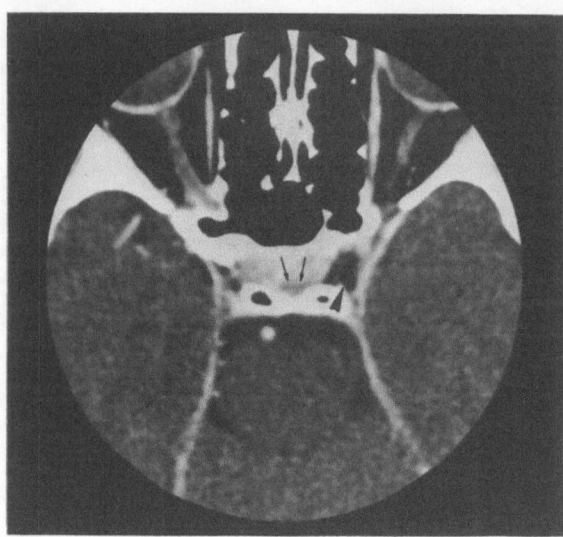

Fig. 9.10. Cushing's disease in a 45-year-old woman. Axial contrast CT scan. Large fat deposits within left cavernous sinus (*arrowhead*). The posterior pituitary is normal (*arrows*)

Chapter 10
Rare Pituitary Adenomas

TSH- and FSH- LH-secreting pituitary adenomas are rather rare. Gonadotropin-producing adenomas may be misdiagnosed as nonsecreting adenomas.

TSH-Secreting Adenomas

TSH-secreting adenomas are classified into two distinct categories (Guinet and Tourniaire):

1. Primarily TSH-secreting adenomas responsible for hyperthyroidism (Fig. 10.1)

2. THS-secreting adenomas secondary to primary hypothyroidism (Fig. 10.2)

Peripheral thyroid insufficiency, more often of congenital origin (thyroid aplasia and ectopy), rarely acquired (thyroidectomy), without replacement therapy, determines at first an adenomatous hyperplasia which secondarily may develop into a macroscopic adenoma. Figure 10.2 illustrates such a macroadenoma secondary to a postsurgical thyroid insufficiency. The parathyroid insufficiency induced by the ablation of parathyroid glands determines the basal ganglia calcifications.

CT appearance of TSH-secreting adenomas are altogether nonspecific. Small adenomas determine a moderate mass effect and appear of lower density than the normal pituitary gland on enhanced CT scans (Fig. 10.1). Macroadenomas look like nonsecreting adenomas (Fig. 10.2).

FSH-LH-Secreting Adenomas

Some adenomas secrete either FSH alone or FSH and LH simultaneously (Trouillas). They have a tendency for rapid growth and generally present a prominent suprasellar extension by the time of their discovery. FSH-LH-secreting adenomas may display an invasive tendency (Fig. 10.3).

Pituitary Oncocytomas

Tumor cells of pituitary adenomas may have the morphologic appearance of oncocytes. CT findings are similar to those of nonfunctioning adenomas (Landolt and Oswald, Rosai).

References

Anniko M, Backlund EO, Lundquvist PG, Samuelsson K, Ritzen M, Tribukait B, Wersall J (1982) Pituitary tumor producing only thyrotropin: a case report. J Oto Rhino Laryngol 44:134–141

Bonneville JF, Dietemann JL (1981) Radiology of the sella turcica. Springer Verlag, Berlin Heidelberg New York

Citrin CM, Davis DO (1977) CT in the evaluation of pituitary adenomas. Invest Radiol 12:27–35

Derome PJ, Jedynak CP, Peillon F (1980) Pituitary adenomas. Biology, physiopathology and treatment. Asclepios Publishers, Paris

Faggiano M, Criscuolo T, Perrone L, Quarto C, Sinisi AA (1983) Sexual precocity in a boy due to hypersecretion of LH and prolactin by a pituitary adenoma. Acta Endocrinol 102:167–172

Grosvalet A, Garel L, Ernest C, Sauvegrain (1980) Adénome hypophysaire secondaire à une hypothyroïdie. Evolution radiologique sous traitement. Ann Radiol 23:159–162

Guinet P, Tourniaire J (1974) L'adénome hypophysaire à cellules thyréotropes. Rev Otoneuroophthalmol 47:253–264

Harris RI, Schatz NJ, Gennarelli T, Savino PJ, Cobbs WH, Snyder PJ (1983) Follicle-stimulating hormone-secreting pituitary adenomas: correlation of reduction of adenoma size with reduction of hormonal hypersecretion after transsphenoidal surgery. J Clin Endocrinol Metab 56:1288–1293

Hill SA, Falko JM, Wilson CB, Hunt WE (1982) Thyrotropin-producing pituitary adenomas. J Neurosurg 57:515–519

Lamberg BA, Pelkonen R, Gordin A, Haltia M, Wahlström T, Paetau A, Leppäluoto J (1983) Hyperthyroidism and acromegaly caused by a pituitary TSH- and GH-secreting tumour. Acta Endocrinol 103:7–14

Landolt AM, Oswald UW (1973) Histology and ultrastructure of an oncocytic adenoma of the human pituitary. Cancer 31:1099–1105

Linquette M, Fossati P (1980) Les adénomes hypophysaires sécrétants. Rev Prat, Paris 30:3017–3040

Mashiter K, Van Nooryen S, Fahlbusch R, Fill H, Skrabal K (1983) Hyperthyroidism due to a TSH-secreting pituitary adenoma: case report, treatment and evidence for adenoma TSH by morphological and cell culture studies. Endocrinol 18:473–483

Miura M, Matsukado Y, Kodama T, Mihara Y (1985) Clinical and histopathological characteristics of gonadotropin-producing pituitary adenomas. J Neurosurg 62:376–382

Racadot J (1980) Adénomes de l'antéhypophyse: histoire naturelle, classification et histopathologie. Rev Prat (Paris) 30:2981–3002

Rosai J (1981) Ackerman's Surgical Pathology. Vol 2, sixth edition. The CV Mosby Company, St. Louis, Toronto, London

Ross LS (1981) Routine hormonal screening of infertile men: is it worthwhile? J Urol 126:756–758

Saeger W, Luedecke DK (1982) Pituitary adenomas with hyperfunction of TSH. Frequency, histological classification, immunocytochemistry and ultrastructure. Virchows Arch 394:255–267

Samaann MA, Osborne BM, Mackay B (1977) Endocrine and morphologic studies of pituitary adenomas secondary to primary hypothyroidism. J Clin Endocrinol Metab 45:903–911

Smallridge RC, Smith CE (1983) Hyperthyroidism due to thyrotropin-secreting pituitary tumors. Diagnostic and therapeutic considerations. Arch Intern Med 143:503–507

Smith CE, Smallridge RC, Dimond RC, Wartofsky L (1982) Hyperthyroidism due to a thyrotropin-secreting pituitary adenoma: studies of thyrotropin and subunit secretion. Arch Intern Med 142:1709–1711

Snyder PF, Sterling FH (1976) Hypersecretion of LH and FSH by a pituitary adenoma. J Clin Endocrinol Metab 42:544

Trouillas P, Claustrat B, Lheritier M, Goutelle A, Tourniaire J (1979) An FSH- and LH-producing pituitary adenoma. 2nd European Workshop on pituitary adenomas. Paris, September 20–22

Trouillas J, Girod C, Sassolas G, Claustrat B, Lheritier M, Dubois MP, Goutelle A (1981) Human pituitary gonadotropic adenoma: histological, immunocytochemical, ultrastructural and hormonal studies in eight cases. J Pathol 136:315–336

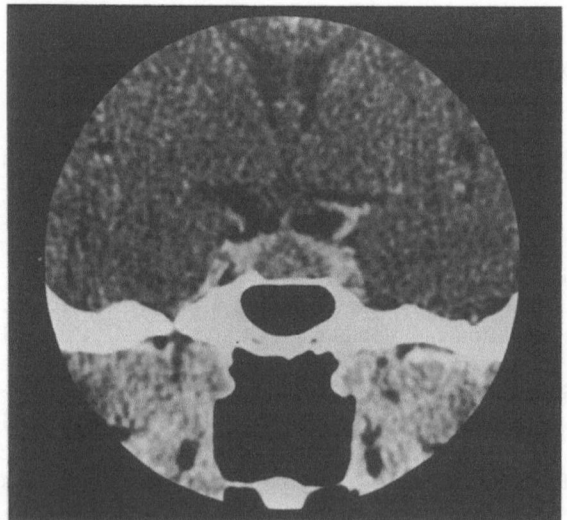
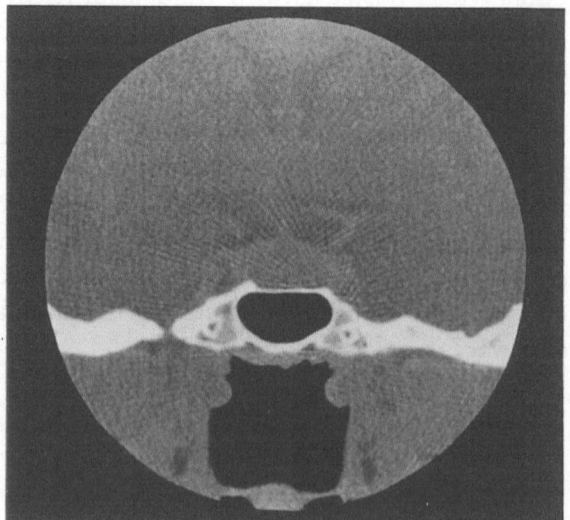

Fig. 10.1a, b. Primarily TSH-secreting adenoma in a 33-year-old woman. **a** Coronal enhanced CT scan. **b** Coronal bone review image. Marked symmetric upward bulging of the sellar diaphragm related to an intrasellar adenoma. Poor enhancement of the pituitary content. The pituitary stalk is not deviated. Thinning of the sellar floor

Fig. 10.2a–c. TSH-secreting adenoma secondary to postoperative hypothyroidism. **a** Plain film, **b** coronal reformatted enhanced CT scan, and **c** axial supratentorial CT scan. **a** The sella turcica is discretely enlarged. **b** Depression of the right part of the sellar floor related to an intrasellar mass which presents a suprasellar extension. **c** Basal ganglion calcifications related to postoperative hypoparathyroidism

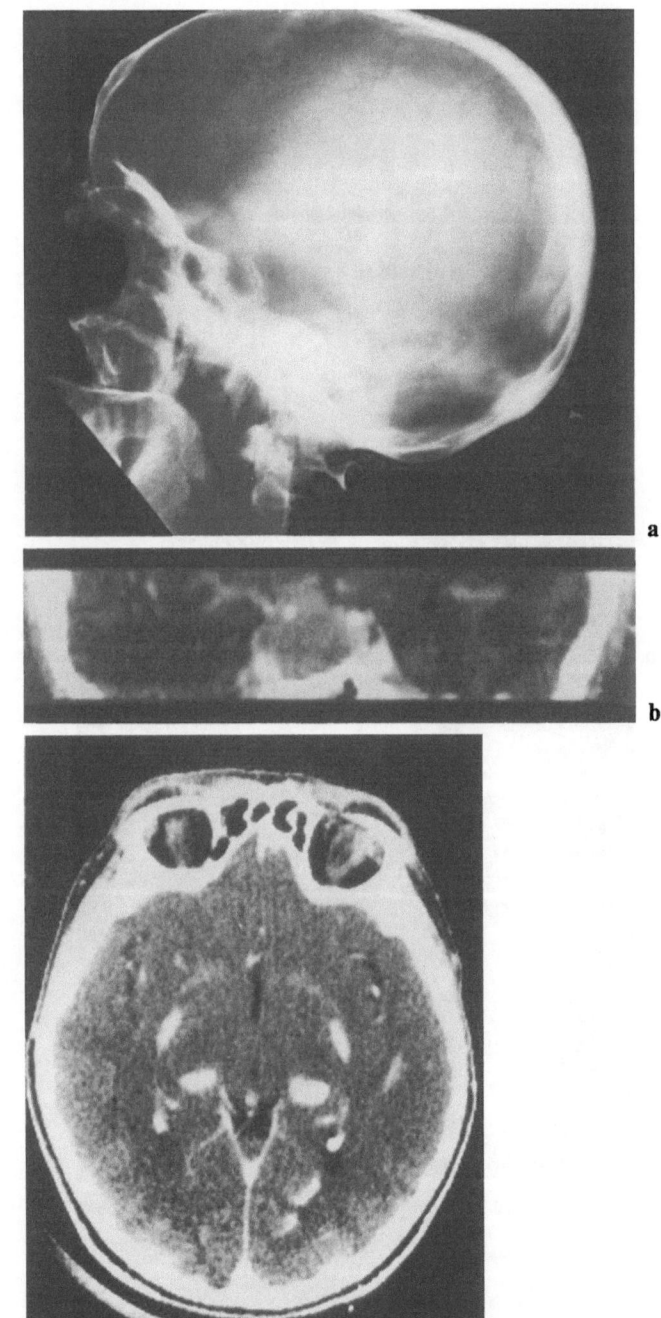

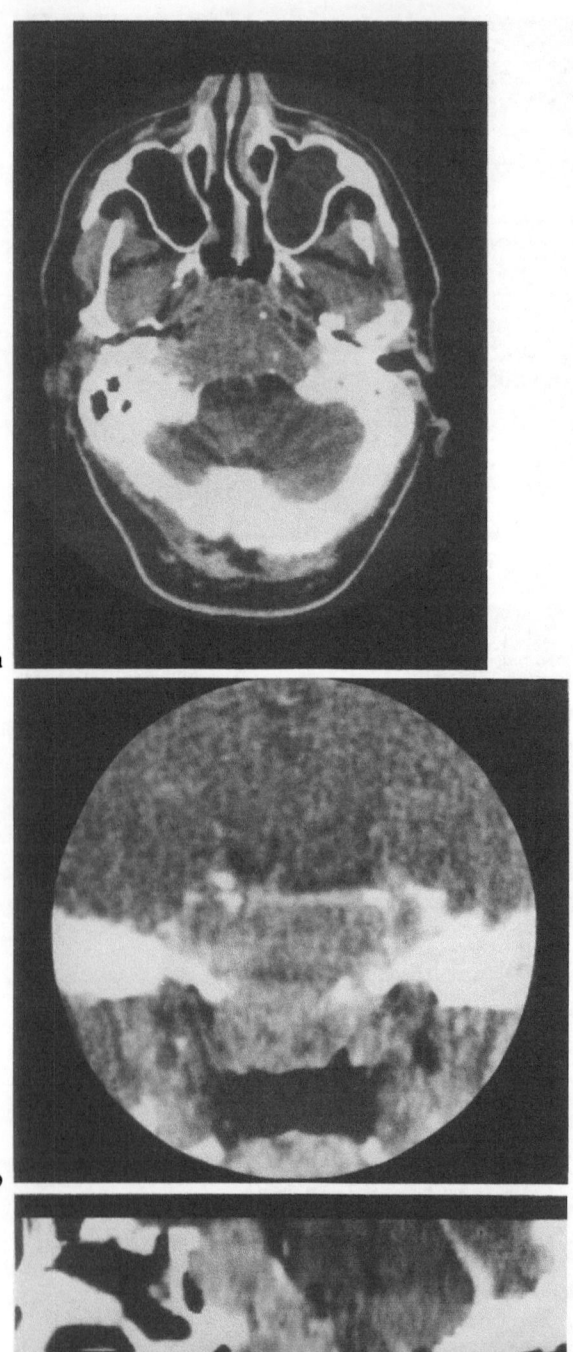

Fig. 10.3a–c. FSH- LH-secreting adenoma in a 50-year-old man. **a** Axial, **b** coronal enhanced CT scans, and **c** sagittal reformatted CT scan. The adenoma presents a suprasellar extension associated to an intrasellar extension. Indeed the sellar floor is destroyed and the tumor is extending to the clivus and the nasopharynx

Chapter 11
Pituitary Adenomas: Spontaneous Evolution – Complications

Spontaneous Evolution of Pituitary Adenomas

Spontaneous Progression

Nonsecreting pituitary adenomas develop less rapidly than GH-secreting pituitary adenomas. ACTH-secreting adenomas of Nelson's syndrome grow rapidly, while ACTH-secreting adenomas of Cushing's disease develop very slowly. Microprolactinomas develop rather slowly. Weiss et al. followed the spontaneous evolution of 27 microprolactinomas over a period of 6 years; they observed an increase in tumoral size in only three of them. They considered that early osteoporosis constitutes a risk of prolonged hyperprolactinemia and that therefore medical treatment is indicated. An annual CT follow-up is recommended (see also Chap. 6).

Rapid Growth of Pituitary Adenomas

Malignant adenomas develop very rapidly by invading the osteomeningeal and the vasculonervous structures of the sellar region. Malignant adenomas may show an intracranial dissemination towards the subarachnoid space.

Growth of prolactinomas may also be accelerated by exogenous and endogenous estrogens (see Chap. 7).

Pituitary Apoplexy

Pituitary apoplexy corresponds to a clinical syndrome characterized by acute headache, nausea, and vomiting, consciousness disorders, fever, stiffness of the neck, visual disorders, and ophthalmoplegia. These symptoms are obviously due to a brutal and important enlargement of the adenoma secondary to an ischemic and/or hemorrhagic necrosis, associated sometimes with an extension of the hemorrhage towards ventricles and subarachnoid space. In fact, this pituitary apoplexy, which is expressed clinically by a dramatic situation sometimes rapidly lethal, is but a rare eventuality (Post, Cardoso). On the other hand, surgical discovery of clinically asymptomatic hemorrhagic and necrosed lesions is frequently encountered, since 10% of these adenomas display such modifications. These necrotic and hemorrhagic lesions may then trigger an evolution towards a cystic adenoma or towards a marked decrease in size of the tumor. Such conditions will lead to an empty sella or a calcification of the adenoma (Fig. 11.1). The pathological modifications may sometimes lead to a spontaneous resolution of the endocrine symptoms (Arisaka, Corkill, Mohr and Hardy, Werner, Findling).

Pituitary apoplexy may occur spontaneously or be favored by several predisposing factors (Cardoso): endogenous estrogens (in pregnancy, for example) (Linquette) or exogenous oral estrogens (contraceptive pill) (Mohr and Hardy); bromocriptine treatment (Mohr and Hardy); anticoagulant treatment (Nourizadeh, Peck); radiotherapy (Weisberg); certain invasive radiological explorations, such as angiography (Reichenthal) and pneumoencephalography (Perpetuo); cranial traumatism (Van Wagenen); or the type of adenoma: several previous observations insist upon the greater frequency of pituitary apoplexy in GH-secreting adenomas (Jacobi); however, recent studies show that hemorrhagic necrosis is more common in nonsecreting adenomas and in prolactinomas than in GH-secreting adenomas (Cardoso, Mohr and Hardy).

The radiologic diagnosis of apoplexy is obviously based on CT scan; however, such investigation may sometimes be difficult due to consciousness disorders and agitation in these patients. Plain X-rays generally reveal modifications in volume and/or in morphology of the sella turcica relative to an intrasellar tumoral process.

The CT findings may be variable and depend upon the type of lesion-hemorrhage, necrosis, or mixed lesion as well as upon the extension and the age of the lesions (Post, Gardeur).

CT scan, thus, allows identification of a spontaneous high-density area within an enlarged pituitary fossa; this high-density area may extend towards the suprasellar region. A subarachnoid or even an intraventricular hemorrhage is sometimes noted (Challa, Kalyanaraman, Patel, and Post). These high-density hemorrhagic areas are sometimes juxtaposed with low-density necrotic and enhancing areas (Post). Intratumoral hemorrhage is sometimes expressed by a horizontal level (Fujimoto) (Fig. 11.2). In case of a strictly ischemic lesion, a centrotumoral low-density area with a ring enhancement is noted. In exceptional cases, pituitary apoplexy may lead to pneumocephalus by rupture of the sellar walls or to an occlusion of the internal carotid artery (Rosenbaum).

The spontaneous evolution of these lesions can only be studied in a few patients; in fact, a surgical cure of the hematoma by a transsphenoidal approach is indicated in most cases. When evolution of the lesion is spontaneously favorable, the hematoma is progressively replaced by a ring enhancement. Such reorganization may bring about formation of a cystic adenoma. Sometimes, late evolution of a hemorrhagic tumoral necrosis ends up in an empty sella turcica (Fig. 11.1).

Spontaneous Regression of Pituitary Adenomas

The clinical symptomatology of certain secreting adenomas may shade off spontaneously. Such evolution is probably the result of necrotico-hemorrhagic phenomena which remain clinally asymptomatic. This classical "recovery" may be accompanied by a calcification of the adenoma (Arisaka, Vaughn). In other cases, spontaneous regression of an adenoma is evoked upon conflict between the importance of osseous deformations and the small size of the adenoma (Fig. 11.3).

Abscess Formation

Usually, intrasellar abscesses develop within a pituitary tumor. An abscess may sometimes reveal the tumor. CT scan shows an intrasellar mass displaying usually a suprasellar extension with a low-density center and ring enhancement (see also Chap. 12). Intrasellar abscesses have also been observed after transsphenoidal pituitary surgery (Enzmann, Nelson).

Evolution Following Medical and Surgical Treatment

(See Chapters 6, 8 and 14).

References

Arisaka O, Hall R, Hughes IA (1983) Spontaneous endocrine cure of gigantism due to pituitary apoplexy. Br Med J 287:1007–1008

Bonneville JF, Dietemann JL (1981) Radiology of the sella turcica. Springer Verlag, Berlin Heidelberg New York

Bonneville JF, Poulignot D, Cattin F, Couturier M, Mollet E, Dietemann JL (1982) Computed tomographic demonstration of the effects of bromocriptine on pituitary microadenoma size. Radiology 143:451–455

Bonneville JF, Cattin F, Poulignot D, Dietemann JL, Couturier M (1983) Value of high-resolution CT in the follow-up of prolactinomas treated with bromocriptine. In: Tolis G (ed) "Human prolactin". Raven Press, New York

Campagnoli C, Belforte L, Massara F, Peris C, Molnatti GM (1981) Partial remission of hyperprolactinemic amenorrhea after bromocriptine-induced pregnancy. J Endocrinol Invest 4:85–91

Cardoso ER, Peterson EW (1984) Pituitary apoplexy: a review. Neurosurgery 14:363–373

Challa VR, Richards F, Davis CH (1981) Intraventricular hemorrhage from pituitary apoplexy. Surg Neurol 16:360–361

Corkill G, Hansson FW, Sobel RA, Keller TM (1980) Apoplexy in a prolactin microadenoma leading to regression of galactorrhea and amenorrhea. Surg Neurol 15:114–115

Cowden EA, Thomson JA (1979) Resolution of hyperprolactinemia after bromocriptine-induced pregnancy. Lancet 1:613

David M, Philippon J, Navarro-Artiles G, Racadot J, Weil BB (1969) Les formes hémorragiques des adénomes hypophysaires: aspects cliniques et étiologiques. Neurochirurgie 15:228–229

Dawson BH, Kothandaram P (1972) Acute massive infarction of pituitary adenomas. A study of five patients. J Neurosurg 37:275–279

Derome PJ, Jedynak CP, Peillon F (1980) Pituitary adenomas. Biology, physiopathology and treatment. Asclepios Publishers, Paris

Domingue JN, Richmond IL, Wilson CB (1980) Results of surgery in 114 patients with prolactin-secreting pituitary adenomas. Am J Obst Gynecol 137:102–108

Emperaire JC, Riemens V, Dubecq JJ, Palmade J, Leuret JPH (1972) Hypophysectomie d'urgence à deux mois de grossesse après induction de l'ovulation. Bordeaux Med 15:1901–1904

Enzmann DR (1984) Imaging of infectious and inflammations of the central nervous system: computed tomography, ultrasound, and nuclear magnetic resonance. Raven Press, New York, pp 52–55

Enzmann DR, Sieling RJ (1983) CT of pituitary abscess. AJNR 4:79–80

Findling JW, Tyrrell JB, Aron DC, Fitzgerald PA, Wilson CB, Forsham PH (1981) Silent pituitary apoplexy: subclinical infarction of an adrenocorticotropin-producing pituitary adenoma. J Clin Endocrinol Metab 52:95–97

Fitz-Patrick D, Tolis G, McGarry EE, Taylor S (1980) Pituitary apoplexy: the importance of skull roentgenograms and computerized tomography in diagnosis. JAMA 244:59–61

Fujimoto M, Yoshino O, Veguchi T, Mizukawa N, Hirakawa K (1981) Fluid blood density level demonstrated by CT in pituitary apoplexy. Report of two cases. J Neurosurg 55:143–144

Galvan G, Frick J, Irnberger T (1981) Bromocriptine-induced cystic tumour regression in advanced prolactinomas. Dtsch Med Wochenschr 106:637–642

Gardeur D, Metzger J (1982) Adénomes hypophysaires intra-sellaires. In: Tomodensitométrie intra-crânienne. Livre V: Pathologie sellaire. Ellipses, Paris, p 41

Gemzell C, Wang CF (1979) Outcome of pregnancy in women with pituitary adenoma. Fertil Steril 31:363–372

Gutin PH, Cushard WG Jr, Wilson CB (1979) Cushing's disease with pituitary apoplexy leading to hypopituitarism, empty sella and spontaneous fracture of the dorsum sellae. J Neurosurg 51:866–869

Hashimoto N, Handa H, Takeuchi J, Ishikawa M, Nakano Y (1982) Collection of gas within a huge chromophobe adenoma. Neuroradiology 23:289–290

Jacobi JD, Fishman LM, Daroff RB (1974) Pituitary apoplexy in acromegaly followed by partial pituitary insufficiency. Arch Intern Med 134:559–561

Kajtar T, Tomkin GH (1971) Emergency hypophysectomy in pregnancy after induction of ovulation. Br Med J 4:88–90

Kalyanaraman UP (1982) Clinically asymptomatic pituitary adenoma manifesting as pituitary apoplexy and fatal third ventricular hemorrhage. Hum Pathol 13:1141–1143

Klibanski A, Neer RM, Beitins IZ, Ridgway EC, Zervas NT, McArthur JW (1980) Decreased bone density in hyperprolactinaemic women. N Engl J Med 303:1511–1514

Kovacs K (1972) Adenohypophysial necrosis in routine autopsies. Endokrinologie 60:309–316

Lamberts SW, Seldenrath HJ, Kwa HG, Birkenhäger JC (1977) Transient bitemporal hemianopsia during pregnancy after treatment of galactorrhea-amenorrhea syndrome with bromocriptine. J Clin Endocrinol Metab 44:180–184

Lamberts SW, Klijn JG, Lange SA, Singh R, Stefanko SZ, Birkenhäger JC (1979) The incidence of complications during pregnancy after treatment of hyperprolactinemia with bromocriptine in patients with radiologically evident pituitary tumors. Fertil Steril 31:614–619

Laws ER Jr, Ebersold MJ (1982) Pituitary apoplexy. An endocrine emergency. World J Surg 6:686–688

Le Pogamp C, Grall JY, Massart C, Ramee A, Toulouse R (1979) Poussée évolutive d'adénome à prolactine au cours d'une grossesse après stérilité traitée. 2 observations. Nouv Presse Med 8:2009–2011

Linquette M, Buvat J, Gauthier A, Gasnault JP, Pagniez T, Decoulx M, Laine E (1977) Apoplexie hypophysaire révélatrice d'un adénome à cellules prolactiniques au cours d'une grossesse permise par la bromocriptine. Nouv Presse Med 6:3525–3531

Majchrzak H, Wencel T, Dragan T, Bialas J (1983) Acute hemorrhage into pituitary adenoma with SAH and anterior cerebral occlusion. J Neurosurg 58:771–773

Marano SR, Sonntag VKH, Speltzler RF (1984) Planum sphenoidale meningioma mimicking pituitary apoplexy: a case report. Neurosurgery 15:859–862

March CM, Kletzky OA, Davajan V, Teal J, Weiss M, Apuzzo MLJ, Marrs RP, Mishell DR Jr (1981) Longitudinal evaluation of patients with untreated prolactin-secreting pituitary adenomas. Am J Obstet Gynecol 139:835–839

McCormick WF, Halmi NS (1970) The hypophysis in patients with "coma dépassé" (respirator brain). Am J Clin Pathol 54:374–383

Mohanty J (1980) Recurrent oculomotor palsy due to hemorrhage in pituitary adenoma. Postgrad Med J 56:54–56

Mohr G, Hardy J (1982) Hemorrhage, necrosis, and apoplexy in pituitary adenomas. Surg Neurol 18:181–189

Nelson PB, Haverkos H, Martinez AJ, Robinson AG

(1983) Abscess formation within pituitary tumors. Neurosurgery 12:331–333

Nourizadeh AR, Pitts FW (1965) Hemorrhage into pituitary adenoma during anticoagulant therapy. JAMA 193:623–625

Parent AD, Bebin J, Smith RR (1981) Incidental pituitary adenomas. J Neurosurg 54:228–231

Patel DV, Shields MC (1979) Intraventricular hemorrhage in pituitary apoplexy. J Comput Assist Tomogr 3:829–831

Peck V, Liberman A, Pinto R, Culliford A (1980) Pituitary apoplexy following openheart surgery. NY State J Med 80:641–643

Perpetuo FO (1976) Pituitary apoplexy after pneumoencephalogram. Arq Neuropsiquiatr 34:298–301

Post JD, David NJ, Glaser JS, Safran A (1980) Pituitary apoplexy: diagnosis by CT. Radiology 134:665–670

Racadot J (1980) Adénomes de l'antéhypophyse: histoire naturelle, classification et histopathologie. Rev Prat (Paris) 30:2981–3002

Reichenthal E, Manor RS, Shalit MN (1980) Pituitary apoplexy during carotid angiography. Acta Neurochir (Wien) 54:251–255

Rilliet B, Mohr G, Robert F, Hardy J (1981) Calcifications in pituitary adenomas. Surg Neurol 15:249–255

Rjosk HK, Fahlbusch R, Von Werder K (1982) Spontaneous development of hyperprolactinaemia. Acta Endocrinol 100:333–336

Robinson B (1983) Intrasellar abscess after transsphenoidal pituitary adenomectomy. Neurosurgery 12:684–686

Rosenbaum TJ, Houser OW, Laws ER (1977) Pituitary apoplexy producing internal carotid artery occlusion: case report. J Neurosurg 47:599–604

Rouge PE, Gaio JM, Hommel M, Halmi S, De Rougemont J, Massot C (1981) Complications neurologiques aiguës des adénomes hypophysaires. Rev Med Int 12:91–94

Rovit RL, Fein JM (1972) Pituitary apoplexy: a review and reappraisal. J Neurosurg 37:280–288

Rudwan MA (1977) Pituitary abscess. Neuroradiology 12:243–248

Selosse P, Mahler C, Klaes RL (1980) Pituitary abscess. J Neurosurg 53:851–852

Symon L, Mohanty S (1982) Hemorrhage in pituitary tumours. Acta Neurochir (Wien) 65:41–49

Van Dalen JW, Greve EL (1977) Rapid deterioration of visual fields during bromocriptine-induced pregnancy in a patient with a pituitary adenoma. Br J Ophthalmol 61:728–733

Van Wagenen WP (1932) Hemorrhage into a pituitary tumor following trauma. Ann Surg 95:625–628

Vaughn TC, Haney AF, Wiebe RH, Kramer RS, Hammond CB (1980) Spontaneous regression of prolactin producing pituitary adenomas. Am J Obstet Gynecol 136:980–984

Von Room E, Van der Vijver JC, Gerratsien G, Hiekster RE, Wattandorf RA (1981) Rapid progression of a suprasellar extending prolactinoma after bromocriptine treatment during pregnancy. Fertil Steril 36:173–177

Weisberg LA (1977) Pituitary apoplexy: association of degenerative change in pituitary adenoma with radiotherapy and detection by cerebral CT. Am J Med 63:109–115

Weiss MH, Teal J, Gott P, Wycoff R, Yadley R, Apuzzo MLJ, Giannotta SL, Kletzky O, March C (1983) Natural history of microprolactinomas: six-year follow-up. Neurosurgery 12:180–183

Werner PL, Shah JH, Kukreja SC, Miller SM, Williams GA (1982) Recurrence of acromegaly after pituitary apoplexy. JAMA 247:2816–2818

Windgrawe SJ, Kay CR, Vessey MP (1980) Oral contraceptives and pituitary adenomas. Br Med J 280:685–686

Yamaji T, Ishibaski M, Kosaka K, Fukushima T, Hori T, Manaka S, Sano K (1981) Pituitary apoplexy in acromegaly during bromocriptine therapy. Acta Endocrinol 98:171–177

Zorub DS, Martinez AJ, Nelson PB, Lam MT (1979) Invasive pituitary adenoma with abscess formation: case report. Neurosurgery 5:718–722

Zumstein V, De Tribolet N (1982) L'apoplexie pituitaire. Rev Med Suisse Romande 102:775–779

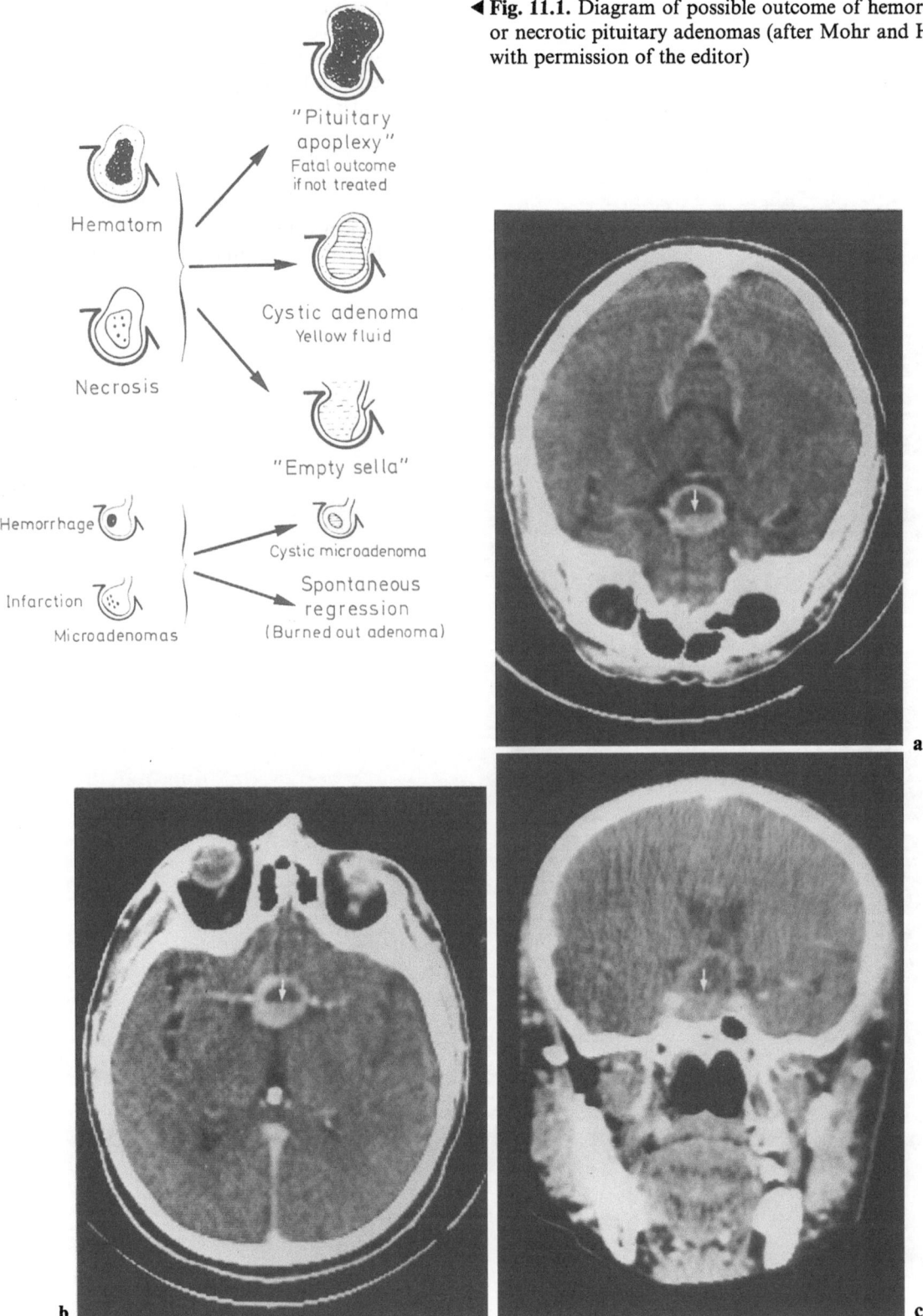

◀ **Fig. 11.1.** Diagram of possible outcome of hemorrhagic or necrotic pituitary adenomas (after Mohr and Hardy, with permission of the editor)

Fig. 11.2a–c. Pituitary apoplexy. Headache and acute bitemporal hemianopia in a 50-year-old man. **a** Axial supine, **b** axial prone, and **c** coronal prone enhanced CT scans. Huge nonfunctioning pituitary adenoma. Note the posteriorly located hemorrhagic component (*arrow*) in the supine axial scan. The contrary is noted in prone position. A horizontal fluid level is present in all the positions (*arrows*)

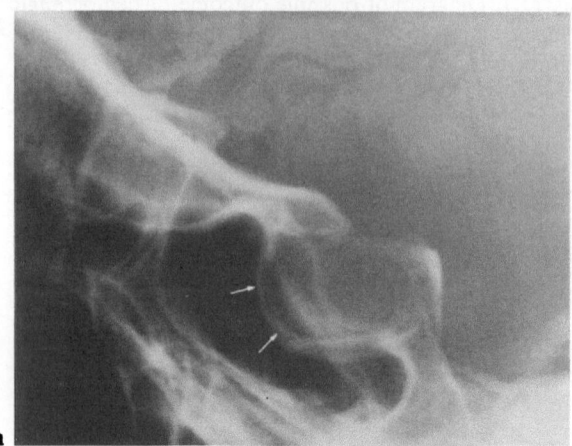

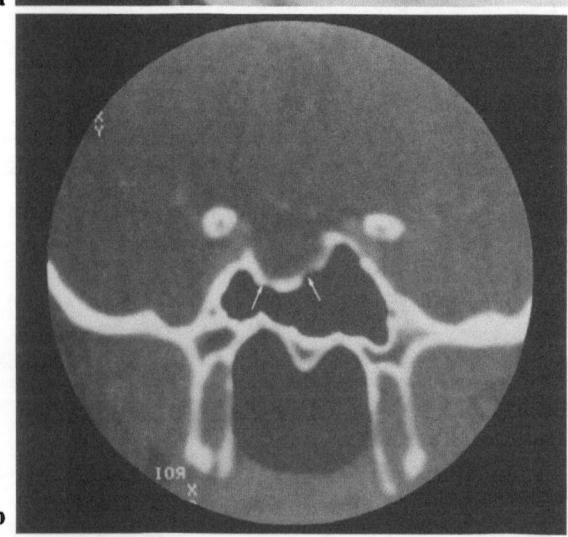

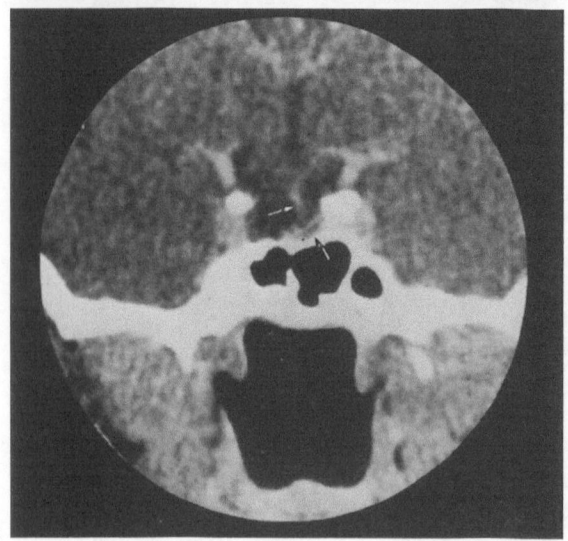

Fig. 11.3a–c. Spontaneous regression of a prolactinoma in a 35-year-old woman. Amenorrhea for 10 years; spontaneous clinical improvement with normalization of prolactinemia. **a** Lateral plain film. Bone changes evoke a typical anteriorly located adenoma (*arrows*). **b** Coronal bone window CT scan confirms the right lateralized depressed sellar floor. Note an irregular and thickened sellar floor (*arrows*). **c** Coronal enhanced CT scan, located 5 mm behind **b**. Partial empty sella. Pituitary stalk (*white arrow*) is oriented to the left part of the pituitary fossa toward the remaining pituitary gland (*black* and *white arrow*)

Chapter 12
Rare Intrasellar Disorders

The CT appearance of most of rare intrasellar lesions is usually nonspecific and surgical investigation is frequently the final means of a reliable diagnosis. However, the confrontation of CT, clinical and biological results sometimes helps to guide the diagnosis; this is particularly the case with lymphocytic adenohypophysis and sarcoidosis (see Table 1).

Table 1. Rare intrasellar lesions

Lesions with low attenuation values	Enhancing lesions	Calcified lesions
Rathke's cleft cysts	Metastases	Craniopharyngioma
Epidermoid and dermoid cysts	Granular cell tumors of the neurohypophysis	Pituitary calculus
Parasitic cysts (cysticercosis hydatidosis)	Gangliocytomas	Sellar spine
Pituitary abscess	Sarcomas and carcinomas of the pituitary gland	Chondroma
Craniopharyngioma	Germinomas	Chordoma
Lipoma	Sarcoidosis	
	Hemangioblastoma	
	Leiomyoma	
	Melanoma	
	Hemangiopericytoma	
	Schwannoma	
	Pituitary hyperplasia secondary to primary end-organ failure or pregnancy	

Hypophyseal Metastases

Hypophyseal metastases are frequent at the terminal phase of certain cancers (breast and bronchial carcinomas mainly), with multiple metastatic dissemination. However, they usually remain asymptomatic. The neurohypophysis seems more frequently affected than the adenohypophysis.

The CT appearance is nonspecific. The sella turcica is enlarged when the metastasis is symptomatic. The tumoral mass raises the diaphragma sellae and is homogeneously enhanced when the tumor is less than 2 cm in diameter. Beyond this size, the tumor is willingly necrotic and displays intratumoral low-density areas. Small metastases of few mm in diameter are recognized as defects in enhancement (Chambers, Gardeur) (Figs. 12.1 and 12.2).

Intrasellar "Cysts"

Cysts of Embryological Origin

Rathke's Cleft Cysts
Cystic lesions are usually of lower density than the brain, but there are exceptions to this rule because epidermoid cysts, cystic craniopharyngiomas, and Rathke's cleft cysts of the same or of higher densities have been described.

On the other hand, all low-density lesions are not necessarily cystic; they may be necrosed adenomas, intrasellar abscesses, and even an empty sella turcica.

Small intrasellar cysts are frequent pathological discoveries. Most of these cysts are tiny and classically asymptomatic. Methodical autopsies reveal cysts varying from 1 to 6 mm in 30% of cases (Muhr).

Several authors insist upon the defects in enhancement that can theoretically be determined by these cysts on a CT investigation, and the problems of differential diagnosis with a micro-adenoma (Chambers). However, in our experience of more than 2300 pituitary CT scans, all realized with sections of 1.5 mm thickness, these small cystic formations are practically never identified and do not raise any problems of differential diagnosis with microadenomas.

These cysts become exceptionally symptomatic. Symptomatic Rathke's cleft cysts of the pituitary gland induce compression of the chiasm and pituitary deficiency. Sometimes only headaches may incite a radiologic investigation. On enhanced CT scans, the cysts appear ring-shaped or egg-shaped. They sometimes appear as lesions with clearly low attenuation value, but others have higher densities and may appear of the same density as the brain. A ring enhancement can be noted around the cyst, raising problems of differential diagnosis with a necrotic adenoma or even with an intrasellar abscess (Dietemann, Turpin) (Figs. 12.3, 12.4, and 12.5).

Epidermoid Cysts

Intrasellar epidermoid cysts derive from embryonic epidermal inclusions and are much less common than suprasellar epidermoid cysts. They usually display CSF densities. However, due to variations in the relative concentrations of keratin and cholesterol, these cysts may display either fat densities, when cholesterol concentration is high, or higher density, when keratin concentration is high (Boggan, Gardeur, Sachsenheimer) (Fig. 12.6).

Craniopharyngiomas

Purely cystic intrasellar craniopharyngiomas are exceptional. They vary from fat density to high density.

Parasitic Cysts

Intrasellar cysticercosis (Prosser) and intrasellar hydatid cyst (Ozgen) are unusual findings. CT scans reveal a mass whose content has approximately CSF density.

Pituitary Abscess

Intrasellar abscesses occur usually within a tumor (adenoma, craniopharyngioma, Rathke's cleft cyst). The abscess usually comes from a neighboring infectious lesion such as a sphenoid sinusitis favored by an immune deficiency; hematogeneous dissemination is exceptionally noted. If the diagnosis of an abscess is delayed, the infection can extend towards the cavernous sinuses and may cause thrombosis of the cavernous sinuses followed by thrombosis of intracranial veins and/or towards the subarachnoid spaces and cause meningitis.

CT scan shows a mass with a low-density center and a marked ring enhancement. A suprasellar extension is noted in most cases. A pronounced destruction of the sellar floor is often present. Signs of sphenoid sinusitis are usually noted (Enzmann).

Craniopharyngiomas

Intrasellar craniopharyngiomas usually appear calcified and purely cystic lesions are exceptional. Mixed-calcified and cystic lesions may be observed (Figs. 12.7 and 12.8).

Tumors of the Neurohypophysis

Primary tumors arising within the neurohypophysis are unusual. Granular cell tumors (Fig. 12.9) and gangliocytomas may be observed. Metastases of the neurohypophysis are much more frequent. Sometimes tumors primarily located within the pituitary stalk extend towards the neurohypophysis; this progression may be observed with metastases or germinomas of the pituitary stalk.

Granular-Cell Tumors

Synonyms for granular-cell type tumors are: choristoma, pituicytoma, myoblastoma, or Abrikossof's tumor. Such asymptomatic tumors, less than 3 mm in diameter are relatively frequently discovered in autopsy series in elderly persons; these lesions may be noted in about 5% of cases (Luse). Symptomatic tumors are very rare and less than 30 such observations are reported in the literature. These tumors are usually revealed by pituitary deficiency sometimes associated to visual field defect.

On CT scans, these symptomatic tumors occupy the whole pituitary fossa with, however, a more pronounced posterior expansion. The suprasellar extension is median and may present polycyclic contours in case of a voluminous tumor (Gardeur). In our observation there was practically no suprasellar extension; the tumor presented homogeneous enhancement (Fig. 12.9).

Gangliocytomas

The CT appearance of these tumors, developed from extremely rare residual nerve cells, has not yet been described.

Primary Malignant Tumors of the Pituitary Gland

Sarcomas and carcinomas may develop in the adenohypophysis. Sarcomas appear almost exclusively after radiotherapy of a pituitary adenoma. Pituitary carcinomas (adenocarcinomas) are degenerated pituitary adenomas. These tumors quickly destroy the sella turcica, invading the sphenoid sinus, the cavernous sinus, the suprasellar cistern, and the petrous apex. Rapid growth and invasion should suggest the diagnosis. CT findings in sarcomas and carcinomas are identical. These tumors are often huge with irregular contours and heterogeneous enhancement because of the presence of numerous areas of necrosis (Martin, Pieterse).

Pituitary "Calculus"

Pituitary "calculus" is an amorphous intrasellar calcification which is usually round, dense, and lying in the anterior part of the sella turcica.

These calcifications probably represent degenerative deposits (Ozonoff and Burrows). The idiopathic "pituitary calculus" should be distinguished from the tumoral calcifications within an adenoma or a craniopharyngioma. To be considered idiopathic, calcifications should be nodular, homogeneous, located in the anterior part of the sella turcica, and lying in a sellar cavity with unmodified contours, in a patient completely free of clinical and biological endocrine disorders (Bonneville and Dietemann, Deramond, Sherman). In all other cases, the possibility of a craniopharyngioma or a calcified adenoma should be evoked. Certain secreting adenomas (prolactin or GH) may present spontaneous involution, probably by necrosis, and secondary calcification (Mohr and Hardy). In young patients, posterior calcifications should first of all suggest a craniopharyngioma.

CT scans confirm the intrasellar calcification and allow study of its contours and homogeneity; CT scans also show the eventual mass effect of this calcified lesion and help the investigation for other anomalies in density within the pituitary gland. The "pituitary calculus" has regular contours, homogeneous density, and is not accompanied by any anomaly within the remaining pituitary gland.

Miscellaneous Intrasellar Disorders

Germinomas

Suprasellar germinomas are usually located in the suprasellar area. Exceptional primary intrasellar germinomas have been described (Banna, Shen). But most of the time, suprasellar germinomas develop in the pituitary stalk and extend towards the pituitary gland. An intrasellar mass with homogeneous enhancement and suprasellar extension is usually noted.

Lymphocytic Adenohypophysitis

Lymphocytic infiltration of the pituitary gland is a rare autoimmune disorder. It usually leads to a pituitary deficiency a few months after delivery; amenorrhea and galactorrhea are frequently present.

CT investigation usually reveals an intrasellar mass with homogeneous enhancement, showing a more or less marked suprasellar extension (Fig. 12.10). These CT findings are altogether nonspecific and are comparable with those of nonsecreting adenomas. When confronted with such a CT appearance, only association with one or several other factors, namely, the circumstances of discovery (secondary to a delivery), presence of serum antibodies to prolactin-secreting cells or, furthermore, presence of another autoimmune disorder (Hashimoto thyroiditis, chronic atrophic lymphocytic gastritis, Addison's disease), may suggest the diagnosis of lymphocytic adenohypophysitis (Quencer, Hungerford).

Sarcoidosis

Sarcoidosis can exceptionally be located within the adenohypophysis. The sarcoidosic granuloma appears as an intrasellar mass with homogeneous enhancement (Del Pozo).

Chordomas and Chondromas

Chordomas and chondromas of intrasellar development are exceptional (Spigos).

Miscellaneous

Hemangioblastomas, leiomyomas, and melanomas may also exceptionally develop within the pituitary gland. A case of hemangiopericytoma (Mangiardi), one of schwannoma (Perone), and one of choriocarcinoma (Yamagami), all three of intrasellar localization, have recently been reported. These three tumors present as homogeneous enhancing masses.

Pituitary Hyperplasia

Apart from pregnancy (see Chap. 7), pituitary hyperplasia has also been described in primary end-organ failure, particularly in primary hypothyroidism.

On CT scans, the pituitary gland appears enlarged, twice or thrice the normal height; a homogeneous enhancement is noted within this mass. With replacement therapy, CT may demonstrate marked decrease in tumor size (Danziger, Floyd, Okuno).

Sheehan's Syndrome

CT may identify a small pituitary gland or an empty sella turcica (Tolis).

Traumatic Pituitary Lesions

Fracture of the sellar floor may be associated with pituitary gland lesions leading to a partial empty sella turcica (Fig. 12.11).

Necrotic and hemorrhagic lesions may also be observed in patients with "coma depassé" (respirator brain) (McCormick).

References

Asa SL, Bilbao JM, Kovacs K, Josse RG, Kreines K (1981) Lymphocytic hypophysitis of pregnancy resulting in hypopituitarism: a distinct clinicopathologic entity. Ann Intern Med 95:166–171

Banna M, Schatz SW, Molot MJ, Groves J (1976) Primary intrasellar germinoma. Br J Radiol 49:971–973

Banna M, Baker HL Jr, Houser OW (1980) Pituitary and parapituitary tumours on CT. Br J Radiol 53:1123–1143

Baskin DS, Townsend JJ, Wilson CB (1982) Lymphocytic adenohypophysitis of pregnancy simulating a pituitary adenoma: a distinct pathological entity. J Neurosurg 56:148–153

Bentson JR, Wilson GH, Helmer E, Winter J (1977) CT in intracranial cysticercosis. J Comp Ass Tomog 1:464–471

Boggan JE, Davis RL, Zorman G, Wilson CB (1983) Intrasellar epidermoid cyst: case report. J Neurosurg 58:411–415

Bonneville JF, Dietemann JL (1981) Radiology of the sella turcica. Springer Verlag, Berlin Heidelberg New York

Bonneville JF, Cattin F, Portha C, Cuenin E, Clere P, Bartholomot B (1985) Computed tomographic demonstration of the posterior pituitary. AJNR 6:889–892

Braun IF, Pinto RS, Epstein F (1982) Dense cystic craniopharyngiomas. AJNR 3:139–141

Brooks BS, El Gammal T, Hungerford GD, Acker J, Trevor RP, Russel W (1982) Radiologic evaluation of neurosarcoidosis. Role of CT. AJNR 3:513–521

Buonaguidi R, Canapicci R, Mimassi N, Ferdeghini M (1984) Intrasellar Cavernous Hemangioma. Neurosurgery 14:732–734

Byrd SE, Winter J, Takahashi M, Joyce P (1980) Symptomatic Rathke's cleft cyst demonstrated on CT. J Comp Ass Tomog 4:411–414

Cabezudo JM, Vaquero J, Garcia de Sola R, Leunda G, Nombela L, Bravo G (1981) CT with craniopharyngiomas: a review. Surg Neurol 15:422–427

Carmel PW, Antunes JL, Chang CH (1982) Craniopharyngiomas in children. Neurosurgery 11:382–389

Chambers AA, Lukin RR, Tomsick TA (1977) Cranial epidermoid tumors: diagnosis by CT. Neurosurgery 1:276–280

Chambers EF, Turski PA, La Masters D, Newton TH (1982) Regions of low density in the contrast-enhanced pituitary gland: normal and pathologic processes. Radiology 144:109–113

Cusick JF, Khang-Cheng HO, Hagen TC, Kun LE (1982) Granular-cell pituicytoma associated with multiple endocrine neoplasia type 2. J Neurosurg 56:594–596

Daniels DL, Williams AL, Thornton RS, Meyer GA, Cusick JF, Haughton VM (1981) Differential diagnosis of intrasellar tumors by CT. Radiology 141:697–701

Danziger J, Wallace S, Handel S, Samaan NB (1979) The sella turcica in primary end organ failure. Radiology 131:111–115

Davis KR, Roberson GH, Taveras JM, New PFJ, Trevor R (1976) Diagnosis of epidermoid tumor by CT. Radiology 119:347–353

Decker RE, Mardayat M, Marc J, Rasool A (1979) Neurosarcoidosis with CT visualization and transsphenoidal excision of a supra- and intrasellar granuloma: case report. J Neurosurg 50:814–816

De Divitiis E, Spaziante R, Cirillo S, Stella L, Donzelli R (1979) Primary sellar chondromas. Surg Neurol 11:229–232

Del Pozo JM, Roda E, Montoya JG, Iglesias JR, Hurtado A (1980) Intrasellar granuloma: case report. J Neurosurg 53:717–719

Deramond H, Revert R, Dietemann JL, Hamza R, Remond A, Trinez G (1983) Incidentally discovered intrasellar calcifications. J Neuroradiology 10:231–241

Derome PJ, Jedynak CP, Peillon F (1980) Pituitary adenomas. Biology, physiopathology and treatment. Asclepios Publishers, Paris

Dietemann JL, Bonneville JF (1985) Radiological Diagnosis of Pituitary Diseases. In: Imura H (ed) The pituitary gland. Raven Press, New York, pp 341–361

Dietemann JL, Balériaux-Waha D, Golabek R, Wackenheim A, Jeanmart L (1979) Aspects tomodensitométri-

ques des kystes épidermoïdes et dermoïdes intracrâniens. A propos de trois observations. J Radiol 60:265–269

Dietemann JL, Lang J, Franck JP, Bonneville JF, Clarisse J, Wackenheim A (1981) Anatomy and radiology of the sellar spine. Neuroradiology 21:5–7

Dietemann JL, Bonneville JF, Buchheit F, Cattin F, Heldt N, Wackenheim A (1983) CT findings in symptomatic Rathke's cleft cysts of the pituitary gland. Report of three cases. Neuroradiology 24:263–267

Dietemann JL, Bonneville JF, Cattin F, Poulignot D (1983) Computed tomography of the sellar spine. Neuroradiology 24:173–174

Doron Y, Behar A, Bellar AJ (1965) Granular cell myoblastoma of the neurohypophysis. J Neurosurg 22:95–99

Earnest FIV, McCullough EC, Frank DA (1981) Fact or artifact: an analysis of artifact in high-resolution CT scanning of the sella. Radiology 140:109–113

Enzmann DR (1984) Imaging of infectious and inflammations of the central nervous system: computed tomography, ultrasound, and nuclear magnetic resonance. Raven Press, New York, pp 52–55

Enzmann DR, Sieling RJ (1983) CT of pituitary abscess. AJNR 4:79–80

Erdheim J, Stumme E (1909) Über die Schwangerschaftsveränderung der Hypophyse. Beitr Pathol Anat 46:1–122

Fitz CR, Wortzmann G, Harwood-Nash DC, Holgate RC, Barry JF, Boldt DW (1978) Computed tomography in craniopharyngiomas. Radiology 127:687–691

Floyd JL, Dorwart RH, Nelson MJ, Mueller GL, Devroede M (1984) Pituitary hyperplasia secondary to thyroid failure: CT appearance. AJNR 5:469–471

Fults D, Kelly DL Jr (1983) A suprasellar atypical teratoma presenting as an intrasellar mass: a case report. Neurosurgery 13:40–43

Gardeur D, Metzger J (1982) Adénomes hypophysaires intra-sellaires. In: Tomodensitométrie intra-crânienne. Livre V: Pathologie sellaire. Ellipses, Paris, p 41

Gardeur D, Nachanakian A, Millard JC, Metzger J, Poisson J, Mashaly R (1978) Neuroradiological aspects and therapeutic incidence of a case of posterior pituitary tumor (choristoma). J Neuroradiology 5:321–326

Gardeur D, Nachanakian A, Van Effenterre R, Zamora G, Metzger J (1979) Analyse tomodensitométrique des craniopharyngiomes. J Radiol 60:51–57

Gross CE, Binet EF, Esguerra JV (1979) Metrizamide cisternography in the evolution of pituitary adenomas and the empty sella syndrome. J Neurosurg 50:472–476

Hungerford GD, Biggs PJ, Levine JH, Shelley BE Jr, Perot PL, Chambers JK (1982) Lymphoid adenohypophysitis with radiologic and clinical findings resembling a pituitary tumor. AJNR 3:444–446

Jedynak CP, Oproiu A, Delalande O, Guiot G (1980) Nonadenomatous intrasellar lesions. In: Derome PJ, Jedynak CP, Peillon F (eds) Pituitary adenomas. Asclepios Publishers, Paris, p 155–169

Kovacs K (1972) Adenohypophysial necrosis in routine autopsies. Endokrinologie 60:309–316

Kovacs K (1973) Metastatic cancer of the pituitary gland. Oncology 27:533–542

La Masters DL, Boggan JE, Wilson CB (1982) Computerized tomography of a sellar spine. Case report. J Neurosurg 57:407–409

Lee BCP (1979) Intracranial cysts. Radiology 130:667–674

Lee KF, Lin JR (1979) Neuroradiology of sellar and juxtasellar lesions. Charles C. Thomas, Springfield, Ill

Luse SA, Kernohan JW (1955) Granular cell tumors of the stalk and posterior lobe of the pituitary gland. Cancer 8:616–622

Mangiardi JR, Flamm ES, Cravioto H, Fisher B (1983) Hemangiopericytoma of the pituitary fossa: case report. Neurosurgery 13:58–62

Martin WH, Cail WS, Morris JL, Constable WC (1980) Fibrosarcoma after high energy radiation therapy for pituitary adenoma. AJR 135:1087–1090

Max MB, Deck MDF, Rottenberg DA (1981) Pituitary metastasis: incidence in cancer patients and clinical differentiation from pituitary adenoma. Neuroradiology 31:998–1002

Mayfield RK, Levine JH, Gordon L, Powers J, Galbraith RM, Rawe SE (1980) Lymphoid adenohypophysis presenting as a pituitary tumor. Am J Med 69:619–623

McCormick WF, Halmi NS (1970) The hypophysis in patients with "coma depassé (respirator brain). Am J Clin Pathol 54:374–383

Mindel JS, Sachdev VP, Kline LB, Sivak MA, Bergman DA, Yang WC, Choi IS, Huang YP (1983) Bilateral intracavernous carotid aneurysm mimicking a prolactin secreting pituitary tumor. Surg Neurol 19:163–167

Mohr G, Hardy J (1982) Hemorrhage, necrosis, and apoplexy in pituitary adenomas. Surg Neurol 18:181–189

Muhr C, Bergström K, Grimelius L, Larsson SG (1981) A parallel study of the roentgen anatomy of the sella turcica and the histopathology of the pituitary gland in 205 autopsy specimens. Neuroradiology 21:55–65

Nelson PB, Haverkos H, Martinez AJ, Robinson AG (1983) Abscess formation within pituitary tumors. Neurosurgery 12:331–333

Nishi Y, Hyodo S, Sakano, Masuda H, Kitamura Y, Shindo H, Sakoda K, Nozumi T Usui T (1984) Pituitary abnormalities detected by high resolution computed tomography with thin slices in primary hypothyroidism and Turner syndrome. Eur J Pediatr 142:25–28

Nudleman KL, Choi B, Kusske JA (1985) Primary pituitary carcinoma: a clinical pathological study. Neurosurgery 16:90–95

Okuno T, Sudo M, Momoi T, Takao T, Ito M, Konishi Y, Yoshioka M, Suzuki J, Nakano Y (1980) Pituitary hyperplasia due to hypothyroidism. J Comp Ass Tomog 4:600–602

Ozanne P, Jedynak CP, Charbonnel B, Derome PJ (1982) Metastases hypophysaires et hypothalamiques. Etude anatomo-clinique de 5 observations. Ann Med Int 133:92–96

Ozgen T, Bertan V, Kansu T, Akalin S (1984) Intrasellar hydatid cyst. J Neurosurg 60:647–648

Ozonoff MB, Burrows EH (1971) Intracranial calcification. In: Newton TH, Potts DG (eds) Radiology of the brain and skull, vol 1, Book 2. Mosby, St. Louis, pp 823–873

Pascaud JL, Vigneu P, Hummel P, Bouchet JB, Rihouet J, Pascaud-Ged E (1982) Absence partielle de carotide interne avec anastomose inter-carotidienne transsellaire. J Radiol 63:37–40

Perone TP, Robinson B, Holmes SM (1984) Intrasellar schwannoma: case report. Neurosurgery 14:71–73

Peyster RG, Hoover E (1984) Computerized tomography in orbital disease and neuro-ophthalmology. Year Book Medical Publishers, Chicago, London

Pieterse S, Dinning TAR, Blumbergs PC (1982) Postirradiation sarcomatous transformation of a pituitary adenoma: a combined pituitary tumor. Case report. J Neurosurg 56:283–286

Pita JC Jr, Shafey S, Pina R (1979) Diminution of large pituitary tumor after replacement therapy for primary hypothyroidism. Neurology 29:1169–1172

Portocarrero CJ, Robinson AG, Taylor AL, Klein I (1981) Lymphoid hypophysitis: an unusual cause of hyperprolactinemia and enlarged sella turcica. JAMA 246:1811

Prosser PR (1978) Intrasellar cysticercosis presenting as a pituitary tumor: successful transsphenoidal cystectomy with preservation of pituitary function. Ann J Trop Med Hyg 27:976–978

Quencer RM (1980) Lymphocytic adenohypophysitis: autoimmune disorder of the pituitary gland. AJNR 1:343–345

Resneck JD, Lederman IR (1981) Traumatic chiasmal syndrome associated with pneumocephalus and sellar fracture. Am J Ophthalmol 92:233–237

Rilliet B, Mohr G, Robert F, Hardy J (1981) Calcifications in pituitary adenomas. Surg Neurol 15:249–255

Rudwan MA (1977) Pituitary abscess. Neuroradiology 12:243–248

Sachsenheimer W, Hamer J (1981) Intrasellar epidermoid with compression of the chiasma. Nervenarzt 52:457–459

Samaann MA, Osborne BM, Mackay B (1977) Endocrine and morphologic studies of pituitary adenomas secondary to primary hypothyroidism. J Clin Endocrinol Metab 45:903–911

Scheithauser BW, Kovacs K, Randall RV (1983) The pituitary gland in untreated Addison's disease. A histologic and immunocytologic study of 18 adenohypophyses. Arch Pathol Lab Med 107:484–487

Schlachter LB, Tindall GT, Pearl GS (1980) Granular cell tumor of the pituitary gland associated with diabetes insipidus. Neurosurgery 6:418–420

Selosse P, Mahler C, Klaes RL (1980) Pituitary abscess. J Neurosurg 53:851–852

Shanklin WM (1951) The incidence and distribution of cilia in the human pituitary with a description of micro-follicular cysts derived from Rathke's cleft. Acta Anat 11:361–381

Shen DY, Guay AT, Silverman ML, Hybels RL, Freidberg SR (1984) Primary intrasellar germinoma in a woman presenting with secondary amenorrhea and hyperprolactinemia. Neurosurgery 15:417–429

Sherman JL, Schnapf DJ, Coker SS (1983) Computed tomography of a pituitary stone. J Comp Ass Tomog 7:1120–1122

Spaziante R, De Divitiis E, Stella L, Cappabianca P, Donzelli R (1981) Benign intrasellar cysts. Surg Neurol 15:274–282

Spigos WST, Khine N (1982) Chordoma of the sellar region. J Comp Ass Tomog 6:154–158

Steinberg GK, Koenig GH, Golden JB (1982) Symptomatic Rathke's cleft cysts. Report of two cases. J Neurosurg 56:290–295

Takeuchi J, Handa H, Keyaki A, Haibara H, Ozaki S (1981) Intracranial angiolipoma. Surg Neurol 15:110–113

Taylon C, Duff TA (1980) Giant cell granuloma involving the pituitary gland. J Neurosurg 52:584–587

Tolis G, Herrera JR (1984) Sheehan's syndrome: in vivo diagnosis with use of computerized axial tomography and pituitary provocative testing. Fertil Steril 41:146–150

Turpin G, Kujas M, Van Effenterre R, Requeda E, Heshmati HM, Racadot J, De Gennes JL (1984) Les kystes mucigènes de l'hypophyse. Trois observations. Presse Med 13:1319–1321

Yamagami T, Handa H, Takeuchi J, Niijima K, Furukawa F (1983) Choriocarcinoma arising from the pituitary fossa with extracranial metastasis: a review of the literature. Surg Neurol 19:469–480

Zorub DS, Martinez AJ, Nelson PB, Lam MT (1979) Invasive pituitary adenoma with abscess formation: case report. Neurosurgery 5:718–722

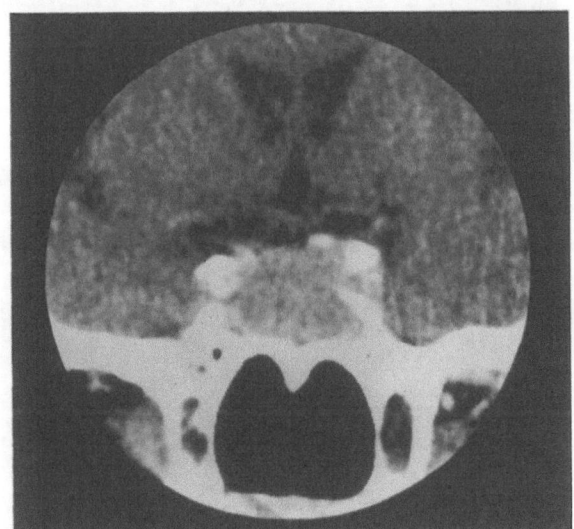
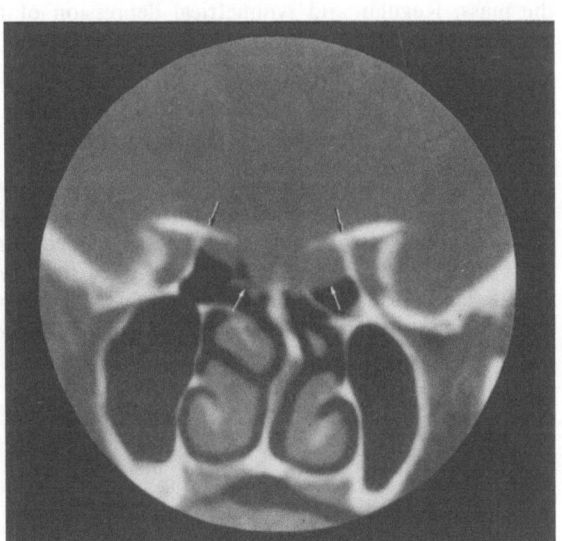

Fig. 12.1 a, b. Pituitary metastasis from a bronchial carcinoma. Coronal enhanced CT scan. **a** Homogeneous enhancement within the enlarged pituitary fossa. Destruction of the sellar floor and tumoral extension to the sphenoid sinus. No suprasellar extension. **b** The most anterior coronal section revealed destruction of the chiasmatic sulcus and planum sphenoidale (*black and white arrows*) with tumoral extension to the posterior ethmoid cells, particularly on the left (*white arrows*)

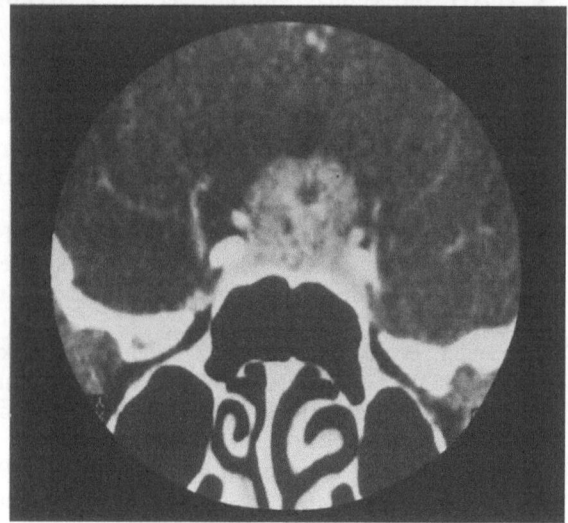

Fig. 12.2. Pituitary metastasis from a breast carcinoma. Postcontrast coronal CT scan. Marked suprasellar extension with heterogeneous enhancement

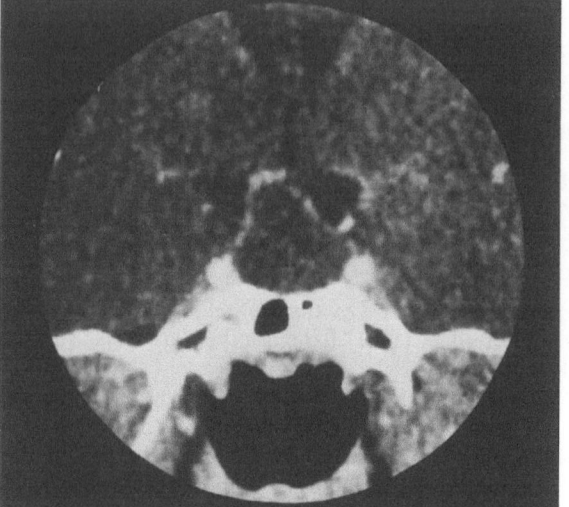

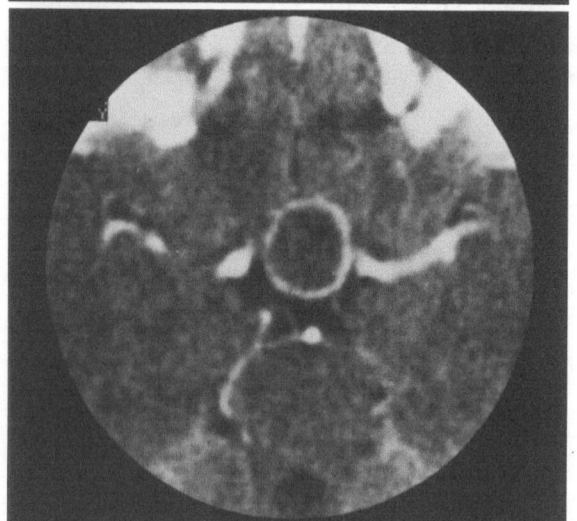

Fig. 12.3a, b. Symptomatic Rathke's cleft cyst in a ▶ 20-year-old woman presenting with headaches and pituitary deficiency. **a** Coronal and **b** axial enhanced CT scans. Low attenuation values are noted within the cyst. Ring enhancement delineates the suprasellar extent of the mass. Regular and symmetrical depression of the sellar floor

Fig. 12.4. Symptomatic Rathke's cleft cyst in a 16-year- ▶ old girl presenting with amenorrhea and galactorrhea. Coronal enhanced CT scan. An isodense rounded intrasellar tumor with suprasellar extent is noted (*white arrows*). The normally enhanced compressed pituitary gland is identified below the cyst (*black* and *white arrows*)

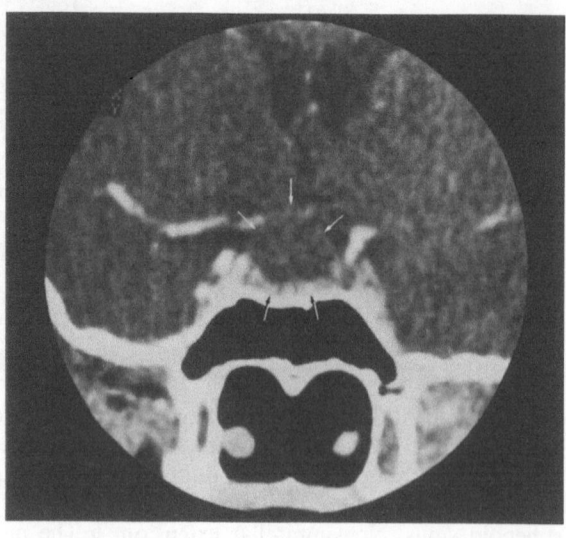

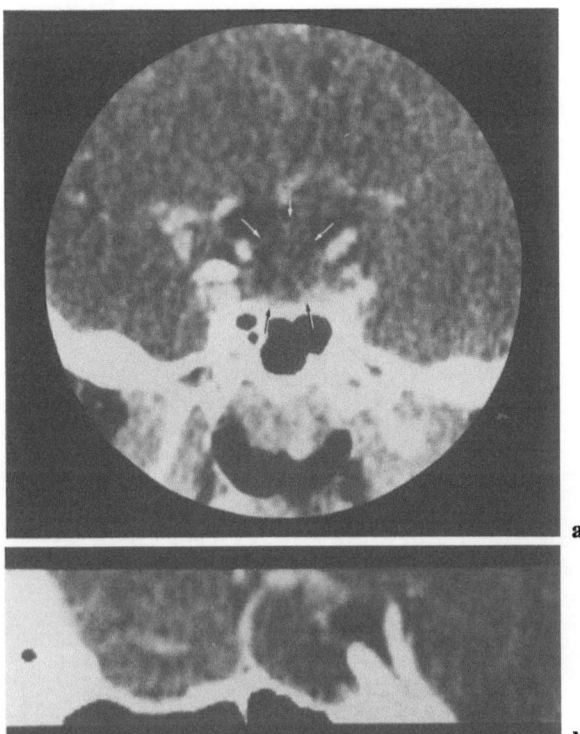

b

Fig. 12.6a, b. Intrasellar epidermoid cyst in an 18-year-old man presenting with delayed growth. **a** Coronal and **b** sagittal reformatted enhanced CT scans. Intrasellar mass with low attenuation values presenting a suprasellar extension (*white arrows*). **a** A thin band of normal pituitary tissue is identified below the epidermoid cyst (*black* and *white arrows*). **b** Peripheral enhancement is noted in the anterior portion of the tumor. Note that the density of the cyst is slightly higher than of the CSF

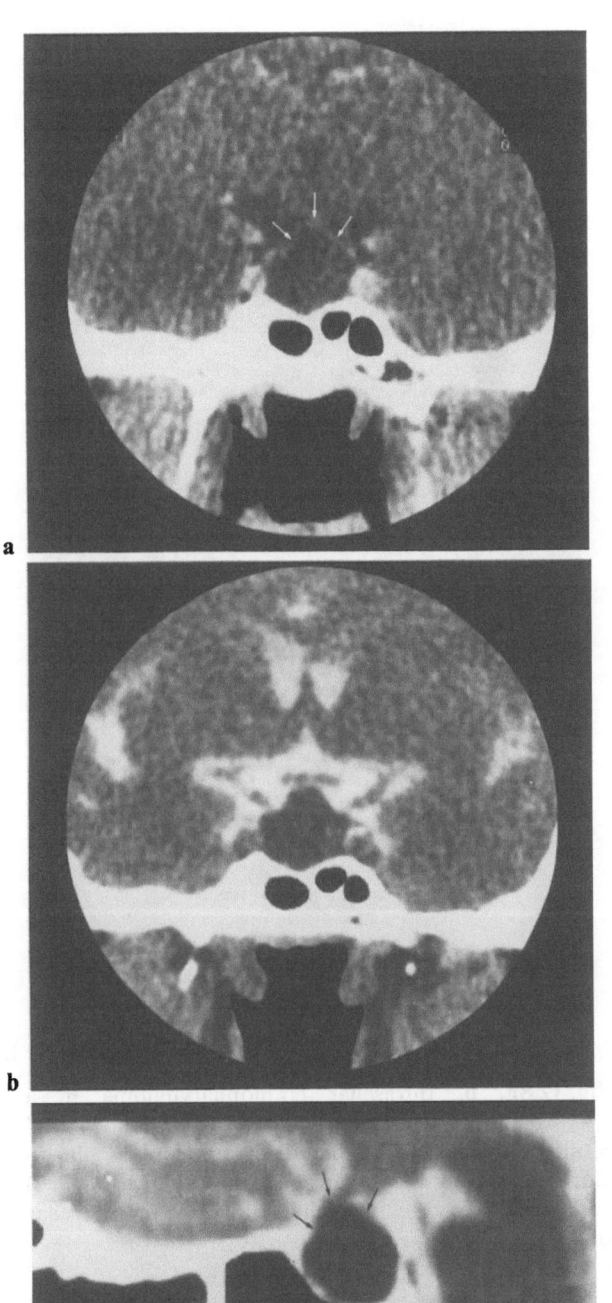

Fig. 12.5a–c. Rathke's cleft cyst. Headache and pituitary deficiency in a 53-year-old woman. **a** Coronal enhanced CT scan. **b** Coronal CT scan with intrathecal contrast media. **c** Sagittal reformatted CT scan after intrathecal contrast. **a** Hypodense intrasellar mass with suprasellar extent (*arrows*). **b, c** Metrizamide cisternography better delineates the upper limit of the mass

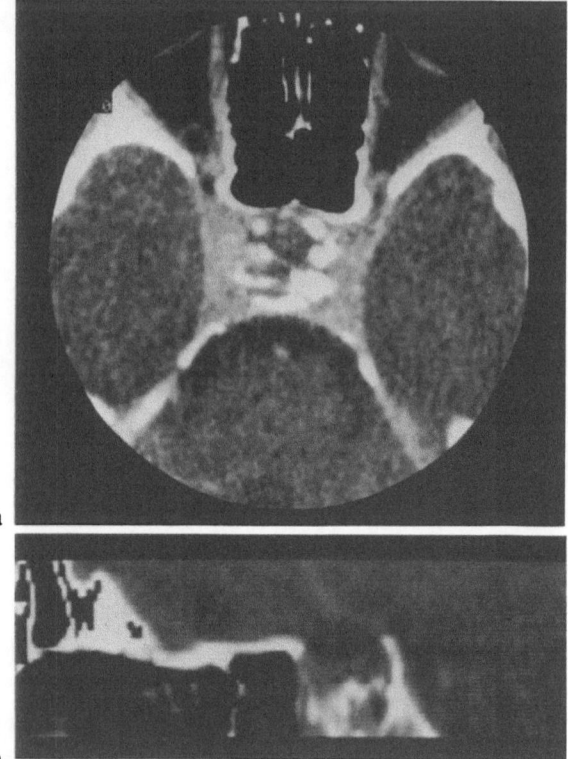

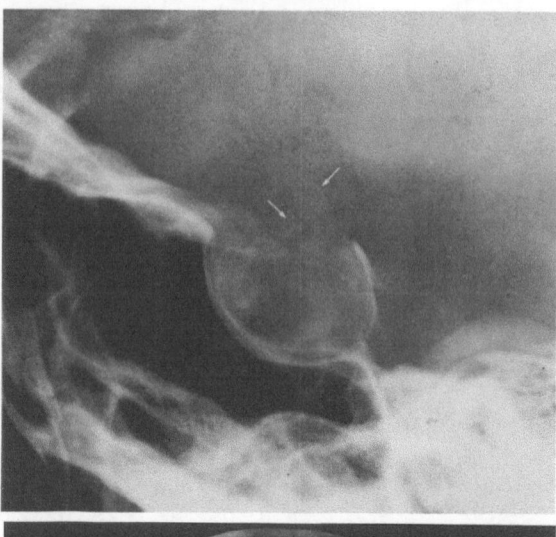

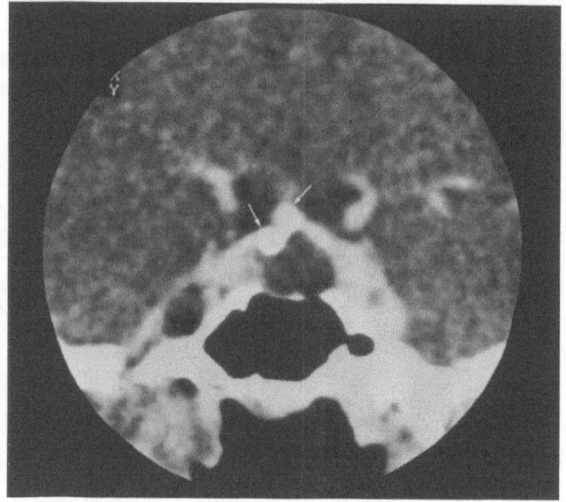

Fig. 12.7 a, b. Intrasellar craniopharyngioma in a 30-year-old man. Incidental discovery. **a** Axial, enhanced CT scan. **b** Sagittal reformatted image (bone window). Calcified intrasellar mass without suprasellar extension

Fig. 12.8 a, b. Intrasellar craniopharyngioma in a 17-year-old girl presenting with headaches and partial pituitary deficiency. **a** Magnified lateral plain film. **b** Coronal enhanced CT scan. Partially calcified (*arrows*) and partially cystic intrasellar tumor. Thinned sellar floor

Fig. 12.9. Granular cell tumor of the neurohypophysis ▶ in a 65-year-old man presenting with pituitary deficiency. Coronal contrast CT scan. An homogeneous enhancing intrasellar mass with small suprasellar extension is noted (*arrows*)

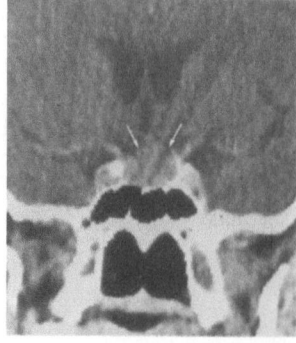

Fig. 12.10. Lymphocytic adenohypophysitis discovered 6 months after delivery in a 25-year-old black woman presenting with pituitary deficiency. Coronal contrast CT scan. A rounded homogeneous enhancing pituitary mass is demonstrated

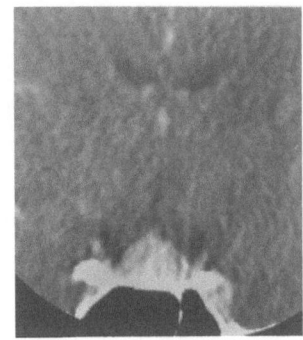

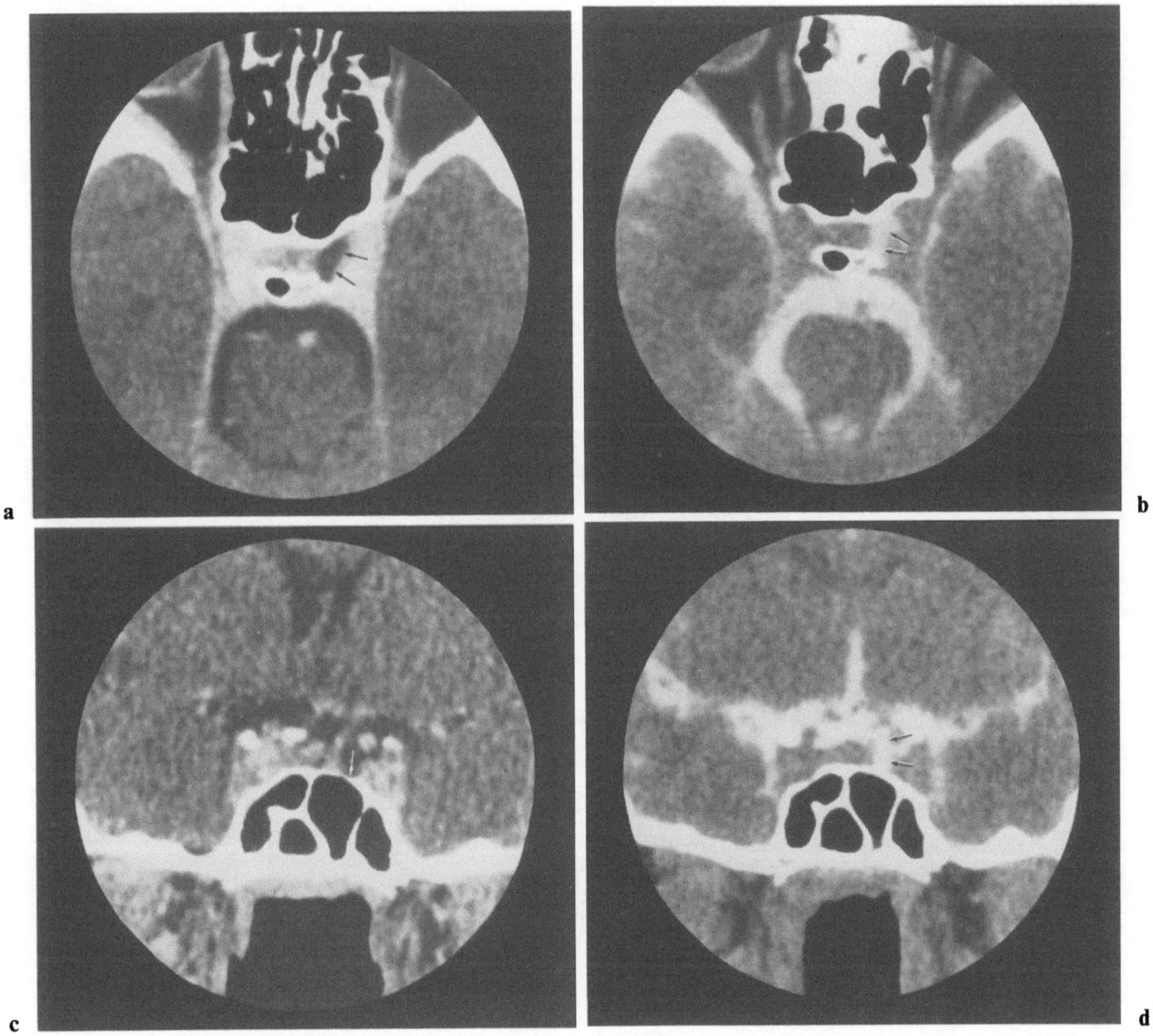

a

b

c

d

Fig. 12.11a–d. Posttraumatic partial empty sella turcica in an 18-year-old man presenting with diabetes insipidus. **a, b** Axial and coronal enhanced CT scans. **c, d** Axial and coronal intrathecally enhanced CT scans. On the left, a pituitary cleft (*arrows*) permits CSF to enter the pituitary fossa

Chapter 13
The Empty Sella Turcica

The empty sella turcica is characterized by an intrasellar herniation of the suprasellar subarachnoid spaces, favored by a dehiscence of the diaphragma sellae. This intrasellar irruption of subarachnoid spaces determines a flattening of the pituitary gland. An empty sella turcica relative to an opening of the diaphragma sellae is termed "primary empty sella turcica;" it should be distinguished from the "secondary empty sella turcica" seen after treatment of an intrasellar tumor (surgery, X-ray therapy, chemotherapy).

Primary Empty Sella Turcica

Dehiscence of the diaphragma sellae is essential for the formation of an intrasellar herniation of the suprasellar cistern. This diaphragma sellae is absent or generously opened in 20% of all individuals. The intrasellar herniation is favored and therefore accelerated on the one hand by suprasellar promoting factors (raised intracranial pressure, posteriorly placed optic chiasm) and on the other hand by hypophyseal factors having in common a reduction of pituitary gland volume (menopause, multiparity, pituitary gland infarction, Sheehan's syndrome, diabetes mellitus, replacement therapy of primary thyroid deficiency, or bromocriptine treatment).

Secondary Empty Sella Turcica

An empty sella turcica may be observed after surgical treatment or X-ray therapy of an intrasellar tumor. Such evolution toward an empty sella has also been observed after prolonged treatment of a microprolactinoma by bromocriptine. An empty sella can exceptionally be seen after rupture of an intrasellar cyst (Fig. 13.12).

In idiopathic primitive pituitary deficiency in children, a very low-heighted pituitary gland (1 to 2 mm) with an empty sella turcica is frequently observed on CT scan.

Clinical Symptoms

Primary empty sella turcica is but rarely symptomatic. Empty sella turcica is discovered in menopaused, frequently multiparous women, presenting headaches and obesity. Visual alterations (visual field defects and, exceptionally, papilledema), endocrine symptoms (pituitary deficiency, hyperprolactinemia, and even more exceptionally diabetes insipidus), or rhinorrhea rarely reveal an empty sella turcica.

CT Findings

CT demonstration of an empty sella turcica needs a technically perfect investigation. Only thin sections (1 or 1.5 mm in thickness, at most) may show CSF densities within the sella turcica. Direct coronal sections best display the content of the pituitary fossa. If direct coronal sections cannot be obtained, as is sometimes the case with old and obese patients, joint or overlapped axial sections of 1 or 1.5 mm thickness should be made in order to obtain coronal and sagittal reformatted images of good quality (Fig. 13.1). Intravenous enhancement is necessary to properly visualize the remaining pituitary gland, recognize an eventual associated microadenoma, and correctly demonstrate the cavernous sinuses and the pituitary stalk.

CSF densities occupy a more or less important volume (Fig. 13.1a). At most, the remaining normal pituitary gland may be difficult to identify, being flattened against the inferior part of the anterior aspect of the dorsum sellae. The whole pituitary fossa, thus, appears occupied by water density (Figs. 13.5 and 13.6). To a lesser degree, a thin band (1–4 mm) of pituitary tissue, whose superior limit still appears concave, may be identified above the sellar floor (Fig. 13.4). Coronal sections often identify a stretched vertical and medial pituitary stalk (Fig. 13.2), surrounded by CSF densities and frequently displaced against the dorsum sellae. Failure to identify the pituitary stalk on coronal and axial sections must lead to prudence and a computerized cisternography should be performed to eliminate any intrasellar or suprasellar cystic lesion (Haughton).

Bony changes are frequently specific. The sella turcica is often enlarged. The most important arachnoidoceles are accompanied by a thinning of the dorsum sellae; these changes are clearly demonstrated in axial sections. Coronal sections visualize changes of the sellar floor depending upon the degree of pneumatization of the sphenoid bone (Bonneville and Dietemann). If pneumatization is absent under the sellar floor, the cortical bone of the sellar floor will show a medial symmetrical depression ("en cuvette") with a spared cortical bone (Fig. 13.1 b). When the sphenoid sinus extends symmetrically under the sellar floor, the latter appears flat and thin (Fig. 13.5). When the sphenoid sinus extends asymmetrically, the floor is depressed toward the lesser pneumatized side (Fig. 13.6). The lateral wall of the pituitary fossa, i.e., the medial wall of the cavernous sinus is often concave. The carotid arteries may extend toward the empty pituitary fossa (Fig. 13.7).

Diagnosis of an empty sella turcica may sometimes be difficult. Water densities measured inside the sella turcica may be erroneous. Bony artifacts sometimes determine intrasellar low-density area on axial sections. These low-density areas are obviously found again in coronal and sagittal reformatted images realized from this series of artifacted axial sections (see Chap. 1). On the other hand, thick axial sections (3–5 mm), may provoke partial volume effect, with false appearance of an empty sella turcica.

Hypodense tumors (cystic craniopharyngiomas, necrosed adenomas, Rathke's cleft cyst) should not be mistaken for an empty sella turcica. Injection of iodinated contrast medium often shows an enhancement on the superior pole of the lesion, especially for necrosed adenoma and Rathke's cleft cyst. The pituitary stalk is either not identified or appears deviated. If any doubt persists, an intrathecally enhanced CT scan may be indicated (Fig. 12.5).

Supratentorial hydrocephalus, as observed in congenital aqueduct stenosis, is accompanied by an expansion of the third ventricle into the sella turcica, thus modifying the latter's morphology. A marked expansion of the ventricles allows a rapid diagnosis of the nature of the intrasellar hypodensity (Fig. 13.8).

In cases of hormonal hypersecretion (prolactin, GH, ACTH), it seems appropriate to investigate for a microadenoma associated with the empty sella turcica. Such investigation may be difficult when the adenoma does not appear clearly as a low attenuation area within the pituitary gland (Gardeur, Smaltino).

When visual symptoms are present, downward displacement of the optic chiasm into the pituitary fossa must be researched (Fig. 13.9). The use of intrathecal contrast medium and sagittal reformatted images can be useful to demonstrate the abnormally displaced optic chiasm (Bursztyn).

Asymmetrical empty sella turcica with deviation of the pituitary may be observed. Indeed the remaining pituitary gland may be located in the lateral part of the pituitary fossa, while the herniated subarachnoid space is occupying the contralateral part of the pituitary fossa (Fig. 13.10). Asymmetrical empty sella turcica may be primary or secondary to the evolution of an intrasellar tumor (see Chap. 11).

In some cases of empty sella turcica, a normal posterior pituitary may be demonstrated behind a remaining anterior pituitary tissue band (Fig. 13.11).

References

Bajraktari X, Bergström M, Brismar K, Goulatia R, Greitz T, Grepe A (1977) Diagnosis of intrasellar external herniation (empty sella) by CT. J Comp Ass Tomog 1:105–116

Banna M, Baker HL Jr, Houser OW (1980) Pituitary and parapituitary tumours on CT. Br J Radiol 53:1123–1143

Bonneville JF, Dietemann JL (1981) Radiology of the sella turcica. Springer Verlag, Berlin Heidelberg New York

Bonneville JF, Poulignot D, Cattin F, Couturier M, Mollet E, Dietemann JL (1982) Apport des méthodes nouvelles dans l'exploration morphologique des tumeurs hypophysaires. Ann Endocrinol (Paris) 43:303–308

Bursztyn EM, Lavyne MH, Aisen M (1983) Empty sella syndrome with intrasellar herniation of the optic chiasm. AJNR 4:167–168

Busch W (1951) Die Morphologie der Sella Turcica und ihre Beziehungen zur Hypophyse. Virchows Arch (Pathol Anat) 320:437–458

Cabanis EA, Van Effenterre R, Iba-Zizen MT (1979) CT in parasellar space-occupying lesions and therapeutic decision. Acta Neurochirurgica (suppl 28) 28:329–333

Daniels DL, Williams AL, Thornton RS, Meyer GA, Cusick JF, Haughton VM (1981) Differential diagnosis of intrasellar tumors by CT. Radiology 141:697–701

Derome PJ, Jedynak CP, Peillon F (1980) Pituitary adenomas. Biology, physiopathology and treatment. Asclepios Publishers, Paris

Dietemann JL, Bonneville JF (1985) Radiological diagnosis of pituitary diseases. In: Imura H (ed) The pituitary gland. Raven Press, New York, pp 341–361

Domingue JN, Wing SD, Wilson CB (1978) Coexisting pituitary adenomas and partially empty sella. J Neurosurg 48:23–28

Drayer BP, Rosenbaum AE, Riegel DB, Bank WO, Deeb ZL (1977) Metrizamide CT cisternography: pediatric applications. Radiology 124:349–357

Gardeur D, Metzger J (1982) Adénomes hypophysaires intra-sellaires. In: Tomodensitométrie intra-crânienne. Livre V: Pathologie sellaire. Ellipses, Paris, p 41

Gross CE, Binet EF, Esguerra JV (1979) Metrizamide cisternography in the evolution of pituitary adenomas and the empty sella syndrome. J Neurosurg 50:472–476

Guibert-Tranier F, Elie G, Guibert JL, Piton J, Caille JM (1980) Selles turcique vides. Diagnostic TDM. J Neuroradiology 7:105–119

Hall K, McAllister VL (1980) Metrizamide cisternography in pituitary and juxtapituitary lesions. Radiology 134:101–108

Hatam A, Bergström M, Greitz T (1979) Diagnosis of sellar and parasellar lesions by CT. Neuroradiology 18:249–258

Haughton VM, Rosenbaum AE, Williams AL, Drayer B (1980) Recognizing the empty sella by CT: the infundibulum sign. AJNR 1:527–529

Hoffman JC Jr, Tindall GT (1980) Diagnosis of empty sella syndrome using Amipaque cisternography combined with CT. J Neurosurg 52:99–102

Kuuliala I, Katevuo K, Ketonen L (1981) Metrizamide cisternography with hypocycloid and CT in sellar and suprasellar lesions. Clin Radiol 32:403–408

Lundberg PO, Osterman PO, Wide L (1981) Serum prolactin in patients with hypothalamus and pituitary disorders. J Neurosurg 55:194–199

Marano GD, Horton JA, Vasquez AM (1981) CT in diabetes insipidus: posterior empty sella. Br J Radiol 54:263–265

Mohr G, Hardy J (1982) Hemorrhage, necrosis, and apoplexy in pituitary adenomas. Surg Neurol 18:181–189

Penley MW, Pribram HFW (1980) Diagnosis of empty sella with small amount of air at CT. Surg Neurol 14:296–301

Peyster RG, Hoover E (1984) Computerized tomography in orbital disease and neuro-ophthalmology. Year Book Medical Publishers, Chicago, London

Pinto RS, Handel SF, Stadhu VK (1979) CT metrizamide cisternography in the recognition of intrasellar cistern. AJR 133:320–321

Roberson GH, Tadmor R, Taveras JM, Kleefield J, Ellis G (1977) CT in metrizamide cisternography. Importance of coronal and axial views. J Comp Ass Tomog 1:241–245

Rozario R, Hammerschlag SB, Post KD, Wolpert SM, Jackson I (1977) Diagnosis of empty sella with CT scan. J Neuroradiology 13:85–88

Sage MR, Lhan ES, Reilly PL (1980) The clinical and radiological features of the empty sella syndrome. Clin Radiol 31:513–519

Sage MR, Blumbergs PC, Fowler GW (1982) The diaphragma sellae: its relationship to normal sellar variations in frontal radiographic projections. Radiology 145:699–701

Sheldon P, Molyneux A (1978) Metrizamide cisternography and CT for the investigation of pituitary regions. Neuroradiology 17:83–87

Smaltino F, Bernini FP, Muras I (1980) CT for diagnosis of empty sella associated with enhancing pituitary microadenoma. J Comp Ass Tomog 4:592–598

Spaziante R, De Divitiis E, Stella L, Cappabianca P, Genovese L (1981) The empty sella. Surg Neurol 16:418–426

Swanson JA, Sherman BM, Van Gilder JC, Chapler FK (1979) Coexistent empty sella and prolactin-secreting microadenoma. Obstet Gynecol 53:258

Taylor S (1982) High resolution computed tomography of the sella. Radiologic clinics of North America 20:207–236

Tolis G, Herrera JR (1984) Sheehan's syndrome: in vivo diagnosis with use of computerized axial tomography and pituitary provocative testing. Fertil Steril 41:146–150

Young WF, Ospina LF, Wesolowski D, Touma A (1981) The primary empty sella syndrome, diagnosis with metrizamide cisternography. JAMA 246:2611–2612

Zull DN, Falko JM (1981) Metrizamide cisternography in the investigation of the empty sella syndrome. Arch Intern Med 141:487–489

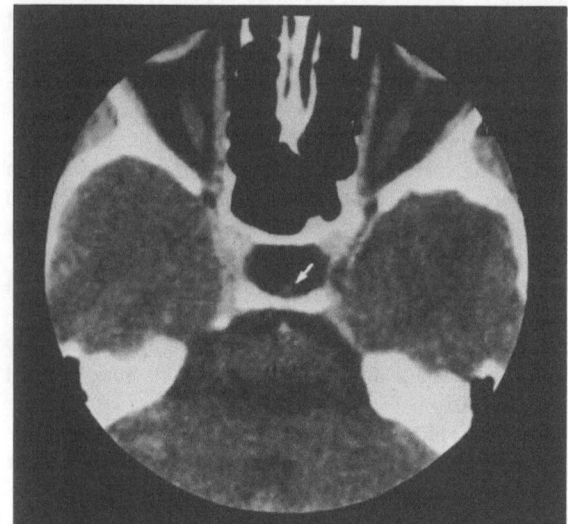

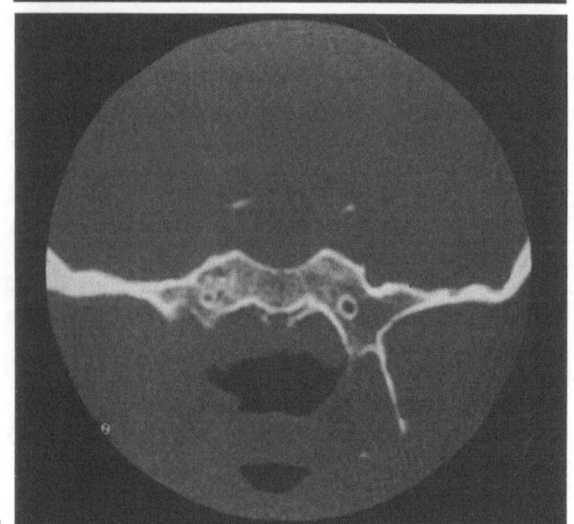

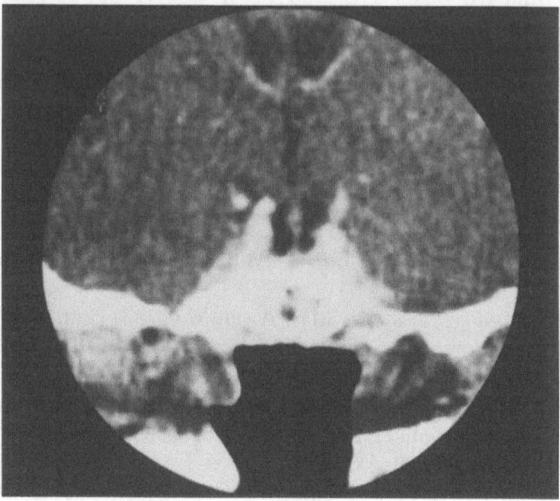

Fig. 13.2. Primary empty sella turcica in a 47-year-old woman presenting with headaches. Coronal enhanced CT scan. The pituitary stalk appears vertical in the midline and elongated

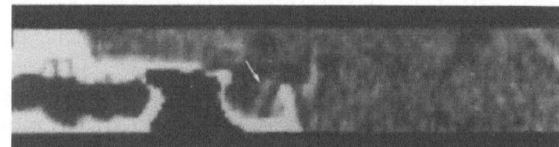

Fig. 13.3. Primary empty sella turcica in a 33-year-old woman presenting with bilateral papilledema. Sagittal enhanced reconstruction. Pituitary stalk (*arrow*) appears clearly in front of the dorsum sellae

Fig. 13.1 a, b. Primary empty sella turcica in a 21-year-old woman presenting with headaches and obesity. **a** Axial enhanced CT scan. **b** Coronal CT scan for bone detail. **a** Enlarged sella turcica; the content appears as water density. The pituitary stalk (*arrow*) is in front of the dorsum sellae. **b** Symmetrical depression of the sellar floor is noted above this nonpneumatized sphenoid body

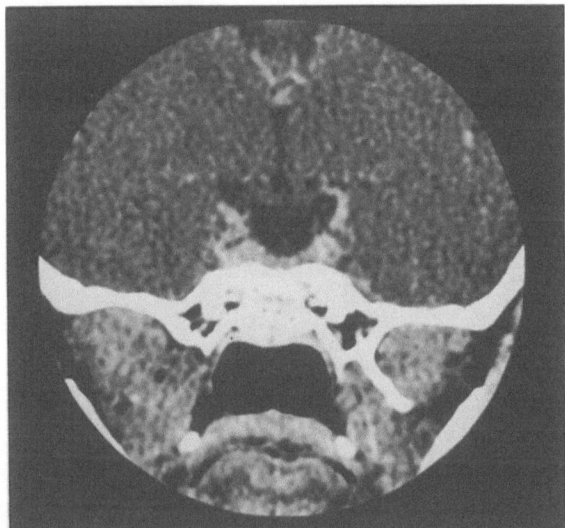

Fig. 13.4. Partially primary empty sella turcica in a 51-year-old man. Coronal enhanced CT scan. Remaining pituitary gland presents an upward concavity. Median symmetrical depression of the sellar floor

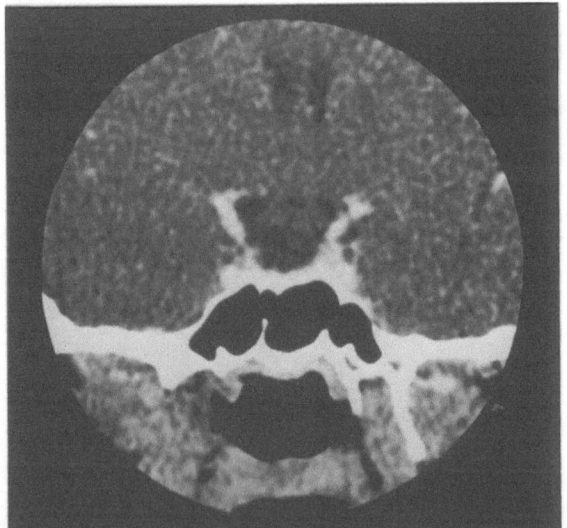

a

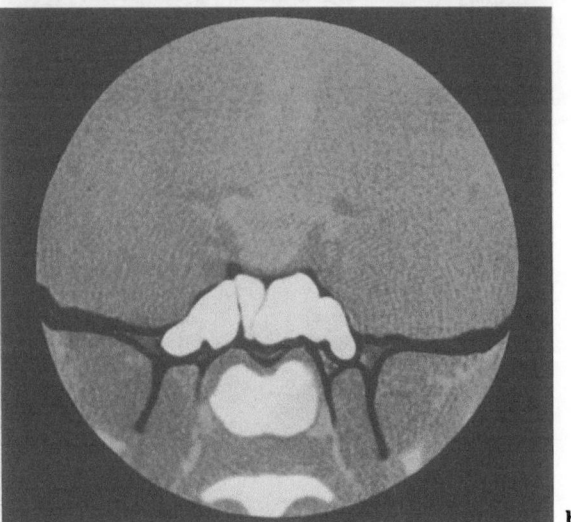

b

Fig. 13.5a, b. Primary empty sella turcica in a 44-year-old woman. Coronal enhanced CT scan. Above a well-pneumatized sphenoid body, the sellar floor appears normal

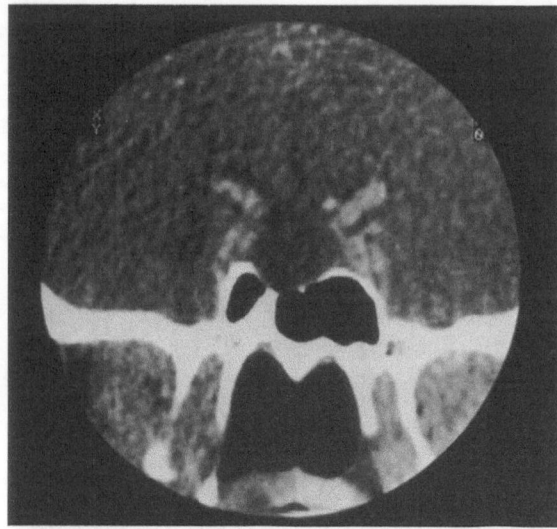

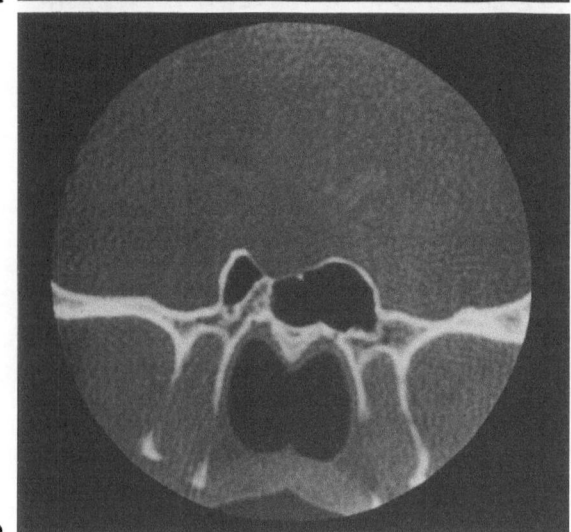

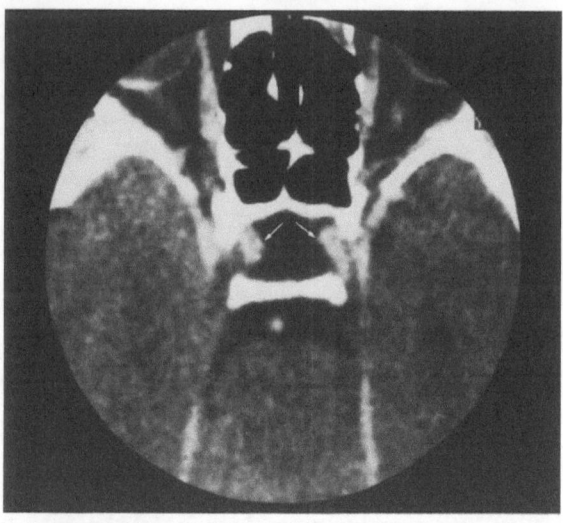

Fig. 13.7. Primary empty sella turcica. Axial enhanced CT scan. Bilateral displacement of the carotid siphons toward the midline (*arrows*)

Fig. 13.6a, b. Primary empty sella turcica. **a** Coronal enhanced CT scan. **b** Coronal CT scan for bone detail. Asymmetrical depression of the sellar floor in a patient with asymmetrical pneumatization of the sphenoid body

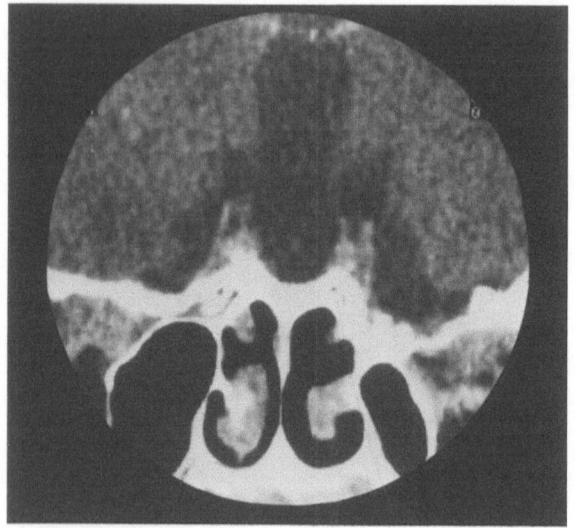

Fig. 13.8. Secondary empty sella turcica. Coronal enhanced CT scan. Empty sella turcica in a patient with marked supratentorial hydrocephalus related to aqueduct stenosis

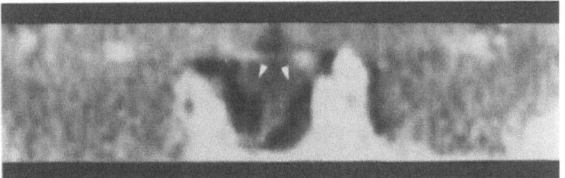

Fig. 13.9. Primary empty sella turcica. Coronal reformatted enhanced CT scan. Downward displacement of the optic chiasm (*arrowheads*) in a patient presenting with visual field defect

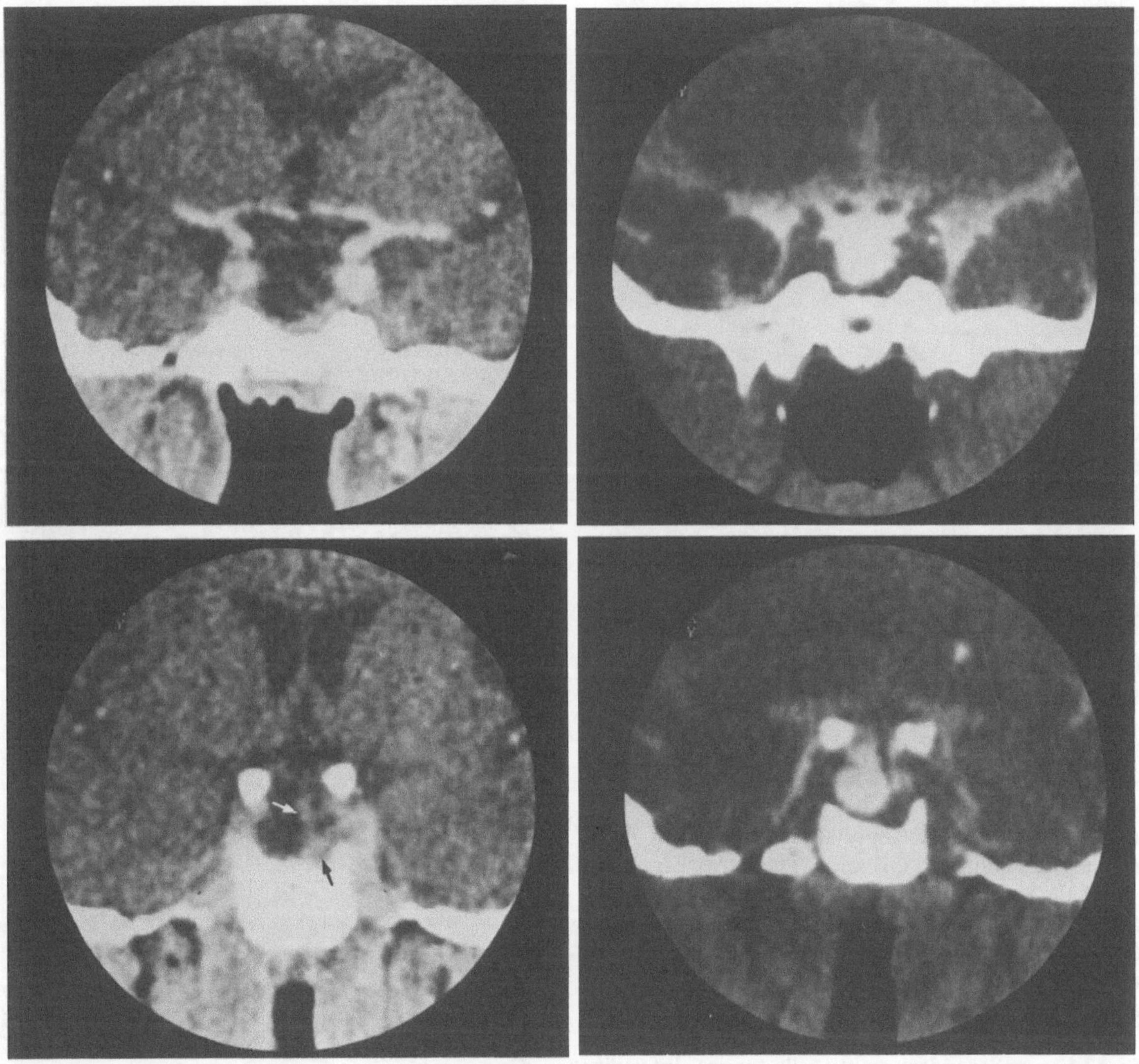

Fig. 13.10 a–d. Secondary empty sella turcica: spontaneous regression of an intrasellar tumor in a 35-year-old man presenting with delayed growth. **a, b** Coronal intravenous enhanced CT scan. **c, d** Coronal CT scans after intrathecal enhancement. Asymmetrical empty sella turcica with left displacement of the distal portion of the pituitary stalk (*white arrow*). The remaining pituitary gland occupies the left part of the pituitary fossa *(black arrow)*

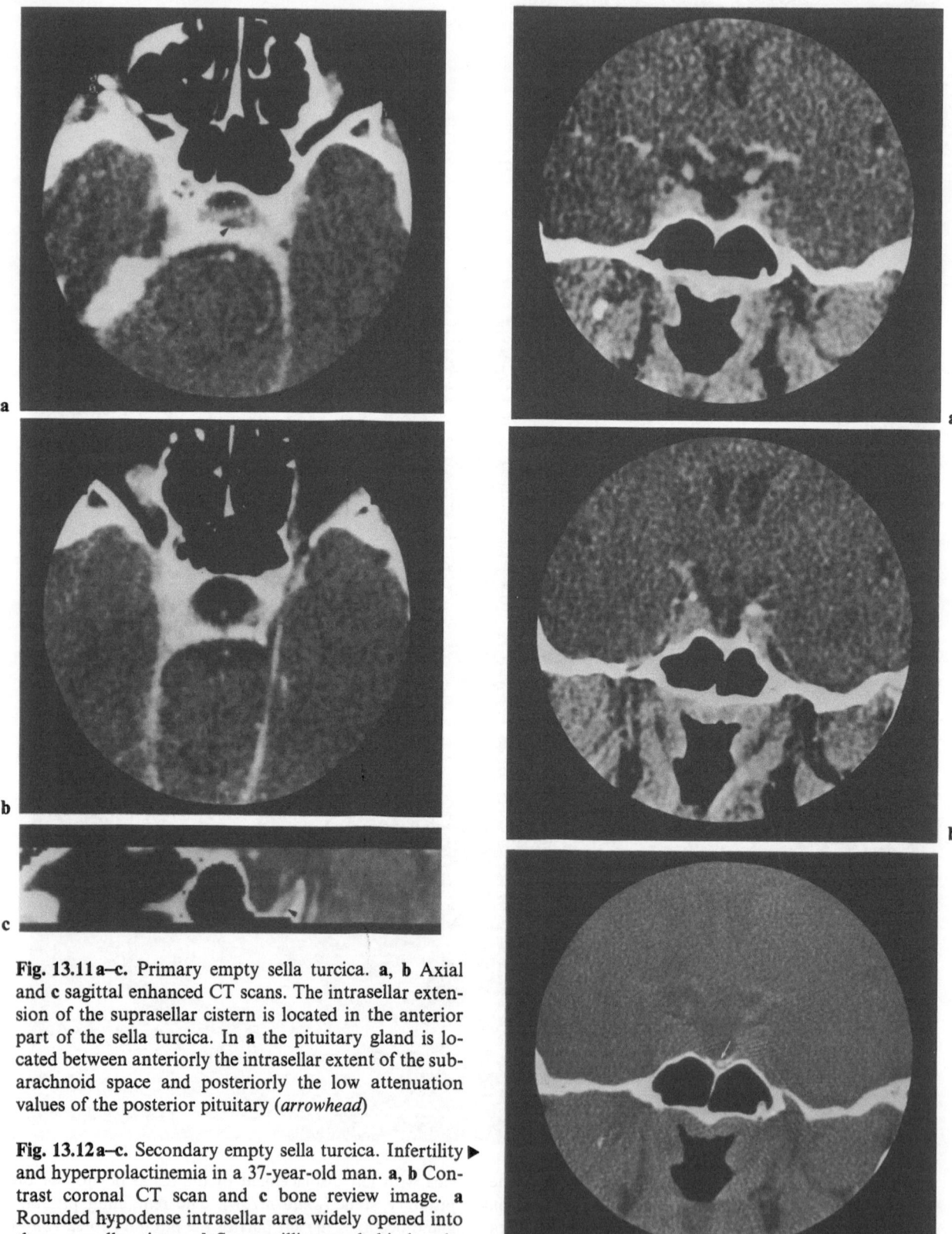

Fig. 13.11a–c. Primary empty sella turcica. **a, b** Axial and **c** sagittal enhanced CT scans. The intrasellar extension of the suprasellar cistern is located in the anterior part of the sella turcica. In **a** the pituitary gland is located between anteriorly the intrasellar extent of the subarachnoid space and posteriorly the low attenuation values of the posterior pituitary (*arrowhead*)

Fig. 13.12a–c. Secondary empty sella turcica. Infertility ▶ and hyperprolactinemia in a 37-year-old man. **a, b** Contrast coronal CT scan and **c** bone review image. **a** Rounded hypodense intrasellar area widely opened into the suprasellar cistern. **b** Some millimeters behind **a**, the elongated pituitary stalk is well shown within the pituitary fossa. Intrasellar curvilinear calcifications (*arrow*) limiting the bottom of the intrasellar extent of the subarachnoid space are in favor of a spontaneously necrosed pituitary cystic tumor

Chapter 14
CT of the Sellar Region After Surgery and/or Radiotherapy

The transsphenoidal approach has become the procedure of choice for removal of intrasellar lesions. The subfrontal approach is reserved for large suprasellar tumors and adenomas with hourglass expansion. Postoperative CT scans can demonstrate surgical changes, complications, residual or recurrent tumors.

Postsurgical Changes

These modifications can be observed on the line of surgical approach, in the surgical cavity, or in the cavernous sinus.

Postsurgical Changes on the Line of Surgical Approach

A more or less extensive hypodensity, generally on the right, is often observed after a subfrontal approach. Bony gaps are almost always observed on the inferior wall of the sphenoid bone and on the intersinusal septa after transsphenoidal approach (Figs. 14.2 and 14.11). Osseous grafts or Silastic brought to reconstitute the sellar floor can also be found inside the sphenoid sinus (Fig. 14.1). Addition of muscles, Gelfoam or Surgicel by the surgeon contribute to increasing the densities measured inside the sphenoid sinus. This hyperdensity due to foreign material remains steady for several years after the operation. However, hyperdensities of the sphenoid sinus and even of the maxillary or ethmoid sinuses due to mucosal thickening or blood disappear spontaneously within a few weeks after surgery.

The Pituitary Region After Surgery

The sellar floor appears more or less well reconstituted. It is often irregular, sometimes denser than before surgery or even bristled with bony spicules pointing inside the pituitary fossa (Fig. 14.3). The persistence of a bony gap can favor a herniation of air from the sinus into the sella turcica (Fig. 14.4).

The content of the sella turcica is variable:

1. The sellar content may be completely or almost completely emptied by the surgeon, having an attenuation value close to 0 Hounsfield unit, communicating widely with the suprasellar subarachnoid spaces. The pituitary stalk, attached to a remnant stump of pituitary tissue, is frequently inclined (Fig. 14.5). A dynamic scan is particularly interesting to distinguish the remaining hypophyseal tissue from the cavernous sinus.

2. The surgeon may make a selective microadenomectomy and the CT scan will show just a defect within the pituitary gland. A normal pituitary gland can even be observed.

3. In the case of a voluminous tumor with suprasellar extent, the surgeon may leave the tumoral shell which remains still for weeks as if it has been suspended over the surgical cavity (Fig. 14.6).

4. Lastly, the surgeon may pack the surgical cavity with muscle, Gelfoam, Surgicel, or fat. In certain cases the attenuation values of the pituitary fossa packed with foreign material may be close to the density of a recurrent tumor; however, the former remains unchanged after injection of contrast material.

We have observed a hypodense image of the cavernous sinus with a convex lateral wall in a patient who presented bleeding during surgery, forcing the surgeon to pack the cavernous sinus to control the bleeding (Fig. 14.7).

Complications

The presence of a small quantity of blood in the surgical cavity is common during the first few days following the operation, but bleeding readily detected by CT scan may prompt a second operation.

Displaced bony fragments threatening the chiasm have also been described. An excess of packing causing visual defects can also be observed. Diagnosis with CT scan is easy if fat has been used for packing. Carotid occlusions and ocular nerve palsies secondary to excessive packing have also been described. If a rhinorrhea occurs after transsphenoidal approach, a CT cisternography may be required.

Residual and Recurrent Tumors

The use of foreign material inside the pituitary fossa may hinder the diagnosis of residual or recurrent tumor. The diagnosis of recurrence is obviously easier if a postsurgical CT scan has been performed some months after the operation. Of course, attenuation values of muscle, Gelfoam, or Surgicel do not change after injection of contrast medium, but recurrent tumors may themselves be poorly enhanced. In an early stage, recurrences are almost always seen as a tissue of the same attenuation value as the initial lesion, often located laterally within the sella (Figs. 14.8, 14.9, and 14.10). Dynamic scan can here bring important data to localize limits of the cavernous sinus and demonstrate eventual lateralized intrasellar recurrent tumors.

Cystic recurrences of pituitary adenomas with ring enhancement are less common (Fig. 14.11).

Certain types of adenomas are known to recur more frequently than others. In our experience, we have observed the most rapid and most voluminous recurrences with FSH- LH-secreting adenomas and with mixed prolactin- and GH-secreting adenomas.

Nonsecreting adenomas can also demonstrate important suprasellar recurrences, but they generally grow more slowly.

X-ray therapy can, on its own, bring about a considerable decrease in tumoral size, giving a CT image of empty sella turcica, as has been observed in several nonoperated acromegalies. On the other hand, in our experience, the pituitary gland generally appears low-heighted in cases of Cushing's disease with no evidence of microadenoma, treated in the past by X-ray therapy. In general, the effect of conventional X-ray therapy on pituitary tumor seems slow and the tumoral volume may sometimes decrease after several years only. However, particle radiotherapy seems more rapidly effective, especially in prolactinomas.

References

Arafah BM (1981) Cure of hypogonadism after removal of prolactin secreting adenoma in men. J Clin Endocrinol 52:91–94

Aubourg PR, Derome PJ, Peillon F, Jedynak CP, Visot A, Le Gentil P, Balagura S, Guiot G (1980) Endocrine outcome after transsphenoidal adenomectomy for prolactinoma: prolactin levels and tumor size as predicting factors. Surg Neurol 14:141–143

Barbarino A, De Marvis L, Menini E, Anile C, Maira G (1979) Prolactin secreting pituitary adenomas: prolactin dynamics before and after transsphenoidal surgery. Acta Endocrinol 91:397–400

Bataini JP (1980) Physiothérapie des adénomes hypophysaires. Rev Prat 30:3063–3068

Bonneville JF, Dietemann JL (1981) Radiology of the sella turcica. Springer Verlag, Berlin Heidelberg New York

Casper RF, Rakoff JS, Quigley ME, Gilliland B, Alksne J, Yen SSC (1980) Changes in pituitary hormones during and following transsphenoidal removal of prolactinomas. Ann J Obstet Gynecol 136:518–523

Ciric I, Mikhael M, Stafford T, Lawson L, Garces R (1983) Transsphenoidal microsurgery of pituitary macroadenomas with long-term follow-up results. J Neurosurg 59:395–401

Danoff BF, Pripstein S, Croce N (1978) Value of CT in delineating suprasellar extension of pituitary adenoma for radiotherapeutic management. Cancer 42:1066

Derome PJ, Jedynak CP, Peillon F (1980) Pituitary adenomas. Biology, physiopathology and treatment. Asclepios Publishers, Paris

Derome PJ, Peillon F, Bara RM, Jedynak CP, Racadot J, Guiot G (1979) Adénomes à prolactine, résultats du traitement chirurgical. Nouv Presse Méd (Paris) 8:577–583

De Schryver A, Vandekerckhove D, Debruyne G (1980) Prolactin secreting pituitary adenoma. Acta Radiol (therapy) 19:169–175

Dolinskas CA, Simeone FA (1985) Transsphenoidal hypophysectomy: Postsurgical CT findings. AJNR 6:45–50

Domingue JN, Richmond IL, Wilson CB (1980) Results of surgery in 114 patients with prolactin-secreting pituitary adenomas. Am J Obst Gynecol 137:102–108

Drayer BP, Wilkins RH, Boehnke M, Horton JA, Rosenbaum AE (1977) Cerebrospinal fluid rhinorrhea demonstrated by metrizamide CT cisternography. Am J Roentgenol 129:149–151

Edwards CRW, Feek CM (1981) Prolactinomas: a question of rational treatment (Editorial). Br Med J 283:1561–1562

Faria MA, Tindall GT (1982) Transsphenoidal microsurgery for prolactin-secreting pituitary adenomas. J Neurosurg 56:33–43

Lee KF, Suh JH, Mazziotta JC (1977) The value of CT in the radiotherapeutic management of juxtasellar tumors. CT 1:111–119

Martin WH, Cail WS, Morris JL, Constable WC (1980) Fibrosarcoma after high energy radiation therapy for pituitary adenoma. AJR 135:1087–1090

McGregor A, Scanlon MF, Hall R (1982) Les adénomes hypophysaires à prolactine. Explorations et traitements. Ann Med Int 133:51–57

Muhr C, Bergström K, Enoksson P, Hugosson R, Lundberg PO (1980) Follow-up study with CT and clinical evaluation 5 to 10 years after surgery for pituitary adenoma. J Neurosurg 53:144–148

Nelson PB, Haverkos H, Martinez AJ, Robinson AG (1983) Abscess formation within pituitary tumors. Neurosurgery 12:331–333

Ory SJ, Land M, Van der Merwe JV, Kramer RS, Hammond CB (1983) Transsphenoidal resection of prolactinomas in seventy patients (Ab). Fertil Steril 39:432

Pieterse S, Dinning TAR, Blumbergs PC (1982) Postirradiation sarcomatous transformation of a pituitary adenoma: a combined pituitary tumor. Case report. J Neurosurg 56:283–286

Robinson B (1983) Intrasellar abscess after transsphenoidal pituitary adenomectomy. Neurosurgery 12:684–686

Rudwan MA (1977) Pituitary abscess. Neuroradiology 12:243–248

Serri O, Rasio E, Beauregard H, Hardy J, Somma M (1983) Recurrence of hyperprolactinemia after selective transsphenoidal adenomectomy in women with prolactinoma. N Engl J Med Aug 4:280

Tucker HS, Grubb SR, Wigand JP, Taylon A, Lankford HV, Blackard WG, Becker DP (1981) Galactorrhea-amenorrhea syndrome: follow-up of forty-five patients after pituitary tumor removal. Ann Intern Med 94:302–307

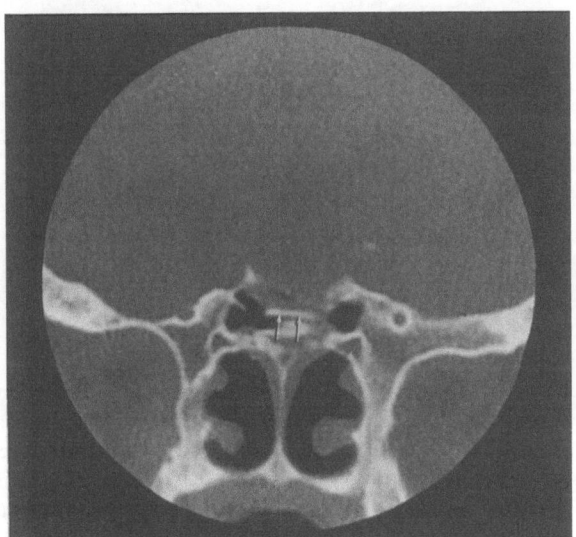

Fig. 14.1. Postoperative sella. Piece of resected bone (*arrows*) is found within the sphenoid sinus. Soft-tissue density represents blood and mucosal thickening

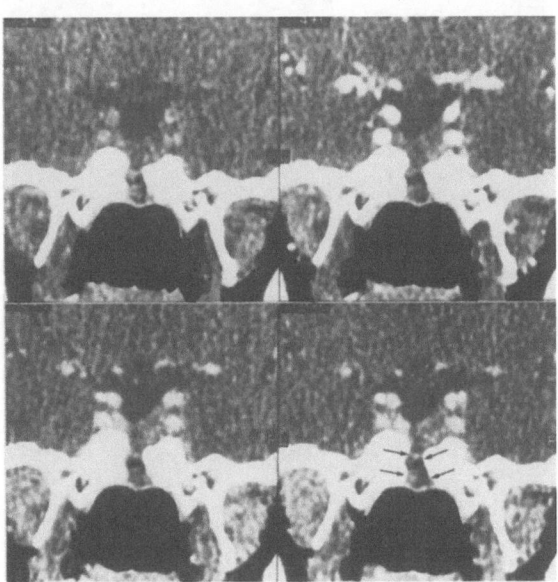

Fig. 14.2. Large bony gap of the sphenoid bone after transsphenoidal surgery. Note that foreign material filling the gap is not enhancing after contrast (*arrows*)

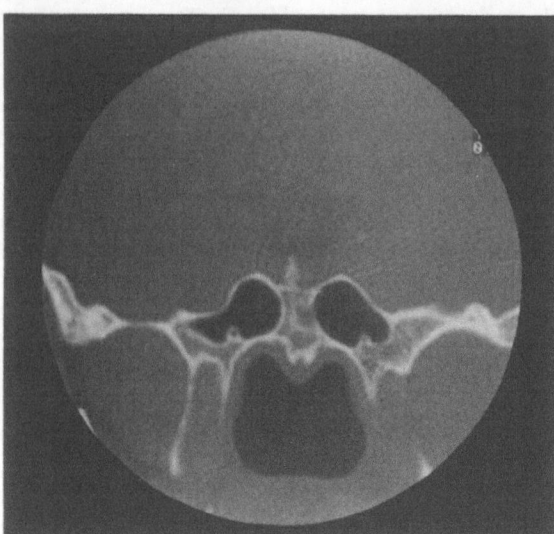

Fig. 14.3. Postoperative sella. A piece of resected bone is pointing towards the pituitary fossa

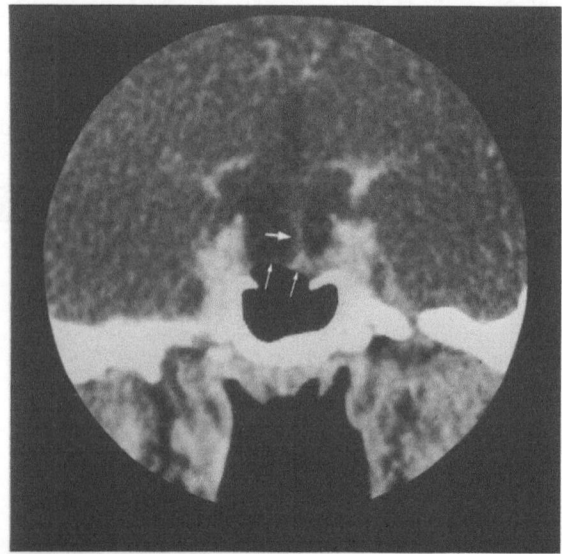

◄**Fig. 14.4a, b.** Intrasellar air herniation after surgery. **a** Contrast-enhanced coronal CT scan. **b** Coronal review image for bone detail. Partially empty sella. The pituitary stalk (*large arrow*) is attached to some residual pituitary tissue located on the left. Air herniation into the pituitary fossa, through a large sellar floor defect (*small arrows*)

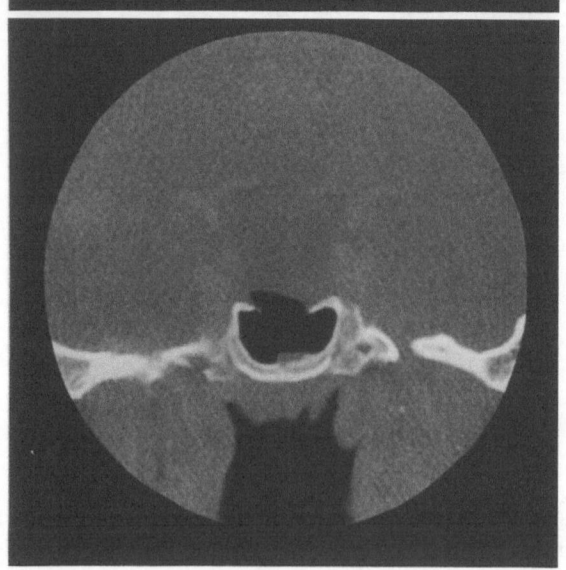

Fig. 14.5. Pituitary fossa after transsphenoidal surgery. Dynamic CT scan. Empty sella. Herniation of the suprasellar cistern into the sella. The pituitary stalk is deviated (*arrow*). Thin pituitary tissue band is demonstrated by dynamic CT along the medial aspect of the cavernous sinus (*arrowhead*)

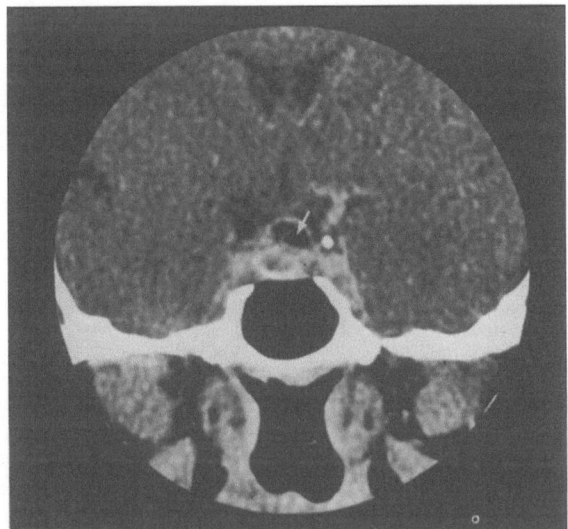

Fig. 14.6. ACTH-secreting adenoma operated 5 years ago. CT follow-up. The pituitary fossa is filled by dense heterogeneous material brought by the surgeon. A large piece of fat (*arrow*) underlined by rim enhancement occupies the bottom of the suprasellar cistern

Fig. 14.7a–c. Packing of cavernous sinus. Selective ▶ transsphenoidal adenomectomy 3 years ago. Persistant hyperprolactinemia. A recurrent prolactinoma is suspected. **a, b** Dynamic CT scan. **c** magnified view of **b**. No enlargement of the pituitary. The sellar diaphragm is flat. The pituitary tuft is on the midline (*arrow*). Defect in enhancement of the left cavernous sinus around the normal carotid artery (*small arrows* in **c**). The lateral wall of the left cavernous sinus is convex laterally. This is *not* an adenomatous invasion of the cavernous sinus: During previous surgery intrasellar bleeding obliged the surgeon to pack the cavernous sinus with tiny pledgets of oxidized cellulose

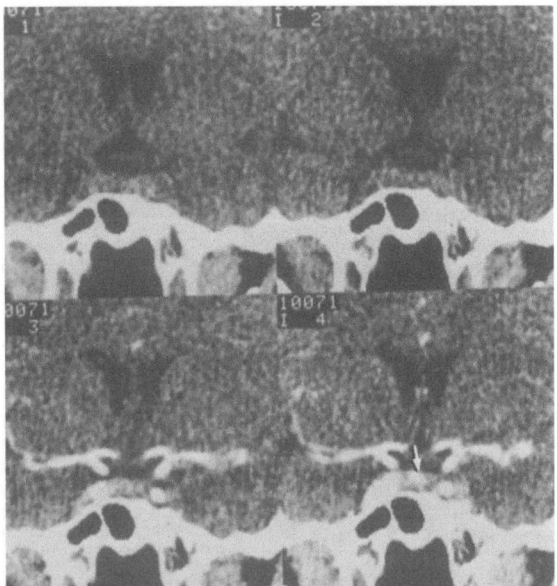

a

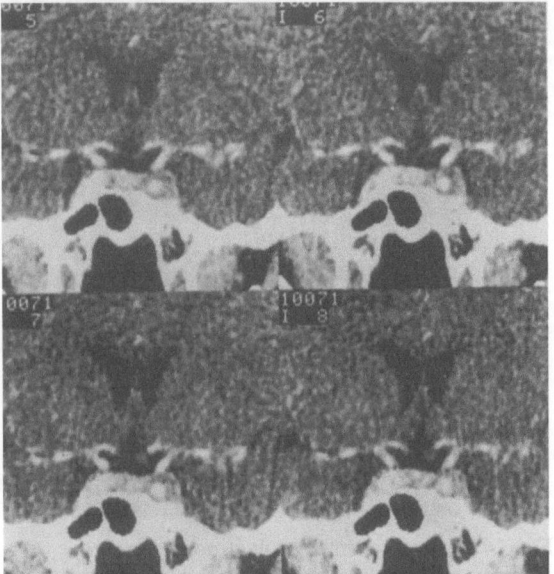

b

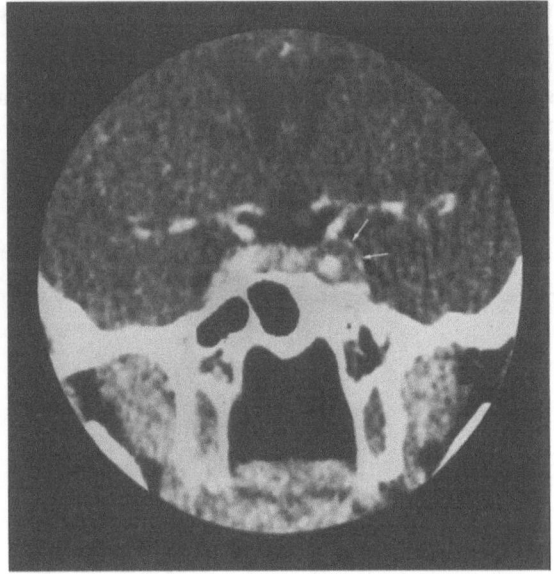

c

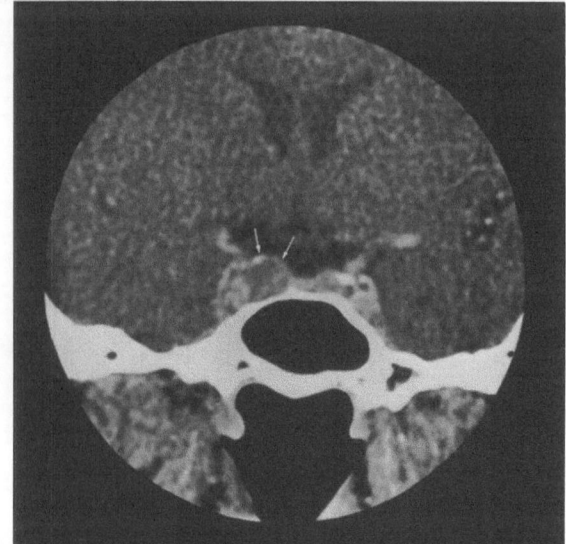

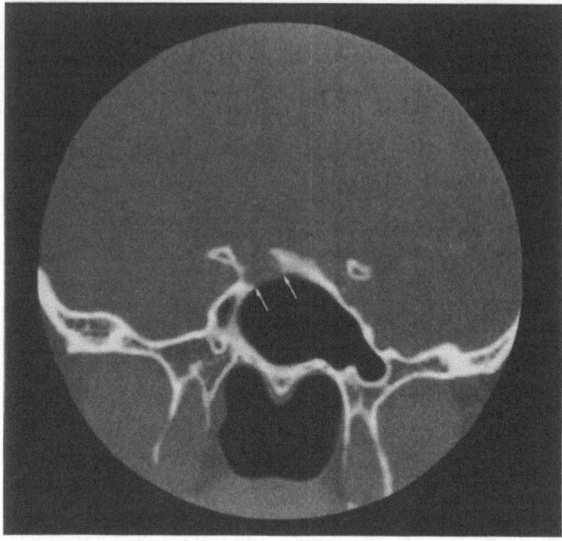

Fig. 14.9. Lateral recurrence of a prolactinoma 10 years after surgery. Note the eroded sellar floor on the left

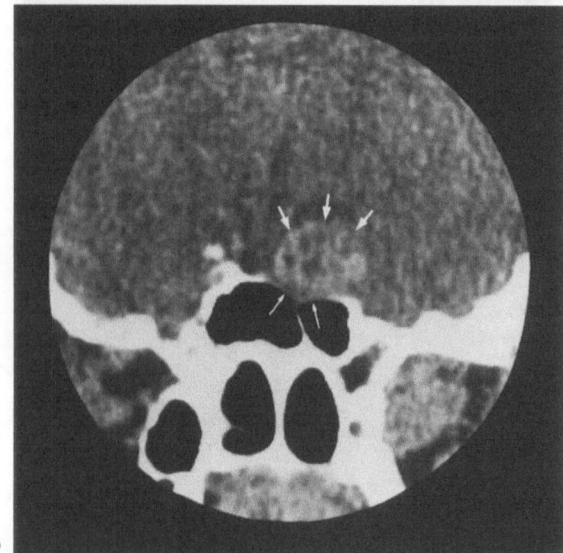

Fig. 14.8 a, b. Recurrent prolactinoma. Previous surgery 5 years ago discovered a right-sided microprolactinoma in this 27-year-old woman. Prolactinemia is again 150 ng/ml. **a** Contrast-enhanced coronal CT scan. A well-limited 7-mm hypodense pituitary adenoma is shown on the right (*arrows*). **b** Bone review image 5 mm more anteriorly: surgical defect of the anterior wall of the sella on the right (*arrows*)

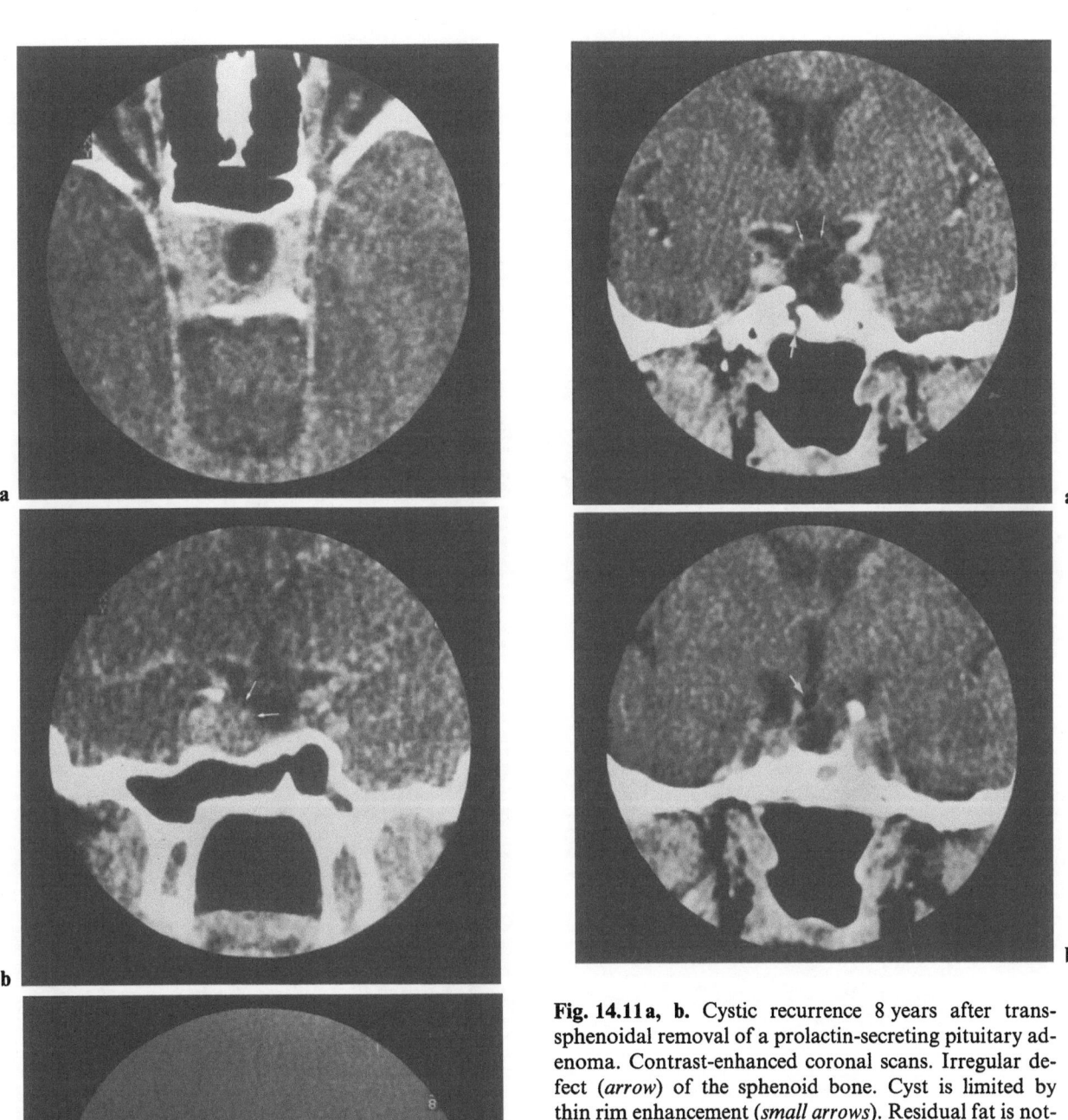

Fig. 14.11a, b. Cystic recurrence 8 years after transsphenoidal removal of a prolactin-secreting pituitary adenoma. Contrast-enhanced coronal scans. Irregular defect (*arrow*) of the sphenoid bone. Cyst is limited by thin rim enhancement (*small arrows*). Residual fat is noted at the top of the cyst (*arrow* in **b**)

◄**Fig. 14.10a–c.** Recurrent pituitary adenoma after surgery. **a** Axial contrast CT scan. **b** Coronal contrast CT scan. **c** Coronal CT scan for bone detail. On the left, free communication between the sella and the suprasellar cistern. On the right, enhancing tissue with convex medial limit represents abnormal secreting tissue (*arrows*); the sellar floor is thin

Chapter 15
The Pituitary Stalk

The pituitary stalk extends from the anterior part of the floor of the third ventricle to the neurohypophysis. The presence of the hypophyseal portal system accounts for the marked enhancement and the constant visualization of the pituitary stalk on CT scans after intravenous enhancement. Modifications in thickness and in topography due to intra- or suprasellar pathological processes will thus be easily visualized.

The Normal Pituitary Stalk

Thin sections of 1 to 2 mm are necessary to visualize correctly and constantly the pituitary stalk on enhanced CT scan. Axial sections perpendicular to the stalk and coronal sections along the stalk axis allow the most precise investigation. The pituitary stalk is but exceptionally perpendicular to the orbitomeatal line. It is more often inclined forwards and downwards.

On axial sections, after injection of contrast medium, the pituitary stalk is clearly identified on the midline, just behind the chiasm and in front of the superior extremity of the dorsum sellae. It appears as a regularly round, homogeneous, hyperdense nodule with a diameter varying from 1 to 2 mm. On computerized cisternography, the diameter of the stalk appears certainly smaller due to partial volume effect, even on very thin sections. The stalk is best visualized when the suprasellar cistern is large (children, cerebral atrophy, and empty sella turcica) (Aubin, Gardeur, Peyster). Above, the hyperdensity of the pituitary stalk is replaced by the hypodensity of the infundibular recess (Wiggli). Below, the hyperdensity of the hypophyseal portal vessels intermingles with the secondary capillary bed which is particularly dense at that point (see Chap. 3) (Xuereb, Bonneville). On coronal

sections, the pituitary stalk appears vertical most of the time. Only an inclination of 5° with respect to the vertical may be accepted as normal (Gardeur). However, constitutional asymmetries of the sella turcica, for example, hemisella turcica, may be accompanied by a greater inclination (see Chap. 13). Sagittal reconstructions are useful to determine the anteroposterior direction of the stalk.

The Abnormal Pituitary Stalk

Primary or secondary tumors or granulomatous lesions can enlarge the pituitary stalk. The pituitary stalk may also be sectioned by a traumatism or deviated by a neighboring tumor.

Enlargement of the Pituitary Stalk

Any pituitary stalk whose diameter exceeds 2–3 mm should be considered pathologic. However, a close study of the shape of the stalk on coronal sections seems even more important than the measurements, especially for stalks whose diameters tend towards the superior limit accepted as normal. In fact, a distortion or a localized enhancement should be retained as abnormal in a clinical context strongly suggesting an affection of the pituitary stalk, especially while investigating a diabetes insipidus. When the expansive lesion is less than 10 mm in diameter, its localization within the pituitary stalk can be clearly established, but when the lesion is more important, it may be difficult to ascertain its origin within the stalk. Enhanced CT scans are identical and nonspecific regardless of whether a primary or secondary tumor or a granulomatous lesion is concerned, showing a hyperdense nodule with a more or less regular outline.

Primary Tumors

Primary tumors are far less common than secondary tumors or nonmalignant lesions. In decreasing order of frequency, one can observe germinomas, granular-cell tumors, and pituicytomas (infundibulomas).

Germinomas. Germinomas of the suprasellar region are identical to those that develop in the pineal region, in the gonads, and in the mediastinum (Burger and Vogel). They develop in children and are revealed by diabetes insipidus, growth retardation, and sometimes visual symptoms or intracranial hypertension. Bipolar germinomas (pineal associated to a suprasellar localization) are not exceptional (Fig. 15.1). Malignant cells are often found in the CSF; an increase in plasmatic level of beta-HCG or of fetoprotein can sometimes be noted. These tumors are radiocurable. Radiologic findings of suprasellar germinomas are often delayed in comparison with the first clinical symptoms; exceptionally, this delay may reach several years. If a CT investigation of a child complaining of diabetes insipidus is normal, repeated examinations should be performed. These include the detection of malignant cells in CSF, the measurement of beta-HCG and delta-foetoprotein levels in plasma, and control CT investigations (Dietemann) (Figs. 15.2 and 15.3).

Granular-Cell Tumors (Choristoma, Abrikossof's Tumor). Granular-cell tumors can develop within the neurohypophysis and the pituitary stalk. They are mainly seen in aged women. Tiny asymptomatic tumors of less than 3 mm in diameter are noted in nearly 5% of methodical autopsies carried out in elderly patients (Doron). Symptomatic tumors are exceptional in intrasellar as well as in suprasellar regions. CT scans reveal a homogeneous tumoral enhancing mass with a more or less regular outline (Gardeur) (Fig. 15.4).

Pituicytomas (Infundibulomas). These tumors, resembling pilocytic astrocytomas, are exceptional (Burger and Vogel).

Metastases

Metastases of the pituitary stalk are relatively frequent, especially in the terminal phase of bronchial and breast carcinoma. Clinical signs (mainly diabetes insipidus) can precede CT findings; other metastatic localizations are often encountered (Peyster and Hoover) (Figs. 15.5, 15.6, and 15.7).

Granulomas

They are mainly represented by histiocytosis X (Hand-Schuller-Christian disease) in children and sarcoidosis and tuberculosis in adults. CT scan reveals an enlarged pituitary stalk or an enhancing nodular suprasellar mass (Fig. 15.8). The authors observed a granuloma related to yersiniosis located within the pituitary stalk (Fig. 15.9).

Deviations of the Pituitary Stalk

Lateralized pituitary gland tumors are often accompanied by a contralateral deviation of the stalk. Such a deviation is sometimes noted in small tumors that do not modify the superior limits of the pituitary gland (Nagagawa).

An asymmetric pituitary gland may be associated with a deviated stalk, the latter being then directed towards the side where the gland is highest (see Chap. 13).

Sections of the Pituitary Stalk

Traumatic rupture of the pituitary stalk is accompanied by an interruption of the capillary bed. Enhanced CT scan may demonstrate this interruption of the hypophyseal portal system of the stalk (Fig. 15.10).

References

Aubin ML, Bentson S, Vignaud J (1978) Tomodensitométrie de la tige pituitaire. J Neuroradiol 51:153–160

Bonneville JF, Cattin F, Moussa-Bacha K, Portha C (1983) Dynamic computed tomography of the pituitary gland: the "tuft sign". Radiology 149:145–148

Brooks BS, El Gammal T, Hungerford GD, Acker J, Trevor RP, Russel W (1982) Radiologic evaluation of neurosarcoidosis. Role of CT. AJNR 3:513–521

Burger PC, Vogel FS (1976) Surgical pathology of the nervous system and its coverings. John Wiley and Sons, New York, London, Sydney, Toronto

Cusick JF, Khang-Cheng HO, Hagen TC, Kun LE (1982) Granular-cell pituicytoma associated with multiple endocrine neoplasia type 2. J Neurosurg 56:594–596

Decker RE, Mardayat M, Marc J, Rasool A (1979) Neurosarcoidosis with CT visualization and transsphenoidal excision of a supra- and intrasellar granuloma: case report. J Neurosurg 50:814–816

Derome PJ, Jedynak CP, Peillon F (1980) Pituitary adenomas. Biology, physiopathology and treatment. Asclepios Publishers, Paris

Dietemann JL, Bonneville JF (1985) Radiological Diagnosis of Pituitary Diseases. In: Imura H (ed) The pituitary gland. Raven Press, New York, pp 341–361

Doron Y, Behar A, Bellar AJ (1965) Granular cell myoblastoma of the neurohypophysis. J Neurosurg 22:95–99

Drayer BP, Rosenbaum AE, Kennerdell JS, Robinson AG, Bank WO, Deeb ZL (1977) CT diagnosis of suprasellar masses by intrathecal enhancement. Radiology 123:339–344

Drayer BP, Rosenbaum AE, Riegel DB, Bank WO, Deeb ZL (1977) Metrizamide CT cisternography: pediatric applications. Radiology 124:349–357

Dupont MG, Gerard JM, Flament-Durand J, Balériaux-Waha D, Mortelmans LL (1977) Pathognomonic aspects of germinomas on CT scan. Neuroradiology 14:209–211

Gardeur D, Metzger J (1982) Adénomes hypophysaires intra-sellaires. In: Tomodensitométrie intra-crânienne. Livre V: Pathologie sellaire. Ellipses, Paris, p 41

Gardeur D, Nachanakian A, Millard JC, Metzger J, Poisson J, Mashaly R (1978) Neuroradiological aspects and therapeutic incidence of a case of posterior pituitary tumor (choristoma). J Neuroradiol 5:321–326

Kuuliala I (1980) The normal suprasellar subarachnoid space in CT. Clinical Radiology 31:155–159

Manelfe C, Louvet JP (1979) CT in diabetes insipidus. J Comp Ass Tomog 3:309–316

Manelfe C, Louvet JP, Boulard C, Regnier C, Rochiccioli P, Bayard F (1978) Hypothalamic-pituitary changes in diabetes insipidus demonstrated by computerised tomography. Lancet 2:1379–1380

Nagagawa Y, Matsumoto K, Fukami T, Takase K (1984) Exploration of the pituitary stalk and gland by high-resolution computed tomography. Comparative study of normal subjects and cases with microadenoma. Neuroradiology 26:473–478

Ozanne P, Jedynak CP, Charbonnel B, Derome PJ (1982) Metastases hypophysaires et hypothalamiques. Etude anatomo-clinique de 5 observations. Ann Med Int 133:92–96

Peyster RG, Hoover ED (1984) CT of the abnormal pituitary stalk. AJNR 5:49–52

Peyster RG, Hoover ED, Adler LP (1984) CT of the normal pituitary stalk. AJNR 5:45–47

Peyster RG, Hoover ED (1984) Computerized tomography in orbital disease and neuro-ophthalmology. Year Book Medical Publishers, Chicago, London

Schlachter LB, Tindall GT, Pearl GS (1980) Granular cell tumor of the pituitary gland associated with diabetes insipidus. Neurosurgery 6:418–420

Stefanko SZ, Talerman A, Mackay WM, Vuzeuski VD (1979) Infundibular germinoma. Acta Neurochirurgica 50:71–78

Steimlé R, Jacquet G, Bourghli A, Pageaut G, Bonneville JF, Abdul Razzak A, Couturier M, Dreyfus T (1980) Suprasellar choristoma. Neurochirurgie 41:49–56

Strand RD, Baker RA, Ordia IJ (1978) Metrizamide ventriculography and CT in lesions about the third ventricle. Radiology 128:405–410

Takeuchi J, Handa H, Otsuka S, Takebe Y (1979) Neuroradiological aspects of suprasellar germinomas. Neuroradiology 17:153–159

Wiggli U, Benz UF (1978) Normal CT anatomy of the suprasellar subarachnoid space. Radiology 128:65–70

Xuereb GB, Prichard MML, Daniel PM (1954) The hypophysial portal system of vessels in man. Q J Exp Physiol 39:219–230

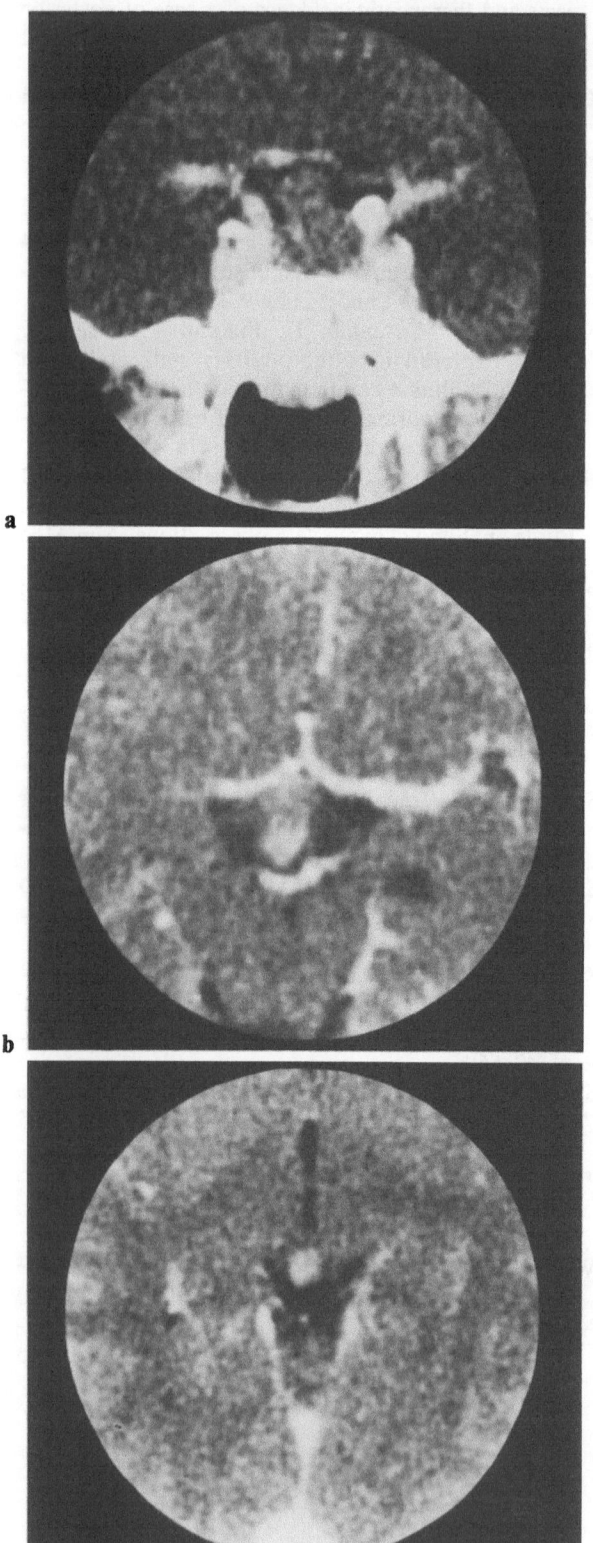

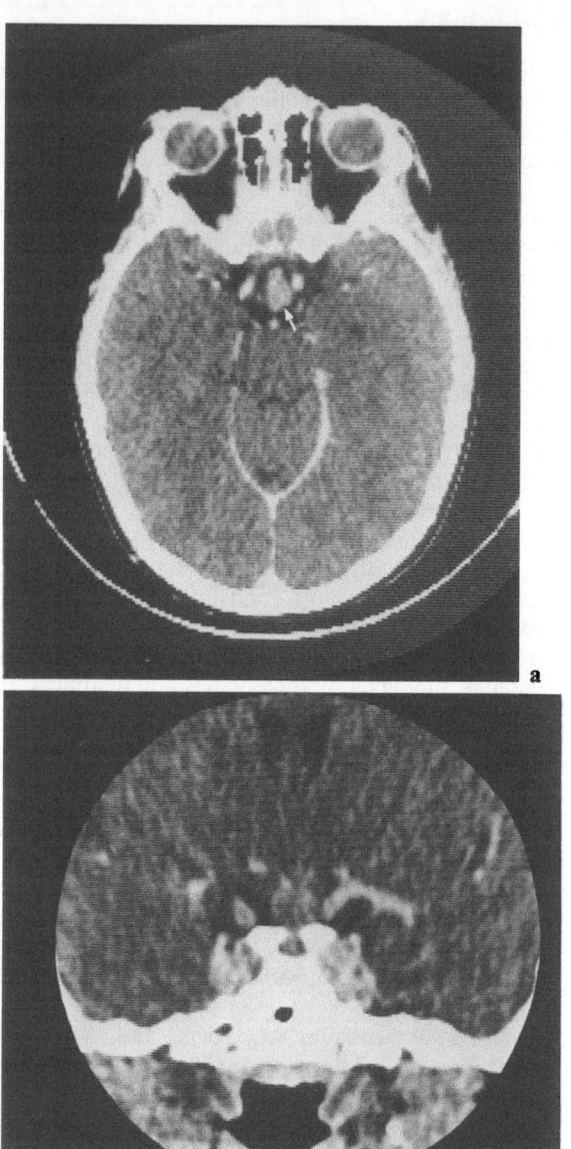

Fig. 15.2a, b. Suprasellar germinoma in an 11-year-old boy presenting with diabetes insipidus for 2 years. A previous CT examination performed 6 months before appeared normal. **a** Axial and **b** coronal enhanced CT scans. Marked homogeneous suprasellar enhancement (*arrow*) related to an oval mass located within the pituitary stalk

◀ **Fig. 15.1a–c.** Germinoma of the neurohypophysis, pituitary stalk, and pineal region in a 12-year-old boy presenting diabetes insipidus. **a** Coronal, **b, c** axial enhanced CT scans. Marked enlargement of the pituitary stalk and neurohypophysis. Homogeneous enhancement of the tumor. Section through the pineal area reveals a second pineal germinoma

Fig. 15.3a–c. Supra- and intrasellar germinoma in a 21-year-old woman with diabetes insipidus and progressive pituitary deficiency developing since 1981. **a** Coronal enhanced CT scan (July 1982). Discrete enlargement of the pituitary stalk is noted associated to marked enhancement. **b** Coronal enhanced CT scan (May 1983). Marked enlargement of the pituitary stalk and sellar content. **c** Coronal enhanced CT scan after radiotherapy (April 1984). The height and density of the pituitary gland appear normal. The pituitary stalk is also quite normal

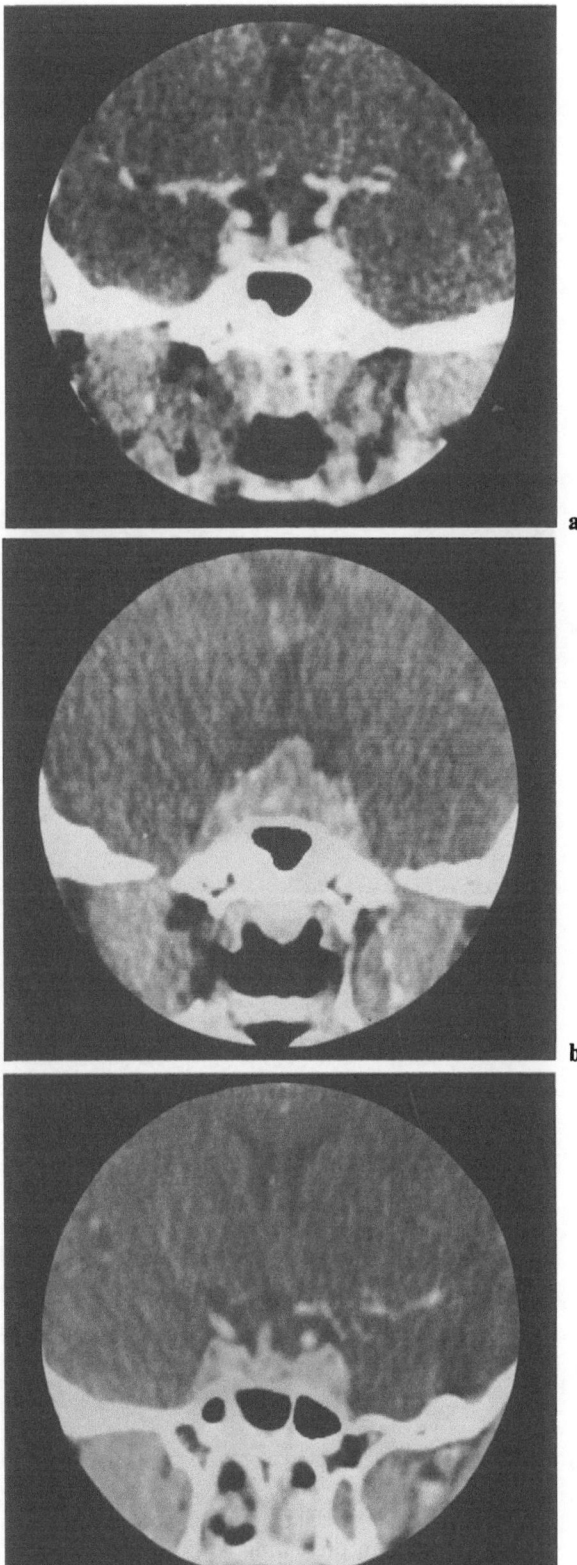

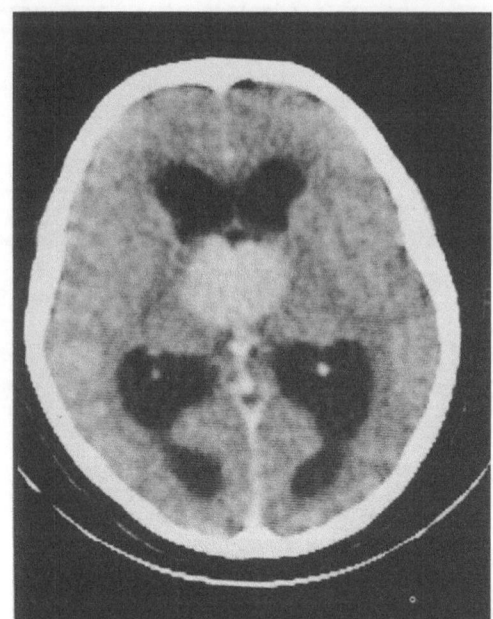

a

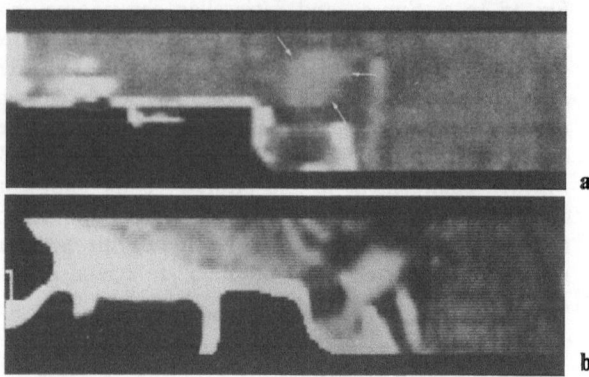

a

b

Fig. 15.5a, b. Pituitary stalk metastasis in two different patients. Sagittal reformatted enhanced CT scans. **a** Testicular cancer and **b** endometrial cancer

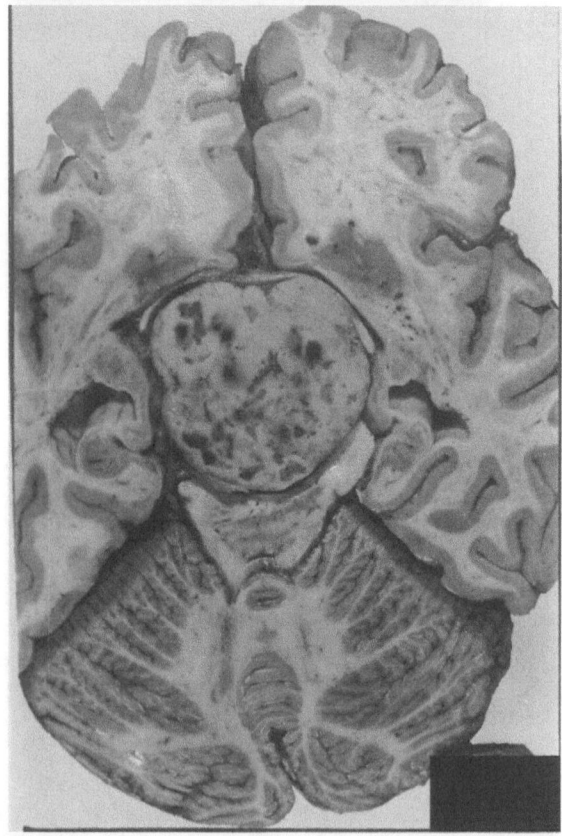

b

Fig. 15.4a, b. Suprasellar granular cell tumor in a 45-year-old woman presenting with progressive mental deterioration. **a** Axial enhanced CT scan. **b** Autopsy specimen of the same patient. The tumor initially developed within the pituitary stalk is extending towards the third ventricle. The tumor is spontaneously hyperdense (not shown) and presents marked homogeneous enhancement

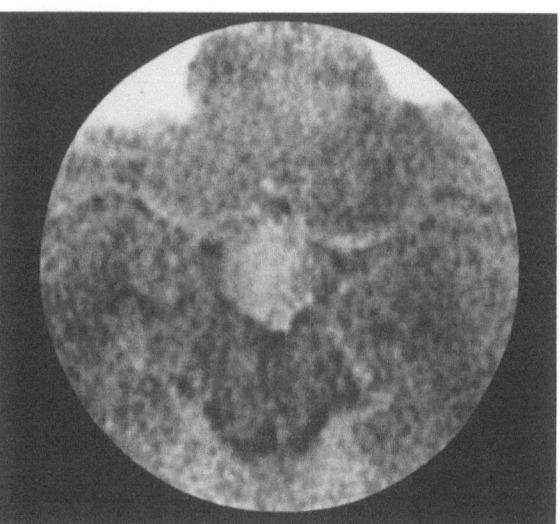

Fig. 15.6. Suprasellar metastasis from a medulloblastoma in an 11-year-old boy. Axial enhanced CT scan

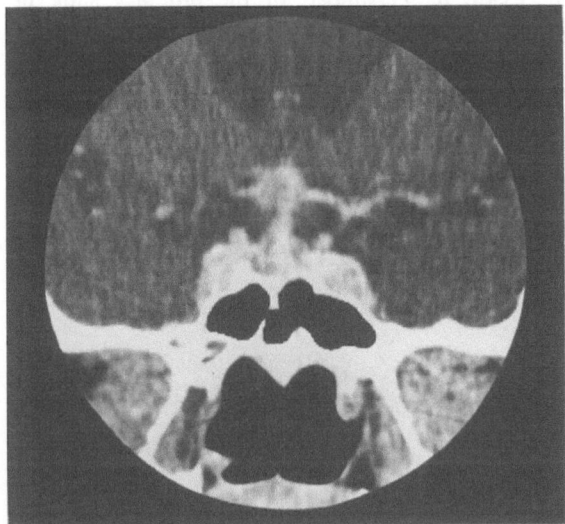

Fig. 15.7. Pituitary gland, pituitary stalk, and hypothalamic metastasis from a bronchial adenocarcinoma in a 60-year-old man presenting with diabetes insipidus. Coronal enhanced CT scan

Fig. 15.8a–c. Sarcoidosis located within the pituitary ▶ stalk. **a** Axial and **b** sagittal reformatted enhanced CT scans. **c** Radiograph of the hand. Enhancing nodular lesion (*arrows*) located within the pituitary stalk. Lytic phalangeal bone lesions

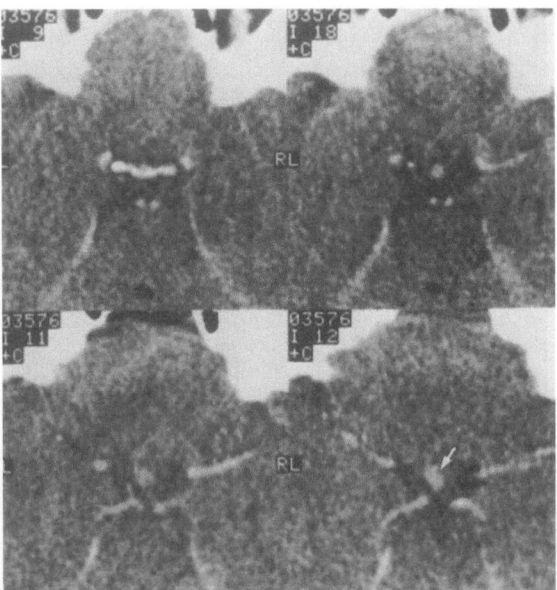

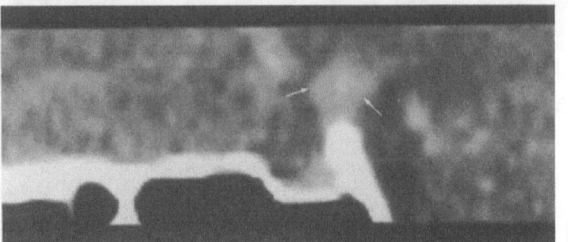

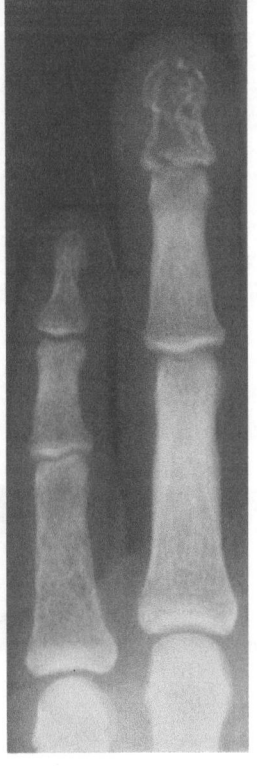

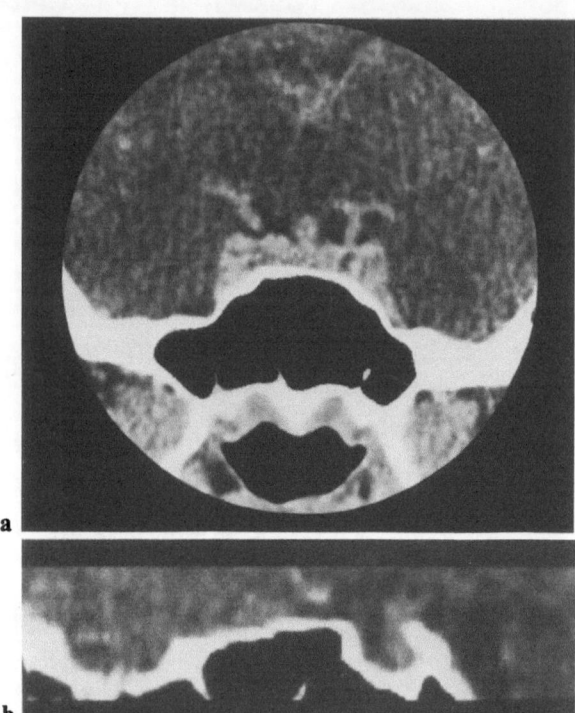

Fig. 15.9a, b. Yersiniosis of the pituitary stalk in a 25-year-old male presenting with diabetes insipidus. **a** Coronal and **b** sagittal reformatted enhanced CT scans. Marked enlargement of the pituitary stalk

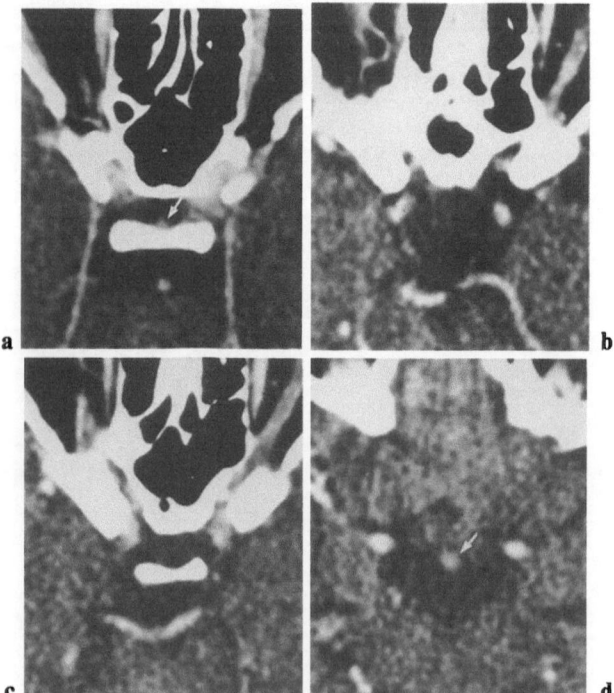

Fig. 15.10a–d. Posttraumatic diabetes insipidus and hyperprolactinemia in a 19-year-old man. Axial enhanced CT scans. The upper and the lower parts of the pituitary stalk only are identified (*arrows*). Probable traumatic rupture of the nonenhanced midpart of the stalk

Chapter 16
Suprasellar Pathology

Craniopharyngiomas

Craniopharyngiomas are benign tumors arising from epithelial remnants of Rathke's pouch. These tumors are mainly seen between the age of 5 and 25 years; however, about 20% of these tumors are encountered after the age of 40.

From a pathological point of view, three elementary tumoral components are present to different degrees: a solid component (usually vascularized), a cystic component, and a calcified component. Suprasellar craniopharyngiomas are either in front of or behind the chiasm, exceptionally intraventricular (within the third ventricle) or intrachiasmatic (Cashion, Duff). The radiologic findings in conventional radiology (Bonneville and Dietemann) as well as in CT scan obviously depend on the topography. The CT scan appearance varies with the degree of each of the three elementary tumoral components (Figs. 16.1, 16.2, and 16.3).

Before injection of contrast medium, the tumor may appear homogeneous, completely calcified, cystic or solid, but most of the time, the tumoral lesion appears heterogeneous with juxtaposition of the three elementary tumoral components (Gardeur).

The cystic component usually appears of low density, with densities ranging from − 20 Hounsfield units to + 20 Hounsfield units. However, spontaneously isodense or even hyperdense cysts are not exceptional. These high attenuation values in nonenhanced CT scans are mainly related to a high intracystic protein concentration (Fitz, Braun).

The calcified component spontaneously appears of high density: either nodular high density areas with irregular outline or linear arciform high-density areas within the wall of a cystic component.

The solid component presents either low attenuation values or appears isodense to the brain.

Intravenous injection of contrast medium determines changes only within the solid component with highly variable enhancement (Fitz, Gardeur, Numaguchi). Enhanced CT scan also allows a better delineation of the tumor from the vessels of the circle of Willis (Figs. 16.1 and 16.2).

CT scans allow a better evaluation of extension of the tumor:

1. An extension towards the foramen of Monro is accompanied by a ventricular hydrocephaly with symmetrical or asymmetrical distention of both lateral ventricles.

2. A subfrontal extension, or towards the posterior fossa, or towards the sphenoid sinus can easily be appreciated.

Meningiomas

Suprasellar meningiomas occur in the meninges of the planum sphenoidale, tuberculum sellae, chiasmatic sulcus, sellar diaphragm, anterior clinoid processes, or dorsum sellae (Fig. 16.4).

These tumors represent about one-third of all intracranial meningiomas and are mainly seen after the age of 40, with a distinct predilection for women. They are spontaneously either isodense or hyperdense; tumoral calcifications may be observed. Intravenous injection of iodinated contrast medium shows a more or less homogeneous enhancement. Low-density areas reflect either a necrosed or a cystic component. A marked cerebral edema affecting the white matter of the frontal lobes is noted in most meningiomas of the planum sphenoidale. CT scans usually do not identify malignant meningiomas (Dietemann).

Whether the CT scan appearance of a suprasellar tumor is typical of a meningioma or not, it is especially the bone changes of the skull that will definitely reinforce the radiologist's conviction towards a meningioma. These bone changes should systematically be looked for on plain films, on axial and coronal sections, or on frontal and sagittal reformatted CT scans (Fig. 16.4):

1. Meningiomas of the planum sphenoidale and of the chiasmatic sulcus are accompanied by a hyperostosis and/or a blistering of the planum (Fig. 16.4). Raised intracranial pressure may also be present (Bonneville and Dietemann).

2. Meningiomas of the tuberculum sellae are often accompanied by hyperostosis.

3. Meningiomas of the anterior clinoid process are accompanied by a hypertrophy of the latter.

4. In meningiomas of the sellar diaphragm, bone changes are often discrete, limited usually to erosion of the dorsum sellae.

Gliomas of the Optic Pathways (Optic Nerve, Chiasm, and Tract)

Gliomas of the optic pathways are rather uncommon tumors, mainly encountered before the age of 20. Von Recklinghausen's neurofibromatosis is present in 30% of cases. These tumors are almost always histologically benign. Malignant gliomas are chiefly seen in adults. Gliomas of the chiasm may extend forwards along the optic nerves, or backwards along the optic tract and even the optic radiations (Daniels, Gardeur, Peyster and Hoover, Savoiardo).

CT investigation of chiasmatic gliomas reveals (Figs. 16.5 and 16.6):

1. *Morphological modifications of the chiasm.* They are specially well displayed on coronal sections. The chiasm becomes globulous and it is admitted that a chiasm more than 6 mm in thickness is suggestive of a tumor.

2. *Modifications in density.* Nonenhanced CT scans reveal an isodense tumor. Tumoral calcifications are rare, but appear more frequently after radiotherapy. A more or less marked enhancement is usually noted.

3. *Sometimes a tumoral extension* towards the optic nerves and/or towards the optic tracts is noted.

4. *Bone changes within the sellar region.* Depression of the chiasmatic sulcus, erosion of the dorsum sellae, and enlargement of one or both optic canals may be noted. These modifications should systematically be looked for, on plain films and on frontal and sagittal reformatted CT scans.

Rare Suprasellar Lesions

Hypothalamic Lesions

Hypothalamic gliomas are usually voluminous at the time of their diagnosis and they cannot be distinguished from gliomas of the optic pathways. Hypo- or isodense before enhancement, these tumors are homogeneously or heterogeneously enhanced after injection of contrast medium.

Hypothalamic metastases appear on enhanced CT scan as nodular hyperdense masses located within the anteroinferior part of the third ventricle. The pituitary stalk may be affected simultaneously. Primary or secondary localizations of lymphomas can also be seen (Lanzieri).

Hamartomas of the tuber cinereum are not true tumoral lesions, but rather congenital malformations formed by an accumulation of normal cells (in this case, an association of cells of the tuber cinereum and glial cells) in an unusual site. Hamartomas are almost always pediculated and attached to the posterior part of the hypothalamus, immediately in front of the mammillary bodies; they can exceptionally be found within the hypothalamus itself. Clinically, hamartomas of the tuber cinereum are usually revealed by isosexual precocious puberty and less commonly by mental changes or seizures. These tumors are usually small, measuring less than 2 cm in diameter. In axial sections, a small hamartoma appears as a small, rounded, isodense, nonenhancing mass, localized between the basilar artery and the pituitary stalk. Direct coronal sections and sagittal reconstructions confirm the pediculated feature of the lesion (Lin, Diebler) (Figs. 16.7, 16.8, and 16.9).

Histiocytosis is classically localized, at least during its initial phase, to the pituitary stalk, but hypothalamic localizations are not exceptional. Suprasellar histiocytoma appears as a mass which is spontaneously of low density, but which is intensely and homogeneously enhanced after injection of contrast medium.

Sarcoidosis can affect the hypothalamus. CT scan reveals a nodular enhancement.

Granular-cell tumors (choristomas, pituicytomas, myoblastomas, Abrikossof's tumors), usually located within the pituitary fossa, have also been described in the suprasellar area (Fig. 16.10).

Dermoid and Epidermoid Cysts (Cholesteatomas), Teratomas, and Lipomas

These are congenital tumors, developing from ecto-, meso-, and/or endodermic embryonic remnants incorporated during closure of the neural tube. Cholesteatomas are relatively more common than teratomas and lipomas. Cholesteatomas are encountered more frequently in adolescents and young adults, while teratomas are observed almost exclusively in children.

Epidermoid Cysts

Epidermoid cysts develop from ectodermic remnants. The epithelial lining is keratinized and squamous; these cysts contain epithelial fragments and cholesterol. They are more often located in the midline. Their wall is but rarely calcified. On CT scans, these cysts appear of a lower density mass with densities neighboring that of CSF. However, depending upon the cholesterol and keratin content of the cyst, these densities may become respectively either negative or frankly positive (above 50 Hounsfield units). Spontaneously high-density cysts cannot be distinguished from a craniopharyngioma (Schubiger). A moderate enhancement is only exceptionally noted within the wall of the cyst.

Dermoid Cysts

Dermoid cysts are derived from ecto- and mesodermic inclusions. The cyst wall is formed of a keratinized stratified epithelium with hair follicles and sebaceous glands. The latter is responsible for a predominently fatty content.

On CT scans, these dermoid cysts almost always have fat attenuation values. Calcifications are common within their walls and they sometimes affect the whole cyst. These cysts may exceptionally burst, either spontaneously or after surgical manipulation; small disseminated low-density adipous areas can then be identified within the cisternal spaces (Dietemann) (Fig. 16.11).

Lipomas

Inclusion of fatty tissue occurs in the midline. Thus, lipomas are located in the corpus callosum, within the suprasellar cisterns, or within the quadrigeminal cistern. Suprasellar lipomas are uncommon and usually small; small peripheric nodular calcifications may be observed (Fig. 16.12).

Teratomas

Developed from ecto-, meso-, and endodermal remnants, these very rare tumors, mainly observed in children, have a very heterogeneous content made of osseous, cartilaginous, dental, pilous, and sebaceous elements. These tumors are localized in the pineal region rather than in the suprasellar cistern (Gardeur, Peyster and Hoover).

Arachnoid Cysts

Suprasellar arachnoid cysts are usually congenital and are mainly observed in children. These liquid cysts may either be strictly suprasellar or may extend towards the sylvian fissure. Their densities are similar to that of CSF. They present neither calcifications nor enhancement (Fig. 16.13).

Typical osseous modifications can sometimes be noted on skull radiographs. These arachnoid cysts should be distinguished from marked dilatation of the third ventricle, as can be observed in congenital aqueduct stenosis (Figs. 16.14 and

16.15). They should also be distinguished from an intraventricular cyst, a cystic craniopharyngioma, or an epidermoid cyst (Armstrong, Hoffmann, Leo).

Cisternography with water-soluble contrast media demonstrates delayed opacification of the arachnoid cyst between the 6th and the 24th hour (Gardeur, Wolpert).

Esthesioneuroblastoma

Esthesioneuroblastoma tumors derive from cells of Locy's ganglion, usually localized in the nasal septum, in the area of the lamina cribosa and crista galli. But these cells can also exceptionally be found in the hypothalamic region, explaining the suprasellar development of these tumors. These tumors may appear calcified and usually enhanced (Burke, Manelfe, Sarwar).

Suprasellar Vascular Abnormalities

Suprasellar Aneurysms

The visualization and appearance of suprasellar aneurysms, developed in the suprasellar region, depend mainly upon their size. In fact, calcifications within the wall and intraaneurysmal thrombosis occur when the aneurysm becomes important in size. Enhancement of the aneurysm depends obviously upon the size of the intraaneurysmal thrombus. Certain entirely calcified and thrombosed aneurysms show no modification after enhancement, thus raising problems of differential diagnosis, namely, with a suprasellar tumor such as a craniopharyngioma.

Miscellaneous Vascular Lesions

Atheromatous arteries may be responsible for compression of the chiasm. CT scans may identify these abnormal vessels. Arteriovenous or purely venous malformations have been described within the chiasm (Fermaglich, Roski).

Suprasellar Arachnoiditis

Arachnoiditis of the chiasmatic cistern is mainly secondary to bacterial meningitis (tuberculosis, pneumococcus meningitis). At an acute phase, the arachnoiditis is expressed by disappearance of the liquid densities in the suprasellar cistern and marked enhancement. At a more advanced phase, disappearance of the liquid densities in the suprasellar cistern persists, but there is no enhancement and calcifications may appear especially in cases of tuberculous meningitis.

References

Anderson FM, Segall HD, Paton WL (1979) Use of CT scanning in supratentorial arachnoid cysts. A report on 20 children and 4 adults. J Neurosurg 50:323–329

Armstrong EA, Harwood-Nash DCF, Hoffman H, Fitz CR, Chuang S, Pettersson H (1983) Benign suprasellar cysts: the CT approach. AJNR 4:163–166

Azar Kia B, Palacios E, Churchill RJ (1977) Diagnosis of sellar and parasellar lesions by CT and other diagnostic modalities. CT 1:249–256

Bamberger C, Rosier J (1980) Examen tomodensitométrique des lésions sellaires et parasellaires. Feuillets de radiologie 20:197–208

Banna M (1976) Arachnoid cysts on computed tomography. AJR 127:979–982

Banna M, Schatz SW, Molot MJ, Groves J (1976) Primary intrasellar germinoma. Br J Radiol 49:971–973

Banna M, Baker HL Jr, Houser OW (1980) Pituitary and parapituitary tumours on CT. Br J Radiol 53:1123–1143

Beal MF, Kleinman GM, Ojemann RG, Hochberg FH (1981) Gangliocytoma of third ventricle: hyperphagia, somnolence and dementia. Neurology 31:1224–1228

Beardsley TL, Brown SVL, Sydnor CF, Grimson BS, Klintwourth GK (1984) Eleven cases of sarcoidosis of the optic nerve. Ann J Ophthalmol 95:62–77

Bentson JR, Wilson GH, Helmer E, Winter J (1977) CT in intracranial cysticercosis. J Comp Ass Tomog 1:464–471

Bonneville JF, Dietemann JL (1981) Radiology of the sella turcica. Springer Verlag, Berlin Heidelberg New York

Braun IF, Pinto RS, Epstein F (1982) Dense cystic craniopharyngiomas. AJNR 3:139–141

Brennan TG, Krishna-Rao CNG, Robinson W (1977) Tandem lesions: chromophobe adenoma and meningioma. J Comp Ass Tomog 1:517–520

Brooks BS, El Gammal T, Hungerford GD, Acker J, Trevor RP, Russel W (1982) Radiologic evaluation of neurosarcoidosis. Role of CT. AJNR 3:513–521

Burke DP, Gabrielsen TO, Knake JE, Seeger JF, Oberman HA (1980) Radiology of olfactory neuroblastoma. Radiology 137:367–372

Cabanis EA, Van Effenterre R, Iba-Zizen MT (1979) CT in parasellar space-occupying lesions and therapeutic decision. Acta Neurochirurgica (suppl 28) 28:329–333

Cabezudo JM, Vaquero J, Garcia de Sola R, Leunda G, Nombela L, Bravo G (1981) CT with craniopharyngiomas: a review. Surg Neurol 15:422–427

Cacciari E, Frejaville E, Cicognani A, Pirazzoli P, Frank G, Balsamo A, Tassinari D, Zappula F, Bergamaschi R, Cristi GF (1983) How many cases of time precocious puberty in girls are idiopathic? J Pediatr 102:357–360

Carmel PW, Antunes JL, Chang CH (1982) Craniopharyngiomas in children. Neurosurgery 11:382–389

Cashion EL, Young JM (1971) Intraventricular craniopharyngioma. Report of two cases. J Neurosurg 34:84–87

Chambers AA, Lukin RR, Tomsick TA (1977) Cranial epidermoid tumors: diagnosis by CT. Neurosurgery 1:276–280

Daniels D, Haughton V, Williams A, Gager W, Berns TF (1980) CT of the optic chiasm. Radiology 137:123–127

Davis KR, Roberson GH, Taveras JM, New PFJ, Trevor R (1976) Diagnosis of epidermoid tumor by CT. Radiology 119:347–353

Decker RE, Mardayat M, Marc J, Rasool A (1979) Neurosarcoidosis with CT visualization and transsphenoidal excision of a supra- and intrasellar granuloma: case report. J Neurosurg 50:814–816

Derome PJ, Jedynak CP, Peillon F (1980) Pituitary adenomas. Biology, physiopathology and treatment. Asclepios Publishers, Paris

Diebler C, Ponsot G (1983) Hamartomas of the tuber cinereum. Neuroradiology 25:93–101

Dietemann JL, Bonneville JF (1985) Radiological Diagnosis of Pituitary Diseases. In: Imura H (ed) The pituitary gland. Raven Press, New York, pp 341–361

Dietemann JL, Balériaux-Waha D, Golabek R, Wackenheim A, Jeanmart L (1979) Aspects tomodensitométriques des kystes épidermoïdes et dermoïdes intra-crâniens. A propos de trois observations. J Radiol 60:265–269

Dietemann JL, Heldt N, Burguet JL, Medjek L, Maitrot D, Wackenheim A (1982) CT findings in malignant meningiomas. Neuroradiology 23:207–209

Drayer BP, Rosenbaum AE, Kennerdell JS, Robinson AG, Bank WO, Deeb ZL (1977) CT diagnosis of suprasellar masses by intrathecal enhancement. Radiology 123:339–344

Drayer BP, Rosenbaum AE, Riegel DB, Bank WO, Deeb ZL (1977) Metrizamide CT cisternography: pediatric applications. Radiology 124:349–357

Duff TA, Levine R (1983) Intrachiasmatic craniopharyngioma. Case report. J Neurosurg 59:176–178

Dupont MG, Gerard JM, Flament-Durand J, Balériaux-Waha D, Mortelmans LL (1977) Pathognomonic aspects of germinoma on CT scan. Neuroradiology 14:209–211

Enzmann DR, Norman D, Mani J, Newton TH (1976) CT of granulomatous basal arachnoiditis. Radiology 120:341–344

Fermaglich J, Kattah J, Manz H (1978) Venous angioma of the optic chiasm. Ann Neurol 4:470–471

Fitz CR, Wortzman G, Harwood-Nash DC, Holgate RC, Barry JF, Boldt DW (1978) Computed tomography in craniopharyngiomas. Radiology 127:687–691

Fults D, Kelly DL Jr (1983) A suprasellar atypical teratoma presenting as an intrasellar mass: a case report. Neurosurgery 13:40–43

Gardeur D, Metzger J (1982) Adénomes hypophysaires intra-sellaires. In: Tomodensitométrie intra-crânienne. Livre V: Pathologie sellaire. Ellipses, Paris, p 41

Gardeur D, Nachanakian A, Van Effenterre R, Zamora G, Metzger J (1979) Analyse tomodensitométrique des craniopharyngiomes. J Radiol 60:51–57

Ghoshhajra K (1981) High-resolution metrizamide CT cisternography in sellar and suprasellar abnormalities. J Neurosurg 54:232–239

Grosvalet A, Ernest C, Diebler C, Sauvegrain J (1981) CT in precocious puberty of central origin. Ann Radiol (Paris) 24:32–38

Gyldensted C, Karle A (1977) CT of intra- and juxtasellar lesions: a radiological study of 108 cases. Neuroradiology 14:5–13

Hall K, McAllister VL (1980) Metrizamide cisternography in pituitary and juxtapituitary lesions. Radiology 134:101–108

Hamer J (1979) The radiological diagnosis of parasellar dermoid cysts with special regard to CT scan. Acta Neurochirurgica (suppl 28) 28:321–322

Handa J, Nakano Y, Heiha A (1977) CT cisternography with intracranial arachnoid cysts. Surg Neurol 8:451–454

Hatam A, Bergström M, Greitz T (1979) Diagnosis of sellar and parasellar lesions by CT. Neuroradiology 18:249–258

Hoffman HJ, Hendrick EB, Humphreys RP, Armstrong EA (1982) Investigation and management of suprasellar arachnoid cysts. J Neurosurg 57:597–602

Kammer KS, Perlman K, Humphreys RP, Howard NJ (1980) Clinical and surgical aspects of hypothalamic hamartoma associated with precocious puberty in a 15-month-old boy. Childs Brain 7:150–157

Kasdon DL, Douglas EA, Brougham MF (1977) Suprasellar arachnoid cyst diagnosed pre-operatively by CT scanning. Surg Neurol 7:299–303

Kitano I, Yoneda K, Yamakawa Y, Fukoi PH, Kinoshita K (1981) Huge cystic craniopharyngioma with unusual extensions. A case report. Neuroradiology 22:38–42

Kokoris N, Rothman LM, Wolintz AH (1980) CT and angiography in the diagnosis of suprasellar mass lesions. Am J Ophthalmol 89:278–283

Kuuliala I (1980) The normal suprasellar subarachnoid space in CT. Clinical Radiology 31:155–159

Kuuliala I, Katevuo K, Ketonen L (1981) Metrizamide cisternography with hypocycloid and CT in sellar and suprasellar lesions. Clin Radiol 32:403–408

Lanzieri CF, Sabato U, Sacher M (1984) Third ventricular lymphoma: CT findings. J Comp Ass Tomogr 8:645–647

Lee BCP (1979) Intracranial cysts. Radiology 130:667–674

Lee KF, Lin JR (1979) Neuroradiology of sellar and juxtasellar lesions. Charles C. Thomas, Springfield, Ill

Lee SH, Delgado TE, Buchheit WA (1977) Intracranial dermoid tumor: diagnosis by CT; a case report. Neurosurgery 1:281–283

Leeds NE, Naidich TP (1977) CT in the diagnosis of sellar and parasellar lesions. Semin Roentgenol 12:121–135

Leo JS, Pinto RS, Hulvat GF, Epstein F, Kricheff II (1979) CT of arachnoid cysts. Radiology 130:675–680

Lin SR, Bryson MM, Gobien CR, Lee YY (1978) Radiologic findings of hamartomas of the tuber cinereum and hypothalamus. Radiology 127:697–704

Lundberg PO, Osterman PO, Wide L (1981) Serum prolactin in patients with hypothalamus and pituitary disorders. J Neurosurg 55:194–199

Maitland CG, Abiko J, Hoyt WF, Wilson CB, Okamura T (1982) Chiasmal apoplexy. J Neurosurg 56:118–122

Manelfe C, Bonafe A, Fabre P, Pessey JJ (1978) CT in olfactory neuroblastoma: one case of esthesioneuroepithelioma and four cases of esthesioneuroblastoma. J Comp Ass Tomog 2:412–420

Marano SR, Sonntag VKH, Speltzler RF (1984) Planum sphenoidale meningioma mimicking pituitary apoplexy: a case report. Neurosurgery 15:859–862

Matthews FD (1983) Intraventricular craniopharyngioma. AJNR 4:984–985

Miller JH, Pena AM, Segall HD (1980) Radiological investigation of sellar region masses in children. Radiology 134:81–87

Mori K, Handa H, Takeuchi J, Hanakita J, Nakano Y (1981) Hypothalamic hamartoma. J Comp Ass Tomog 5:519–521

Moseley IF, Sanders MD (1982) Computerized tomography in neuro-ophthalmology. Chapman and Hall, London

Murali R, Epstein F (1979) Diagnosis and treatment of suprasellar arachnoid cyst. Report of 3 cases. J Neurosurg 50:515–518

Naidich TP, Pinto RS, Kushner MJ, Lin JP, Kricheff II, Leeds NE, Chase NE (1976) Evaluation of sellar and parasellar masses by computed tomography. Radiology 120:91–99

Nakstad PHJ, Skalpe IO (1981) CT in the evaluation of the supraclinoid arteries in suprasellar pituitary gland tumors. Acta Radiol (Diagn) 22:399–402

Numaguchi Y, Kishikawa T, Ikeda J, Fukui M, Kitamura K, Tsukamsto Y, Masuo K, Maatsuura K (1981) Neuroradiological manifestations of suprasellar pituitary adenomas, meningomas and craniopharyngiomas. Neuroradiology 21:67–74

Ozanne P, Jedynak CP, Charbonnel B, Derome PJ (1982) Metastases hypophysaires et hypothalamiques. Etude anatomo-clinique de 5 observations. Ann Med Int 133:92–96

Peyster RG, Hoover ED (1984) CT of the abnormal pituitary stalk. AJNR 5:49–52

Peyster RG, Hoover ED (1984) Computerized tomography in orbital disease and neuro-ophthalmology. Year Book Medical Publishers, Chicago, London

Post JD (1981) La tomodensitométrie de l'orbite et de la région juxtasellaire chez l'enfant. J Neuroradiology 8:171–194

Raymond LA, Tew J (1978) Large suprasellar aneurysms imitating pituitary tumour. J Neurol Neurosurg Psychiatry 41:83–87

Roski RA, Gardner JH, Spetzler RF (1981) Intrachiasmatic arteriovenous malformation. J Neurosurg 54:540–541

Salcman M (1979) Correlation of absorption coefficients with intracranial fluid protein concentrations and specific gravities. Neurosurgery 5:16–20

Sarwar M (1979) Primary sellar-parasellar esthesioneuroblastoma. AJR 133:140–141

Sato M, Ushio Y, Arita N, Mogami H (1985) Hypothalamic hamartoma: report of two cases. Neurosurgery 16:198–206

Savoiardo M, Harwood-Nash DC, Tadmor R, Scotti G, Musgrave MA (1981) Gliomas of the intracranial anterior optic pathways in children. Radiology 138:601–610

Schubiger O, Valavanis A, Gessagaga E (1983) Dense supra-sellar epidermoid cyst. A case report. Neuroradiology 24:269–271

Shen DY, Guay AT, Silverman ML, Hybels RL, Freidberg SR (1984) Primary intrasellar germinoma in a woman presenting with secondary amenorrhea and hyperprolactinemia. Neurosurgery 15:417–429

Strand RD, Baker RA, Ordia IJ (1978) Metrizamide ventriculography and CT in lesions about the third ventricle. Radiology 128:405–410

Strother CM, Sackett JF, Appen RE (1977) Anatomic considerations for CT of the optic chiasm. Arch Neurol 34:713–714

Stuart CA, Neelon FA, Lebovitz HE (1978) Hypothalamic insufficiency: the cause of hypopituitarism in sarcoidosis. Ann Intern Med 88:589–594

Takeuchi J, Handa H, Otsuka S, Takebe Y (1979) Neuroradiological aspects of suprasellar germinomas. Neuroradiology 17:153–159

Taveras JM, Wood EH (1976) Diagnostic neuroradiology. 2nd Ed Williams and Wilkins, Baltimore

Taylor S (1982) High resolution computed tomography of the sella. Radiologic clinics of North America 20:207–236

Vignaud J, Aubin ML, Bories J (1979) Apport de la tomodensitométrie à l'exploration de la région sellaire et supra-sellaire. Rev Neurol 135:41–50

Visot A, Rougerie J, Derome PJ (1980) Gliomes optochiasmatiques. Neurochirurgie 26:181–192

Volpe BT, Foley KM, Howieson J (1978) Normal CT

scans in craniopharyngioma. Ann Neurol 3:87–89

Wiggli U, Benz UF (1978) Normal CT anatomy of the suprasellar subarachnoid space. Radiology 128:65–70

Wiggli U, Benz UF, Müller HR (1977) Chiasmasyndrome – computertomographische Diagnostik. Klin Mbl Augenheilk 170:290–296

Wirt TC, Hester RW (1978) Suprasellar arachnoid cysts. Surg Neurol 9:322

Wolpert SM, Scott RM (1981) Value of metrizamide CT cisternography in the management of cerebral arachnoid cysts. AJNR 2:29–36

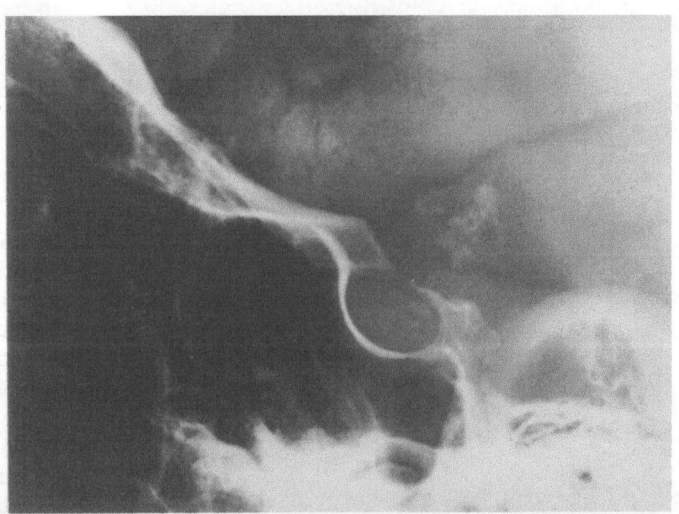

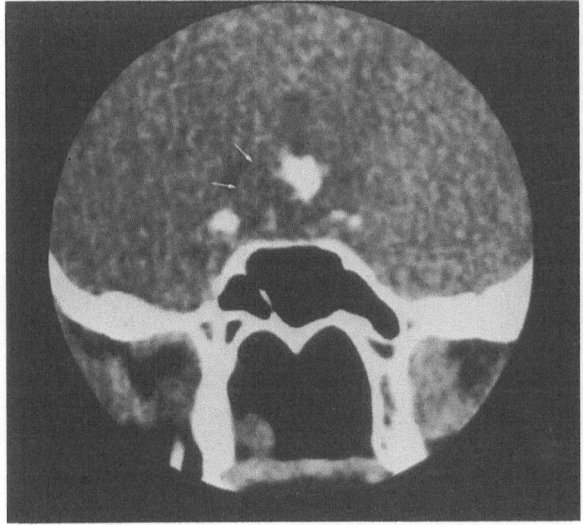

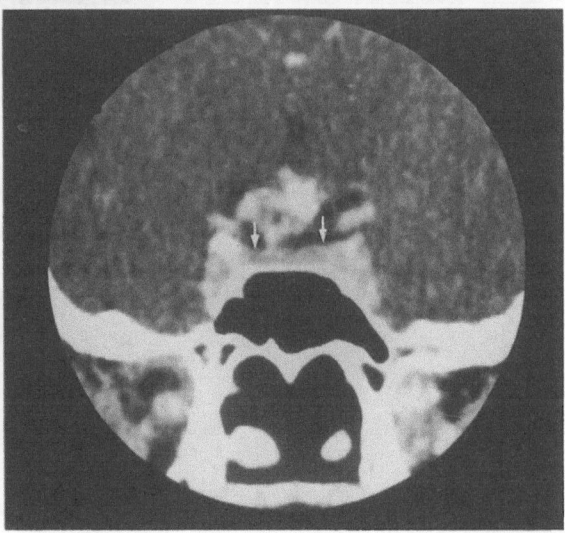

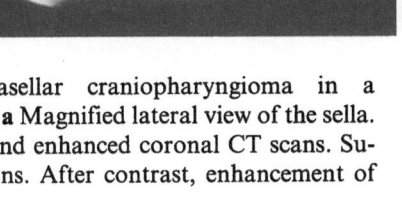

Fig. 16.1a–c. Suprasellar craniopharyngioma in a 55-year-old woman. **a** Magnified lateral view of the sella. **b, c** Nonenhanced and enhanced coronal CT scans. Suprasellar calcifications. After contrast, enhancement of the solid component of the tumor, which was isodense before contrast (**b** *small arrows*). **c** The pituitary gland below the tumor appears quite normal (*arrows*)

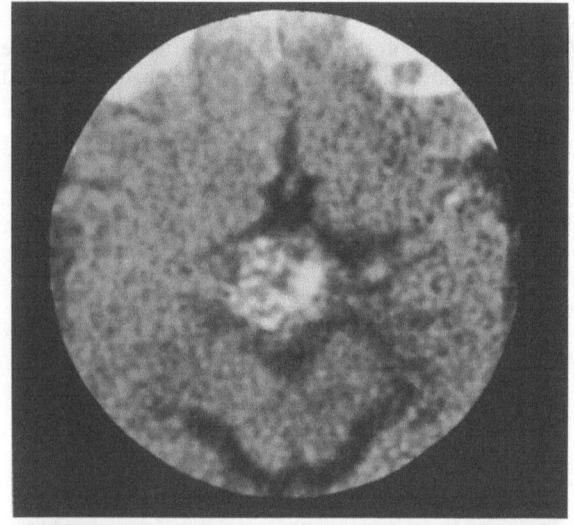

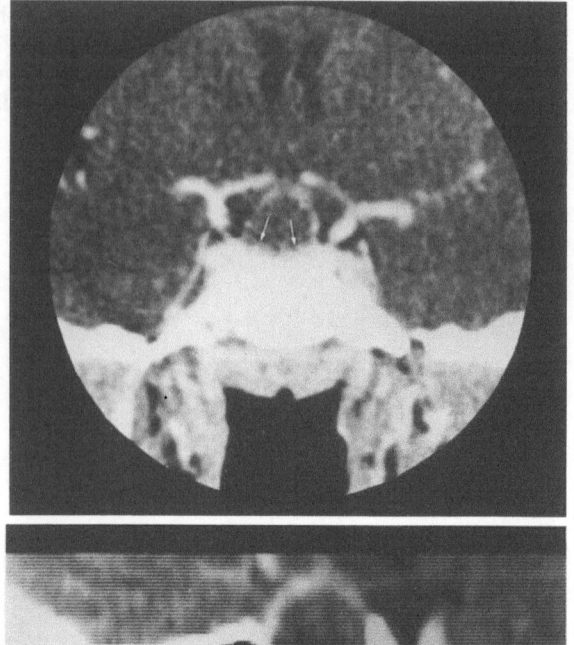

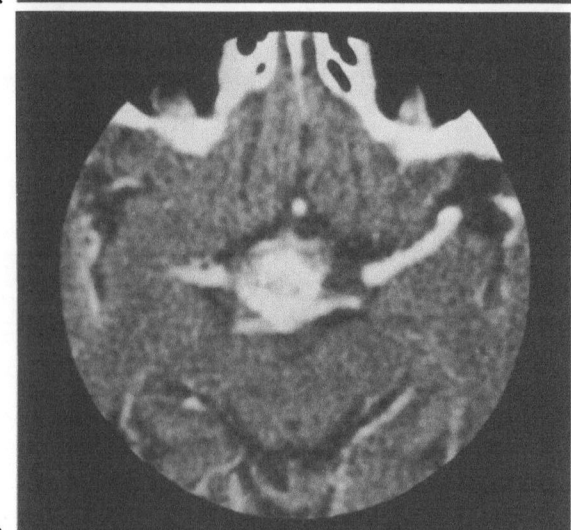

Fig. 16.3a, b. Suprasellar craniopharyngioma in a 17-year-old boy presenting with delayed growth and hyperprolactinemia. **a** Coronal and **b** sagittal reformatted enhanced CT scans. Cystic intra- and suprasellar mass with ring enhancement. The pituitary gland is compressed (*arrows*)

◀**Fig. 16.2a–c.** Suprasellar craniopharyngioma in a 66-year-old man. **a** Nonenhanced axial, **b** enhanced axial, and **c** coronal enhanced CT scans. Nodular calcifications in the suprasellar area. After contrast, marked enhancement of the tumor. The pituitary gland itself is normal

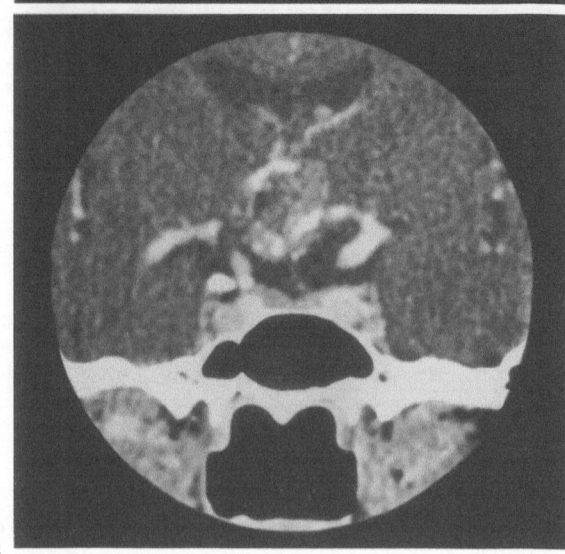

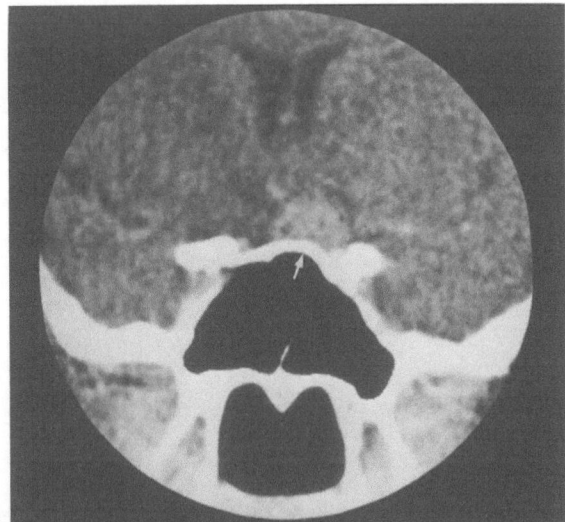

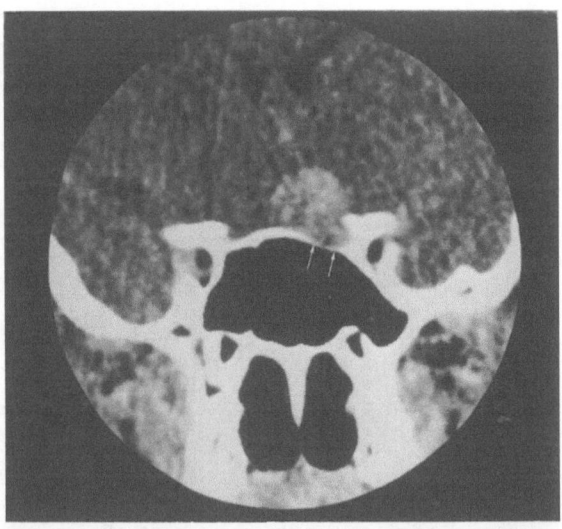

Fig. 16.5. Intracranial optic nerve glioma in a 15-year-old boy with isolated left visual loss. Coronal enhanced CT scan. Rounded homogeneous hyperdense mass developed in the intracranial portion of the left optic nerve. The endocranial foramen of the optic canal appears enlarged and the chiasmatic sulcus is eroded (*arrows*). (Compare with Fig. 16.4a)

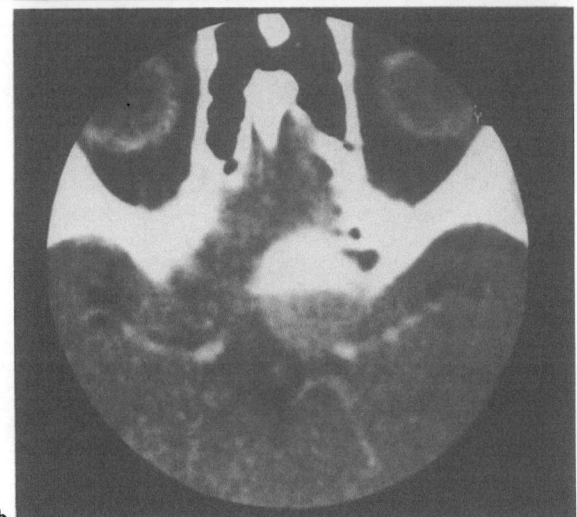

Fig. 16.4a, b. Meningioma of the planum sphenoidale and chiasmatic sulcus in a 35-year-old man presenting with left progressive visual loss. **a** Coronal and **b** axial enhanced CT scans. The meningioma arising from the left part of the planum sphenoidale and chiasmatic sulcus is determining an upward convexity of both osseous structures (blistering) (*arrow*), better seen in coronal section than in axial section

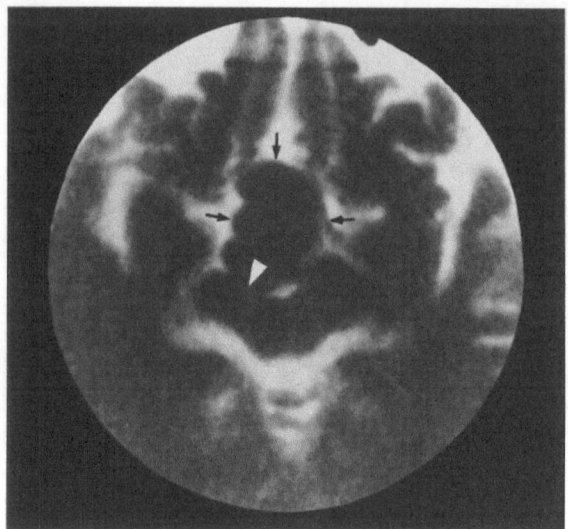

Fig. 16.6. Optic chiasm glioma in a 72-year-old man. Axial intrathecal enhanced CT scans. Polycyclic suprasellar mass (*arrows*) arising from the optic chiasm extending posteriorly towards the right optic tract and compressing the right cerebral peduncle (*arrowhead*)

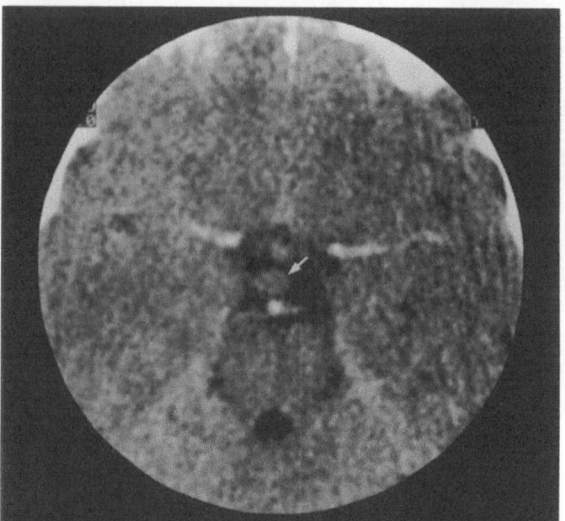

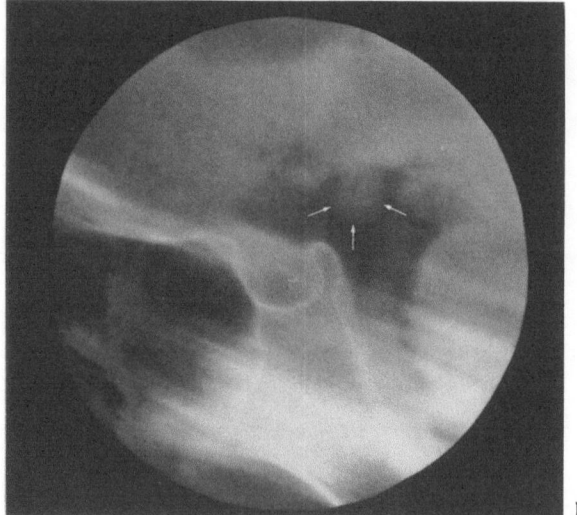

Fig. 16.8a, b. Hypothalamic hamartoma in an 18-year-old man with delayed growth. **a** Axial enhanced CT scan and **b** midsagittal pneumotomoencephalography. **a** The axial CT scan demonstrates a small isodense oval mass located between the pituitary stalk and the basilar artery (*arrow*). **b** The hamartoma appears clearly on pneumoencephalography (*arrows*)

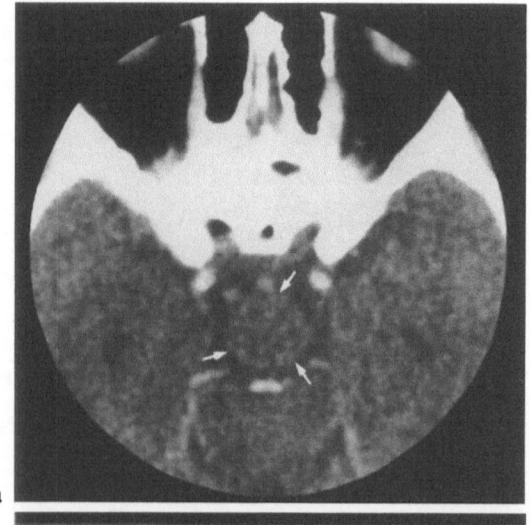

Fig. 16.7a, b. Hypothalamic hamartoma in a 15-year-old girl presenting with precocious puberty from the age of 2. **a** Axial and **b** sagittal reformatted enhanced CT scans. Rounded isodense suprasellar mass (*arrows*) located behind the optic chiasm and extending into the peduncular cistern

Fig. 16.10a–d. Hypothalamic choristoma in a 38-year- ▶ old hyperprolactinemic woman (165 ng/ml). **a** Lateral magnified view of the sella. Marked demineralization of the dorsum sellae. **b** Sagittal reformatted image, **c** axial and **d** coronal enhanced CT scans. Large homogeneously enhanced suprasellar mass. Note that the coronal section only demonstrates a normal pituitary gland below the tumor (*arrows*)

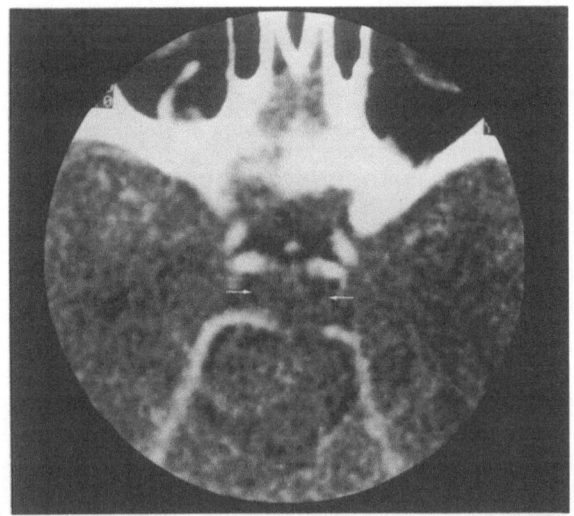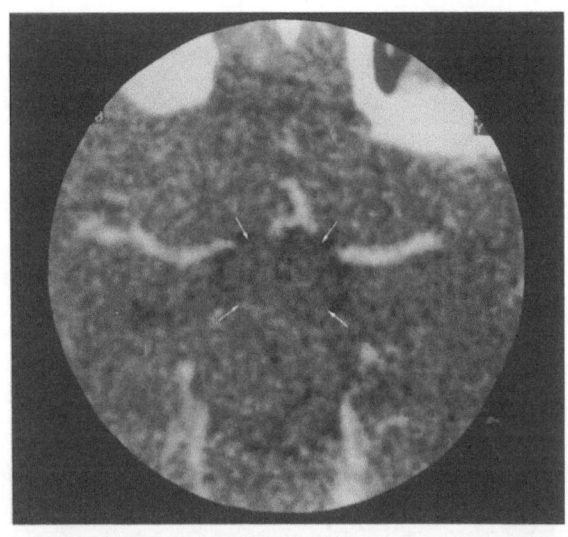

Fig. 16.9 a, b. Hypothalamic hamartoma in a 6-year-old boy presenting with precocious puberty. Axial enhanced CT scans. Isodense supra- and retrosellar mass (*arrows*) developed between the optic chiasm and the cerebral peduncles

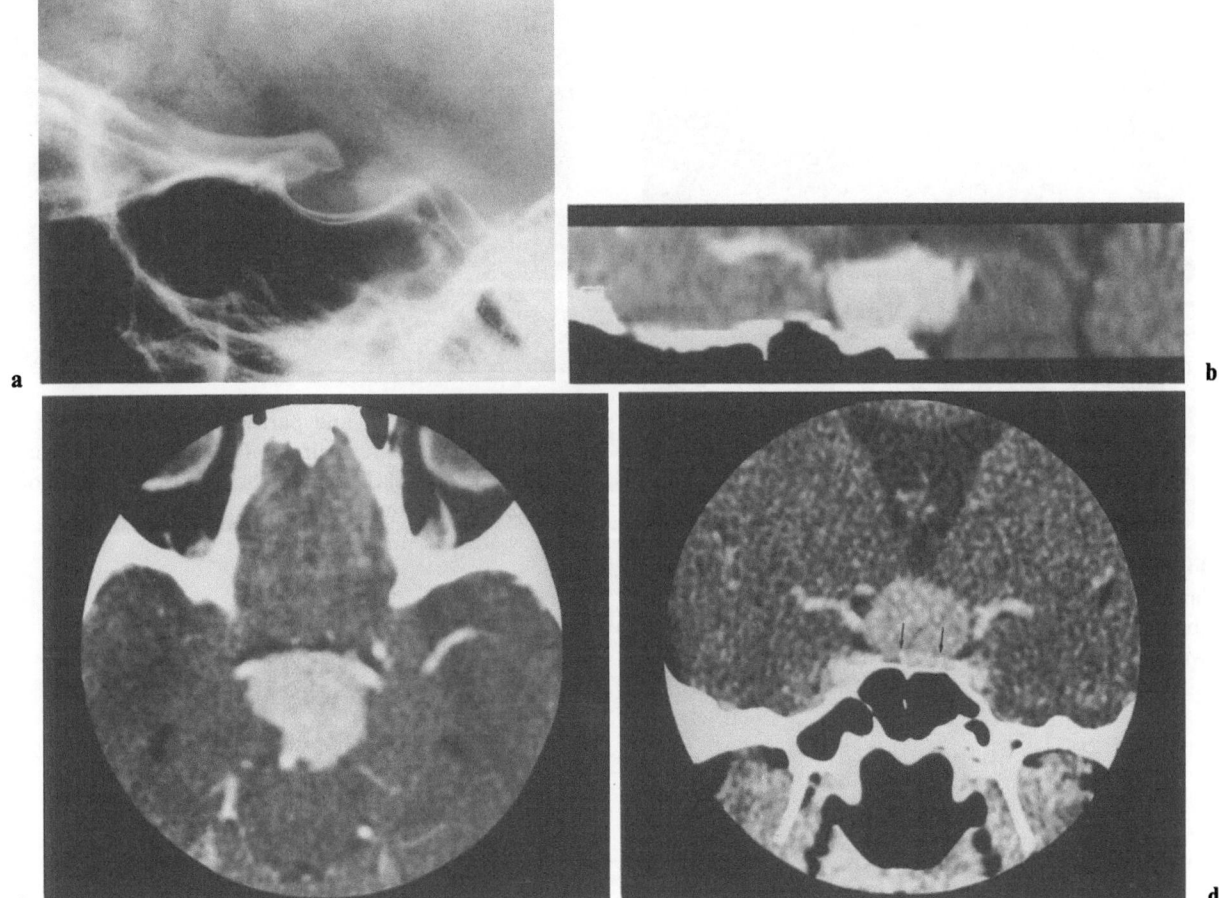

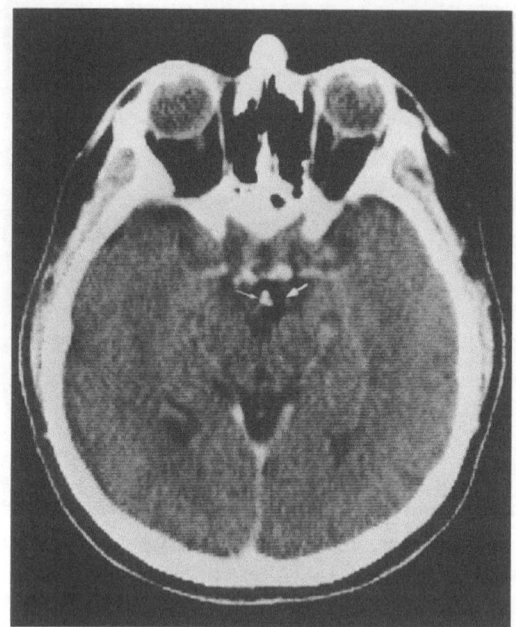

Fig. 16.11. Retro- and suprasellar dermoid cyst. A nodular calcification (*small arrow*) and fatty density values (*large arrow*) are noted within the suprasellar tumor

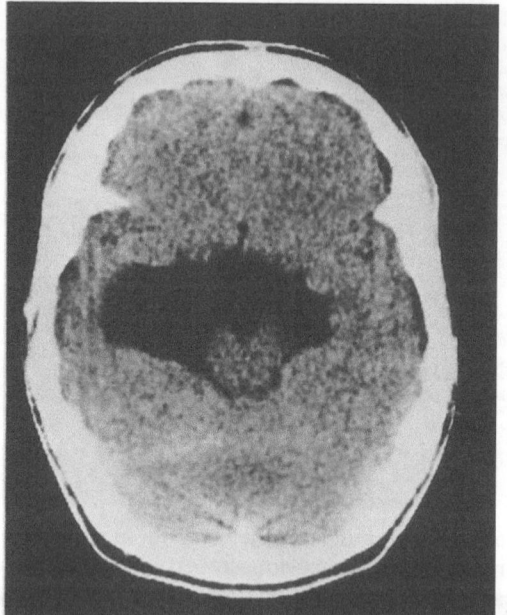

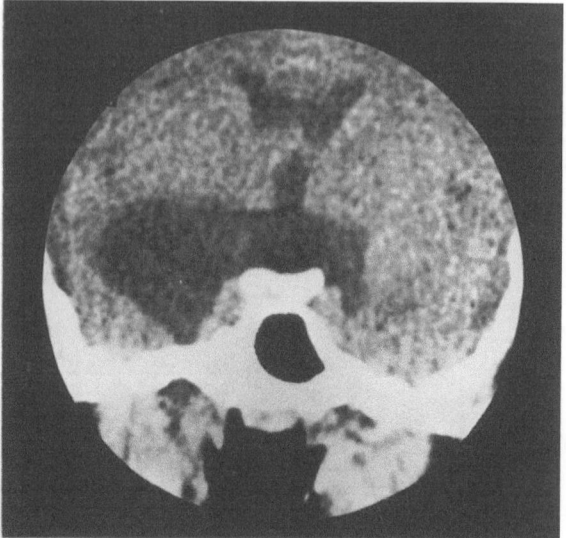

Fig. 16.13a, b. Supra- and parasellar arachnoid cyst in a 14-year-old boy presenting with temporal seizures. **a** Axial and **b** coronal CT scans. The suprasellar arachnoid cyst is extending towards the right temporal fossa

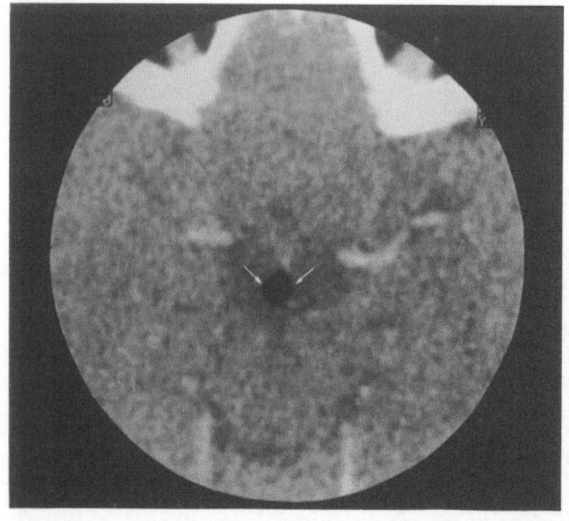

Fig. 16.12. Suprasellar lipoma. Small rounded suprasellar mass (*arrows*) located between the pituitary stalk and the cerebral peduncles. Negative fatty density values within the mass

Fig. 16.14. Intrasellar extension of a dilated third ventricle in a 29-year-old woman with aqueduct stenosis presenting with obesity and pituitary deficiency. Enhanced coronal CT scan. Marked supratentorial hydrocephalus; the third ventricle is extending towards the sella turcica and compresses the pituitary gland (*arrows*)

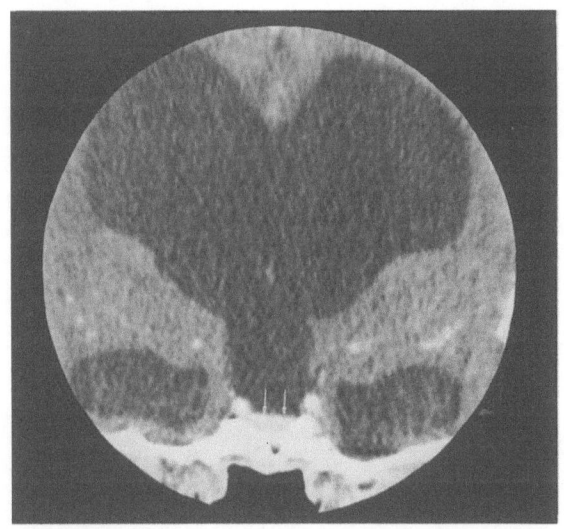

Fig. 16.15a, b. Supratentorial hydrocephalus related to a posterior fossa tumor (Hand-Schuller-Christian disease) in an 18-year-old woman presenting with diabetes insipidus. **a** Lateral plain film and **b** coronal enhanced CT scan. Enlargement of the sella turcica related to marked dilatation of the third ventricle. The pituitary gland appears normal and the pituitary stalk shortened

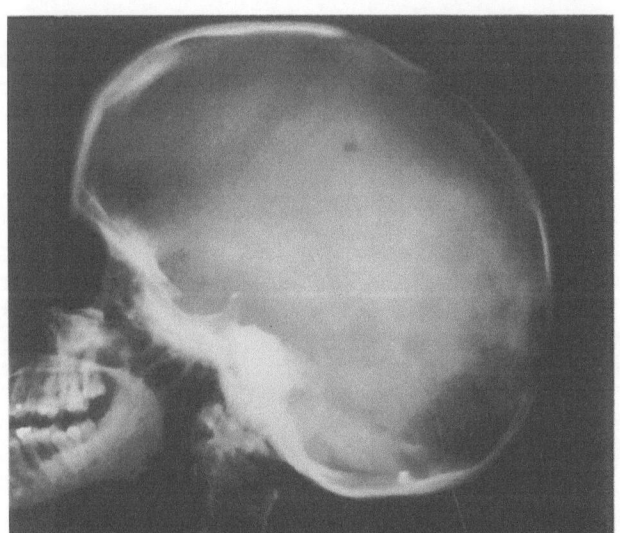

a

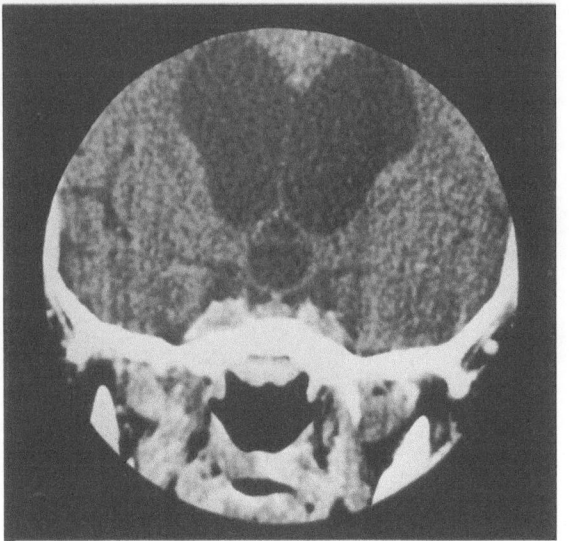

b

Chapter 17
Parasellar Pathology

Lesions developed within the cavernous sinus frequently determine ocular palsies sometimes associated with involvement of the first and/or second branch of the fifth cranial nerve. If a laterally developed pituitary lesion may be responsible for such a cavernous sinus syndrome, on the other hand, a cavernous lesion (mainly an aneurysm of the internal carotid artery) may also extend towards the pituitary fossa and affect the pituitary gland.

Intracavernous aneurysms of the internal carotid artery and meningiomas of the cavernous sinus are by far the most common lesions. Parasellar neurinomas, metastases of the cavernous sinus, and infectious lesions are rather rare. Tumors of the sphenoid body and sphenoid sinus and even juvenile pharyngeal fibromas may also involve the cavernous sinus.

Common Cavernous Lesions

Intracavernous Aneurysms of the Internal Carotid Artery

These vascular malformations are particularly common in women more than 50 years old and are responsible for 20% of cavernous sinus syndromes (Peyster and Hoover).

A symptomatic intracavernous aneurysm is usually large enough to give characteristic CT modifications (Fig. 17.1), the malformations often being giant:

1. The cavernous sinus is usually enlarged, especially transversally.

2. The carotid sulcus appears thinned and depressed towards the midline; the dorsum sellae is laterally eroded.

3. The superior orbital fissure may be enlarged.

4. Calcifications of the wall of the aneurysm are often noted.

5. An intense and homogeneous enhancement is usual; this enhancement affects only the non-thrombosed part of the aneurysm (Fig. 17.2).

6. Bilateral intracavernous aneurysms may be observed (Mindel).

Meningiomas of the Cavernous Sinus

Developed within the wall of the cavernous sinus, these tumors enlarge the cavernous sinus and show a posterior tumoral expansion towards the free edge of the tentorium cerebelli. After enhancement, this tumoral expansion gives a triangular "swallow-tail" image (Figs. 17.3 and 17.4). This sign is, however, not pathognomonic. Nodular tumoral calcifications are but rarely observed.

Uncommon Cavernous Lesions

Metastases of the Cavernous Sinus

Involvement of the cavernous sinus is secondary either to an osseous metastasis or a primary metastasis of the cavernous sinus (Gardeur). These primary metastases enlarge the cavernous sinus in the same way as a meningioma; a more or less homogeneous enhancement is noted (Kline).

Parasellar Neurinomas

These rare tumors develop usually from the branches of the trigeminal nerve and exceptionally from the ocular nerves. At the time of their diagnosis, these neurinomas are often large tumors; the cavernous sinus is enlarged and a more or less homogeneous enhancement is noted (Fig. 17.5).

Cavernous Carotid Fistulas

Communication between the high-pressure internal carotid system and the low-pressure cavernous venous system may result:

1. From a traumatic rupture of internal carotid artery.

2. From a spontaneous rupture of an internal carotid artery whose wall has been weakened (Ehlers-Danlos syndrome, atheromatous disease)

3. From a spontaneous rupture of an aneurysm of the internal carotid artery along its intracavernous portion

To these etiologies should be added dural fistulas of the cavernous sinus which have, to a lesser extent, the same clinical consequences.

When a pulsating proptosis exists, CT scan usually shows a dilated superior ophthalmic vein (Fig. 17.6). When the outflow from the fistula is important, a bilateral enlargement may be noted. Distension of intramuscular veins together with an edema are expressed by diffused swelling of extraocular muscles. The cavernous sinus may appear distended and the superior orbital fissure is sometimes enlarged.

Tolosa-Hunt's Syndrome

The Tolosa-Hunt's syndrome is characterized by a painful ophthalmoplegia due to a nonspecific granulomatous infiltration of the anterior part of the cavernous sinus and the region of the superior orbital fissure. CT scan is often normal, but in a few cases an enlarged cavernous sinus may be noted. In fact, the diagnosis is based, on the one hand, on orbital venography which shows an obstruction of the superior ophthalmic vein at the level of the superior orbital fissure and/or the cavernous sinus, and, on the other hand, on the rapid therapeutic response to steroids.

Chondromas and Chondrosarcomas

The parasellar region is an usual location for chondromas and chondrosarcomas which develop from embryonic cartilaginous remnants. CT scans can identify a calcified parasellar mass destroying laterally the sella turcica, the dorsum sellae, and the sellar floor (Fig. 17.7) (Lee, Reuter and Weber).

Nasopharyngeal, Sphenoid Sinus, and Sphenoid Bone Tumors

Tumors developed within the nasopharyngeal region, the sphenoid sinus, and the sphenoid body may, of course, invade the cavernous region (Carmody, Spigos) (Fig. 17.8).

References

Azar Kia B, Palacios E, Churchill RJ (1977) Diagnosis of sellar and parasellar lesions by CT and other diagnostic modalities. CT 1:249–256

Bamberger C, Rosier J (1980) Examen tomodensitométrique des lésions sellaires et parasellaires. Feuillets de radiologie 20:197–208

Banna M, Baker HL Jr, Houser OW (1980) Pituitary and parapituitary tumours on CT. Br J Radiol 53:1123–1143

Bonneville JF, Dietemann JL (1981) Radiology of the sella turcica. Springer Verlag, Berlin Heidelberg New York

Cabanis EA, Van Effenterre R, Iba-Zizen MT (1979) CT in parasellar space-occupying lesions and therapeutic decision. Acta Neurochirurgica (suppl 28) 28:329–333

Carmody RF, Rickles DJ, Johnson SF (1983) Giant cell tumor of the sphenoid bone. J Comp Ass Tomog 7:370–373

Clifford-Jones RE, Ellis CJK, Stevens JM, Turner A (1982) Cavernous sinus thrombosis. J Neurol Neurosurg Psychiatr 45:1092–1097

De Divitiis E, Spaziante R, Cirillo S, Stella L, Donzelli R (1979) Primary sellar chondromas. Surg Neurol 11:229–232

Diaz F, Latchow R, Duvall AJ, Quick CA, Erickson DL (1978) Mucoceles with intracranial and extracranial extensions. Report of two cases. J Neurosurg 48:284–288

Dietemann JL, Heldt N, Burguet JL, Medjek L, Maitrot

D, Wackenheim A (1982) CT findings in malignant meningiomas. Neuroradiology 23:207–209

Gore RM, Weinberg PE, Kim KS (1980) Sphenoid sinus mucoceles presenting as intracranial masses on CT. Surg Neurol 13:375–379

Hamer J (1979) The radiological diagnosis of parasellar dermoid cysts with special regard to CT scan. Acta Neurochirurgica (suppl 28) 28:321–322

Handa J, Nakano Y, Heiha A (1977) CT cisternography with intracranial arachnoid cysts. Surg Neurol 8:451–454

Harbour RC, Trobe JD, Ballinger WE (1984) Septic cavernous sinus thrombosis associated with gingivitis and parapharyngeal abscess. Arch Ophthalmol 102:94–97

Harris FS, Rhoton AL (1976) Anatomy of the cavernous sinus. A microsurgical study. J Neurosurg 45:169–180

Hatam A, Bergström M, Greitz T (1979) Diagnosis of sellar and parasellar lesions by CT. Neuroradiology 18:249–258

Hayman LA, Evans RA, Hinck VC (1979) Rapid high dose (RHD) contrast CT of perisellar vessels. Radiology 131:121–123

Hori T, Muraoka K, Hokama Y, Takami M, Saito Y (1982) A growth-hormone-producing pituitary adenoma and an internal carotid artery aneurysm. Surg Neurol 18:108–111

Jing BS, Goepfert H, Close LG (1978) CT of paranasal sinus neoplasms. Laryngoscope 88:1485–1503

Khangure MJ, Apsimon HT (1981) Some pitfalls in the diagnosis of pituitary tumors: the importance of carotid angiography. Surg Neurol 16:300–308

Kline LB, Galbraith JG (1981) Parasellar epidermoid tumor presenting as painful ophthalmoplegia. J Neurosurg 54:113–117

Kline LB, Acker JD, Post JD (1982) CT evaluation of the cavernous sinus. Ophthalmology 89:374–385

Lee KF, Lin JR (1979) Neuroradiology of sellar and juxtasellar lesions. Charles C. Thomas, Springfield, Ill.

Leeds NE, Naidich TP (1977) CT in the diagnosis of sellar and parasellar lesions. Semin Roentgenol 12:121–135

MacPherson P, Anderson DE (1981) Radiological differentiation of intrasellar aneurysms from pituitary tumours. Neuroradiology 21:177–183

Mindel JS, Sachdev VP, Kline LB, Sivak MA, Bergman DA, Yang WC, Choi IS, Huang VP (1983) Bilateral intracavernous carotid aneurysm mimicking a prolactin secreting pituitary tumor. Surg Neurol 19:163–167

Moseley IF, Sanders MD (1982) Computerized tomography in neuro-ophthalmology. Chapman and Hall, London

Naidich TP, Pinto RS, Kushner MJ, Lin JP, Kricheff II, Leeds NE, Chase NE (1976) Evaluation of sellar and parasellar masses by computed tomography. Radiology 120:91–99

Osborn AG, Johnson L, Roberts TS (1979) Sphenoidal mucoceles with intracranial extension. J Comp Ass Tomog 3:335–338

Parsons C, Hodson N (1979) CT of paranasal sinus tumors. Radiology 132:641–645

Pascaud JL, Vigneu P, Hummel P, Bouchet JB, Rihouet J, Pascaud-Ged E (1982) Absence partielle de carotide interne avec anastomose inter-carotidienne transsellaire. J Radiol 63:37–40

Peyster RG, Hoover E (1984) Computerized tomography in orbital disease and neuro-ophthalmology. Year Book Medical Publishers, Chicago, London

Post JD (1981) La tomodensitométrie de l'orbite et de la région juxtasellaire chez l'enfant. J Neuroradiology 8:171–194

Post JD, Mendez DR, Kline LB, Acker JD, Glaser JS (1985) Metastatic disease to the cavernous sinus: clinical syndrome and CT diagnosis. J Comp Ass Tomog 9:115–120

Price H, Danzinger A (1980) CT findings in mucoceles of the frontal and ethmoidal sinuses. Clin Radiol 31:169–174

Quencer RM, Stokes NA, Wolfe D, Page LK (1979) Melanotic nerve sheath tumor of the gasserian ganglion and trigeminal nerve. AJR 133:142–144

Reich NE, Zelch JV, Alfidi RJ, Meaney TF, Duchesneau PM, Weinstein MA (1976) CT in the detection of juxtasellar lesions. Radiology 118:333–335

Resneck JD, Lederman IR (1981) Traumatic chiasmal syndrome associated with pneumocephalus and sellar fracture. Am J Ophthalmol 92:233–237

Reuter K, Weber AL (1981) Parasellar chondrosarcoma in a patient with Ollier's disease. Neuroradiology 22:151–154

Rhoton AL, Harris FS, Renn WH (1977) Microsurgical anatomy of the sellar region and cavernous sinus. Clin Neurosurg 24:54–85

Rothfus WE, Curtin HD (1984) Extraocular muscle enlargement: a CT review. Radiology 151:677–681

Schubiger O, Valavanis A, Hayek J (1980) Neuroma of the cavernous sinus. Surg Neurol 13:313–316

Som PM, Shugar JMA (1980) The CT classification of ethmoid mucoceles. J Comp Ass Tomog 4:199–203

Spigos WST, Khine N (1982) Chordoma of the sellar region. J Comp Ass Tomog 6:154–158

Taylor S (1982) High resolution computed tomography of the sella. Radiologic clinics of North America 20:207–236

Trobe JD, Glaser JS, Post JD (1978) Meningiomas and aneurysms of the cavernous sinus. Arch Ophthalmol 96:457–467

Umansky F, Nathan H (1982) The lateral wall of the cavernous sinus, with special reference to the nerves related to it. J Neurosurg 50:228–234

Villani R, Ducati A, Betinelli A (1979) Sphenoid sinus mucoceles: clinical and radiological features. Acta Neurochirurgica (suppl 28) 28:433–437

Wing SD, Anderson RE, Osborn AG (1980) Dynamic cranial CT; preliminary results. AJNR 1:135–139

Yoshizawa T, Tomono Y, Note T, Maki Y (1982) Juxtasellar mycotic abscess. Surg Neurol 21:49–52

Zilkha A, Daiz AS (1980) CT in carotid cavernous fistula. Surg Neurol 14:325–329

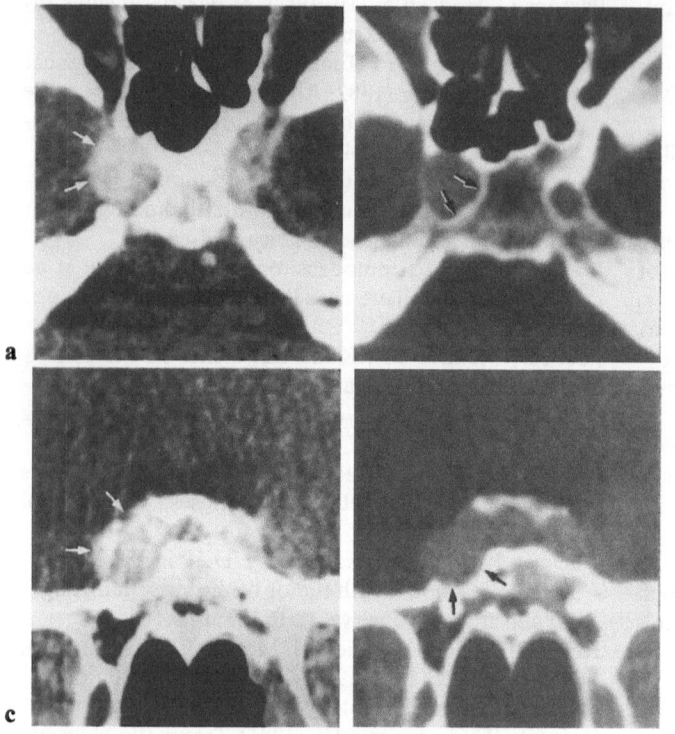

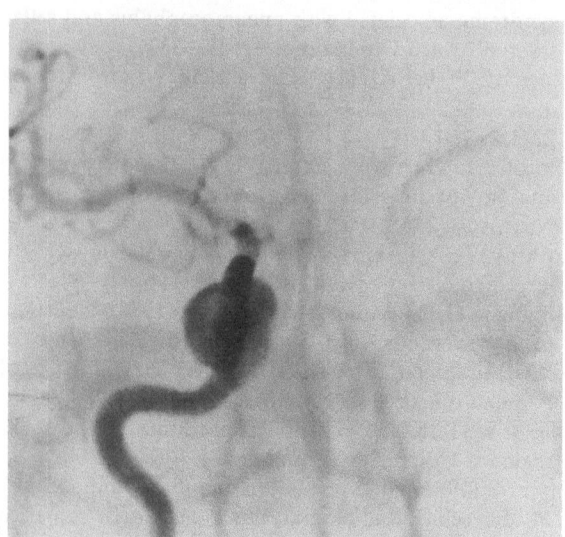

Fig. 17.1 a–e. Right intracavernous carotid aneurysm in a 36-year-old woman presenting with right sixth nerve palsy. **a** Axial, **b** axial with large window, **c** coronal, **d** coronal with large window enhanced CT scans, and **e** right carotid angiography. Enlargement of the right cavernous sinus (*white arrows*) and deepening of the carotid sulcus (*black arrows*). Without bolus injection of contrast media density values are similar both in the aneurysm and in the cavernous sinus

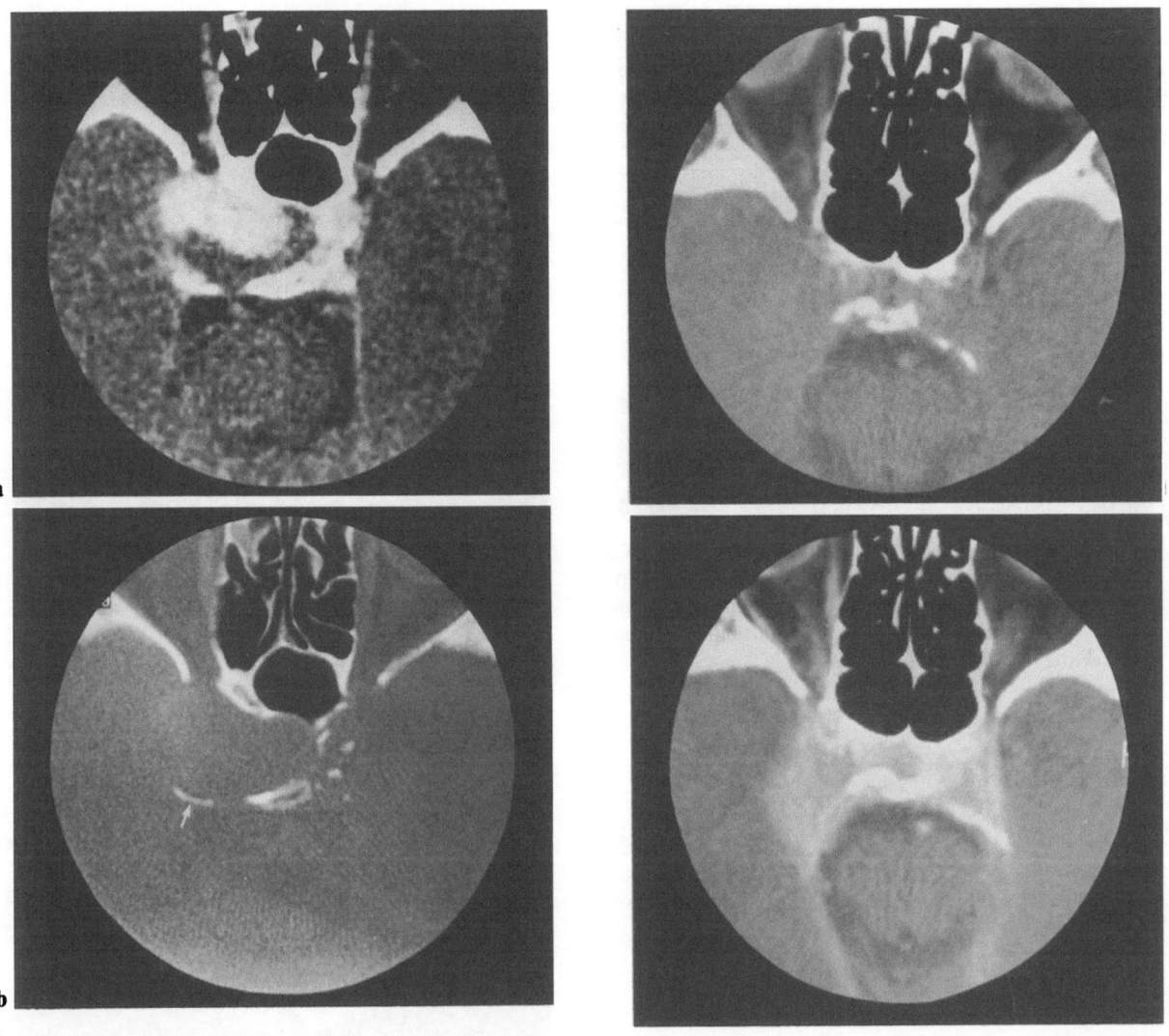

Fig. 17.2a, b. Right giant intracavernous carotid aneurysm in a 60-year-old man presenting with right third nerve palsy. Axial enhanced CT scans. Giant partially thrombosed right intracavernous aneurysm. Marked erosion of the right part of the dorsum sellae is noted. Small calcifications are present within the aneurysm wall (*arrow*). Note also the presence of calcifications within the left internal carotid wall

Fig. 17.3a, b. Meningioma of the right cavernous sinus in a 73-year-old woman presenting with right third nerve palsy. **a** Nonenhanced and **b** enhanced axial CT scans. **a** Without enhancement, the tumor appears quite isodense. **b** Marked enhancement after contrast and extension toward the tentorium

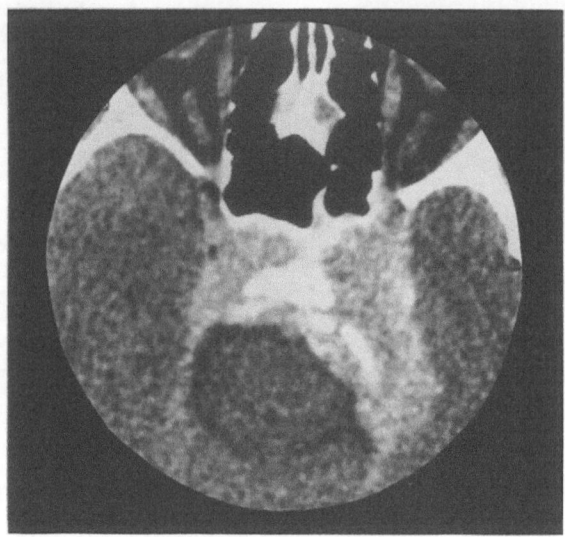

Fig. 17.4. Meningioma of the left cavernous sinus in a 55-year-old man presenting with left sixth nerve palsy. Enhanced axial CT scan. Marked enlargement and homogeneous enhancement of left cavernous sinus. Extension toward the tentorium

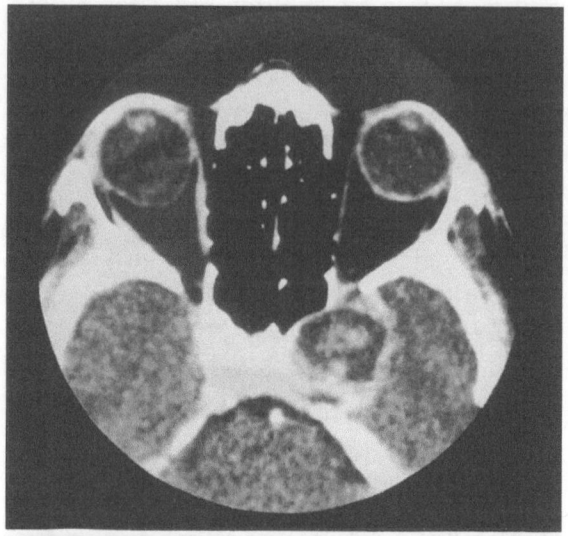

a

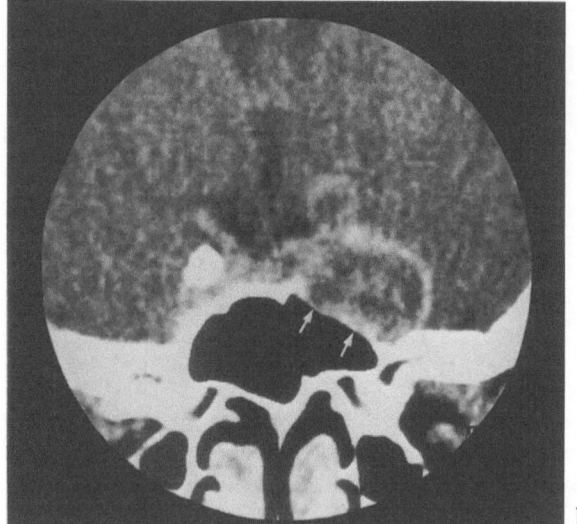

b

Fig. 17.5a, b. Left intracavernous fifth nerve neurinoma in a 54-year-old woman complaining of left facial pain. **a** Axial and **b** coronal enhanced CT scans. Marked enlargement of the left cavernous sinus associated with bone erosion (*arrows*). Heterogeneous enhancement within the tumor

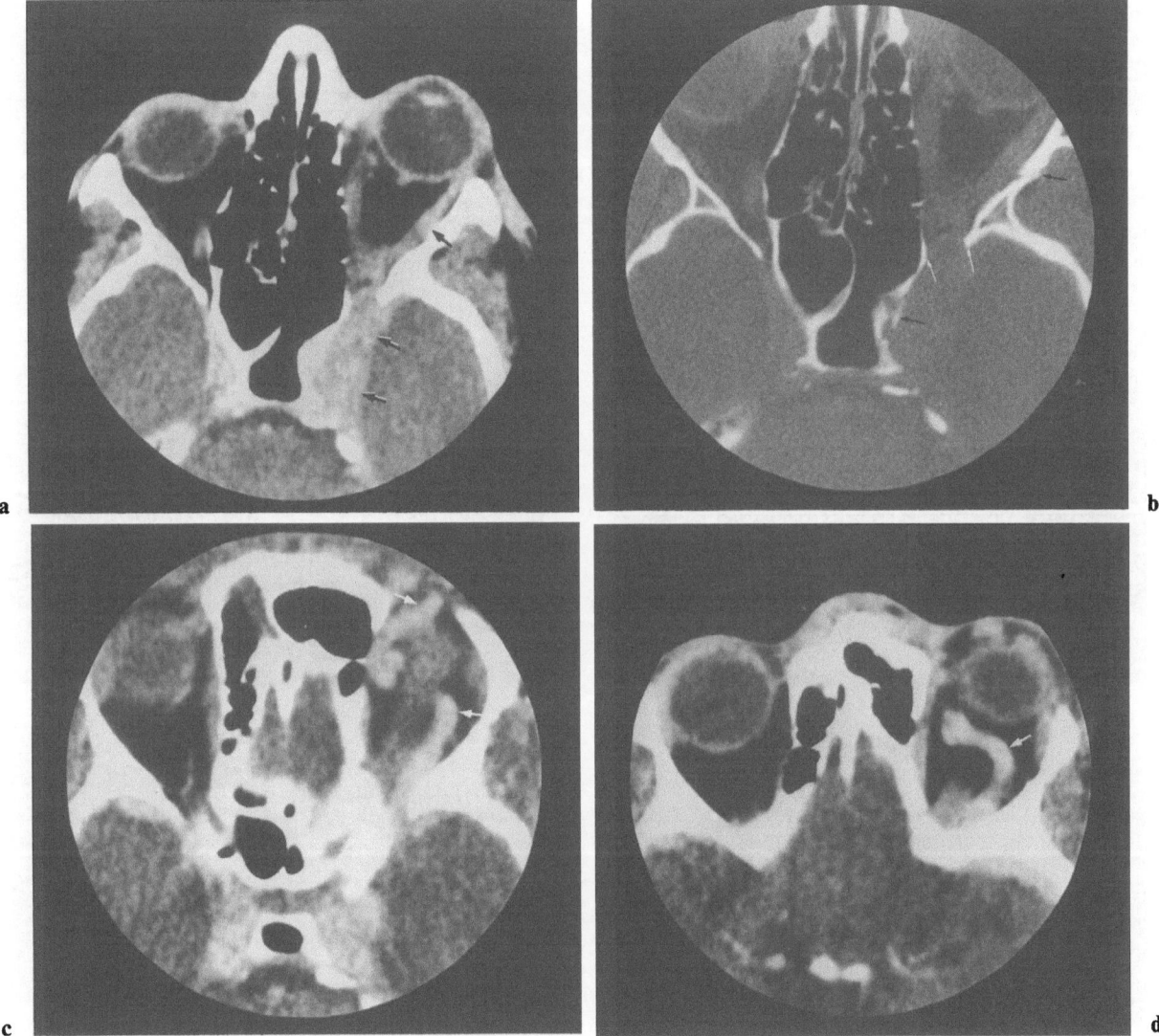

Fig. 17.6a–d. Left carotid-cavernous fistula following trauma (*small black arrows*) in a 21-year-old man. Axial enhanced CT scans. The left cavernous sinus appears enlarged (*black* and *white arrows*), as is the superior orbital fissure (*small white arrows*). Marked left proptosis is present. The left superior orbital vein is greatly dilated and appears tortuous (*large white arrows*). The left medial and lateral recti are enlarged (*large black arrow*)

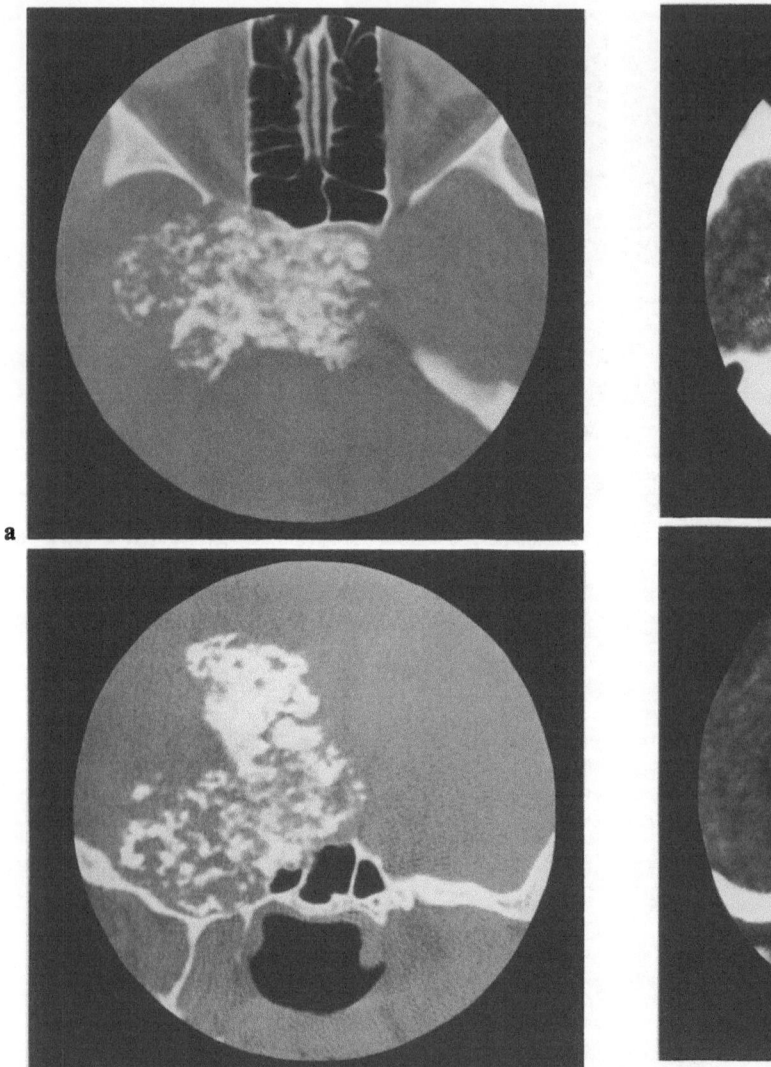

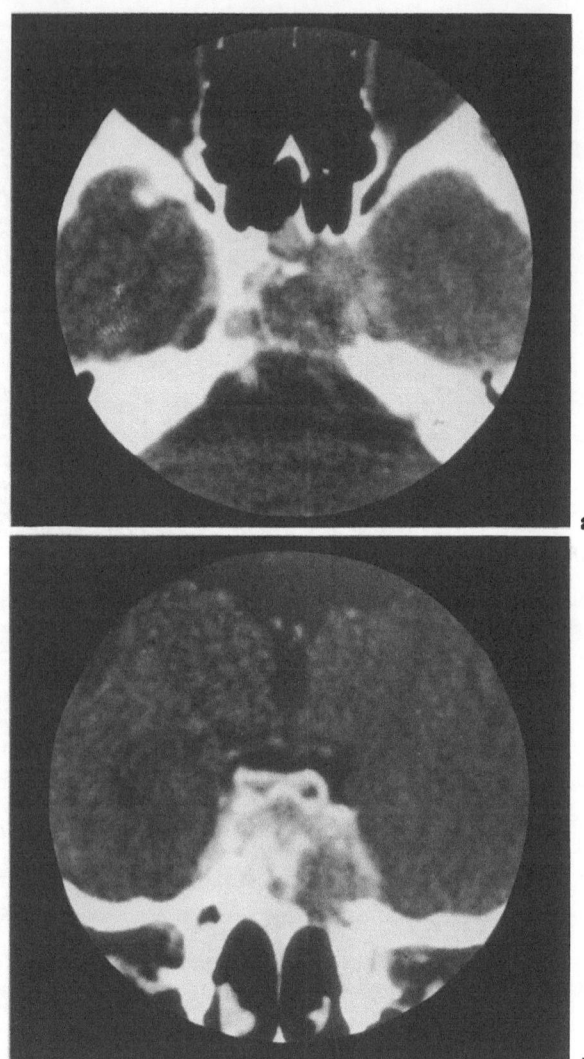

Fig. 17.7a, b. Recurrent right parasellar chondroma in a 27-year-old male presenting with headache and right ophthalmoplegia. **a** Axial and **b** coronal nonenhanced CT scans. Huge right calcified parasellar mass with destruction of the right part of the sella turcica and extension to the right cavernous sinus and the right temporal fossa. Enlargement of the right superior orbital fissure

Fig. 17.8a, b. Sphenoid myeloma in a 60-year-old man presenting with left ophthalmoplegia and facial pain. **a** Axial and **b** coronal enhanced CT scans. Marked destruction of the sphenoid bone. Myeloma is extending to the sella turcica, left cavernous sinus, and sphenoid sinus

Chapter 18
Picture Problems

This chapter has a question and answer format. Cover the right-hand side of each page before checking your answer.

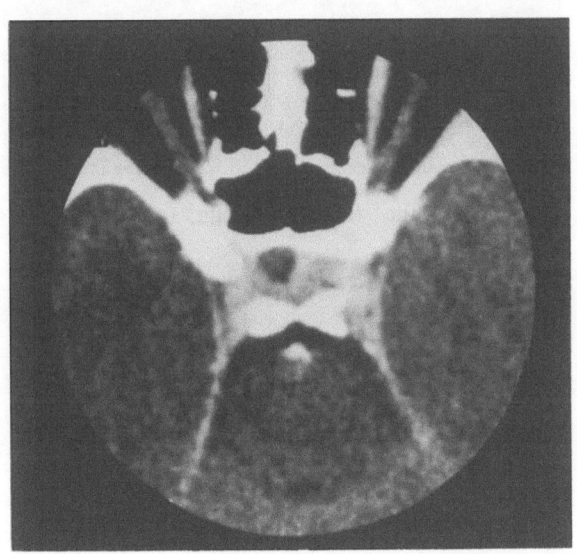 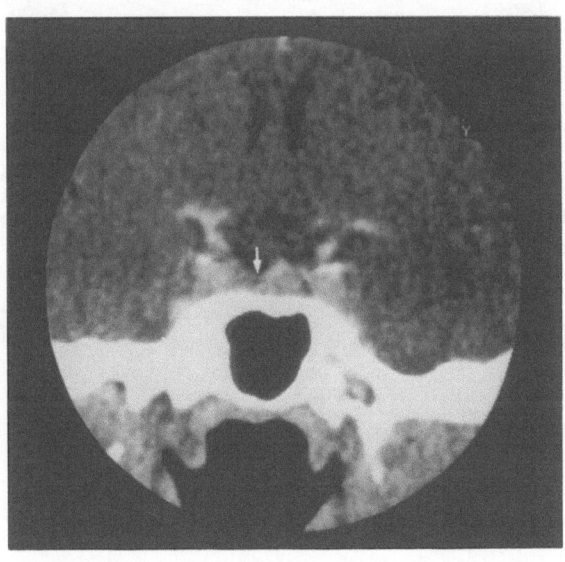

Fig. 18.1. Q: Low-dense area at the anterior part of the pituitary. Is it a pituitary adenoma?

A: On coronal section, demonstration of an asymmetrically depressed sellar diaphragm, allowing intrasellar CSF herniation on one side (*arrow*)

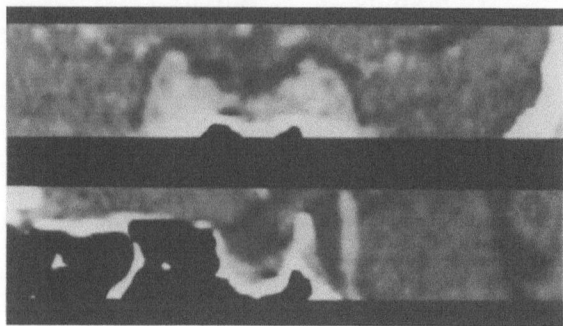

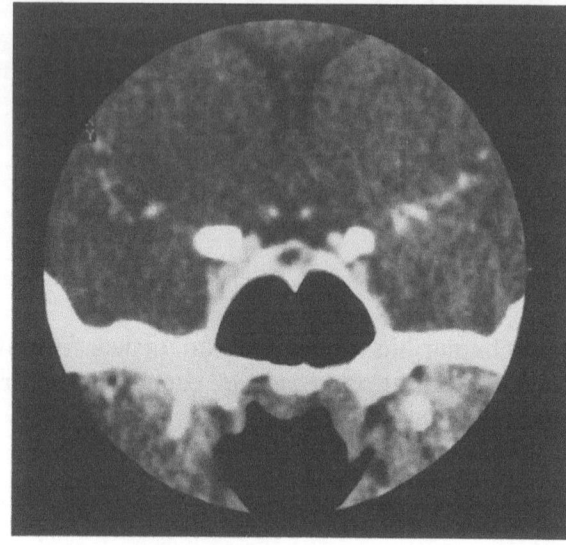

Fig. 18.2. Q: When a low-dense area is demonstrated on reformatted images...

A: Pay attention to possible streak artifacts (see also Figs. 1.10 and 1.11)

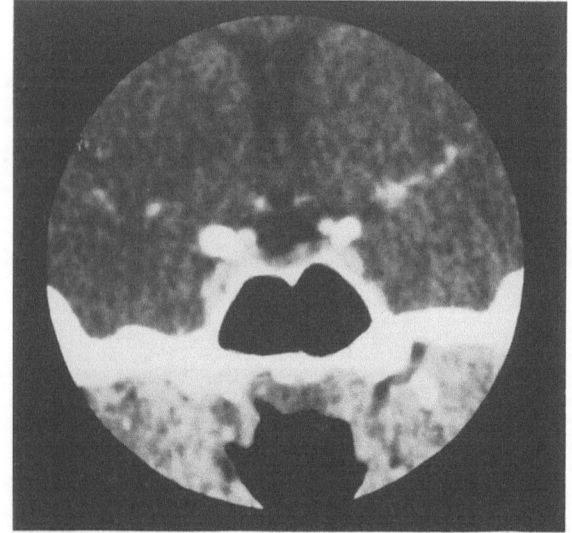

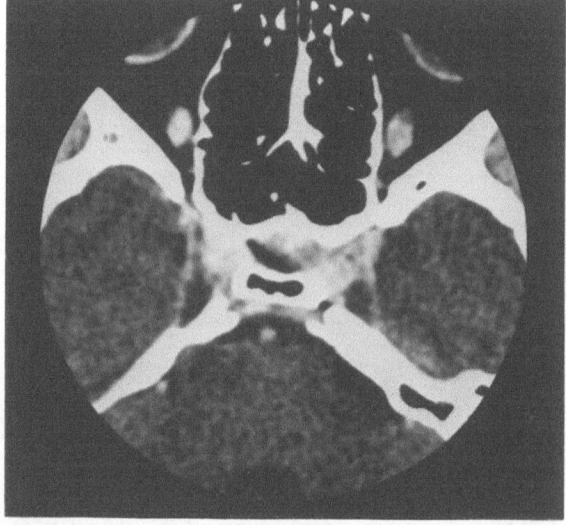

Fig. 18.3. Q: When a low-dense area is demonstrated on the most anterior coronal section...

A: Remember the "tuberculum sign" (see also Fig. 1.14)

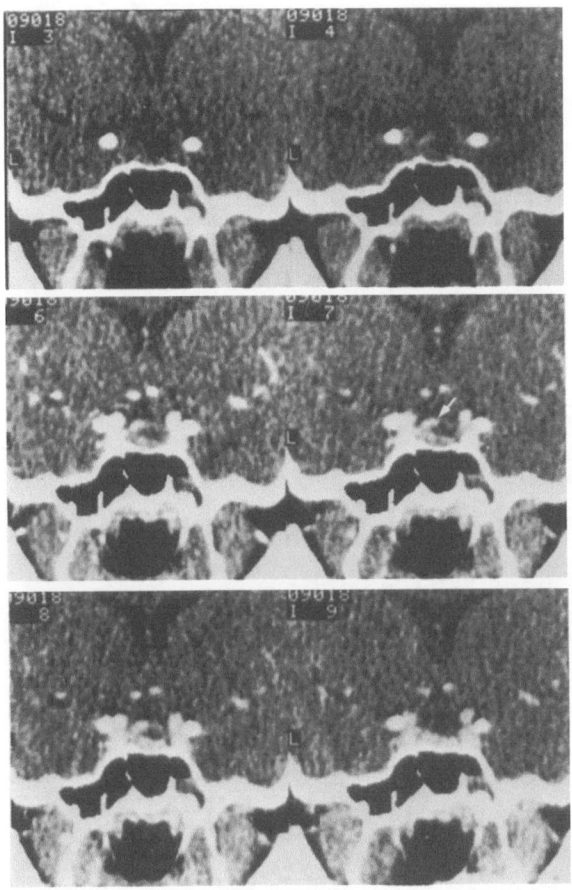

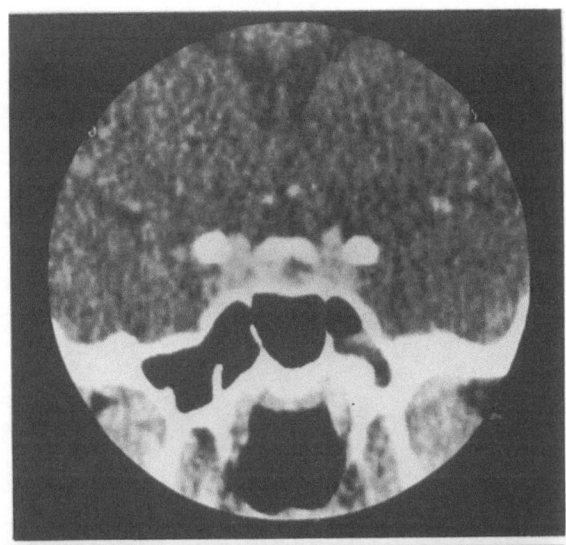

Fig. 18.4. Q: Demonstration by dynamic CT scan of a transient unilateral hyperdensity (*arrow*) above the pituitary. Partial ring enhancement of a tumor? Vascular malformation? But the image is no longer visible lower right (10 s after lower left)

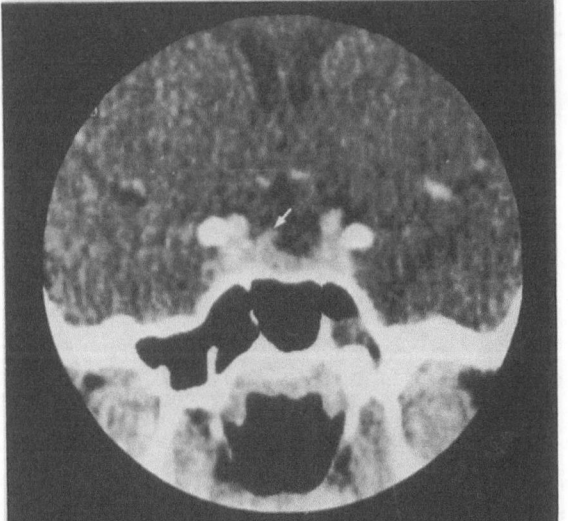

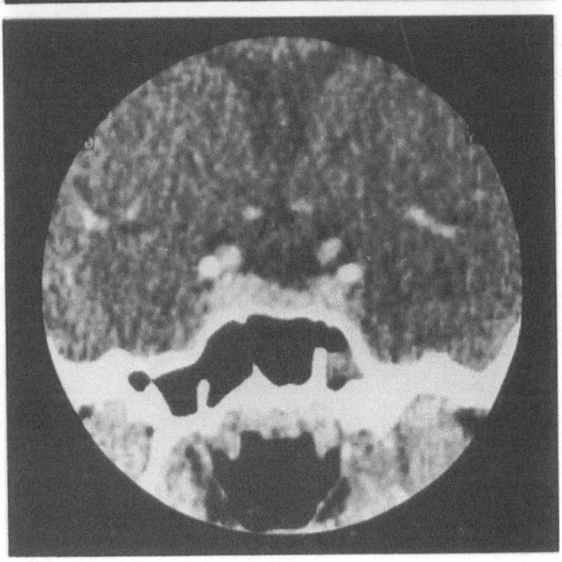

A: The patient's head was moving during dynamic CT ▶ scan, so that the right part of the tuberculum (*arrow*) was not continuously visible during all the dynamic CT scan. Here are three consecutive 1.5 mm sections from front to back for explanation

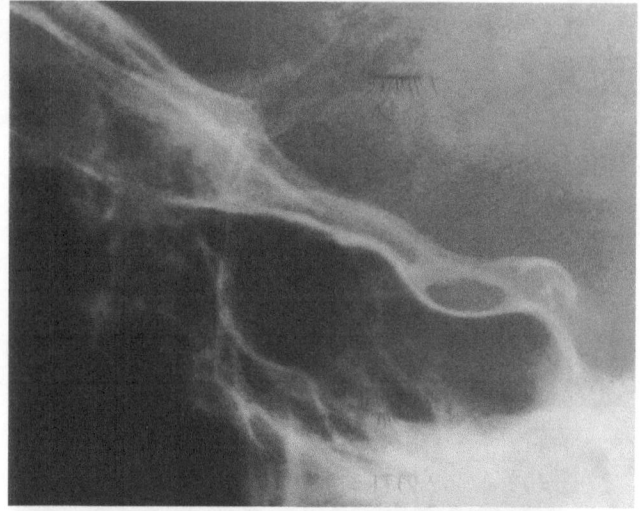

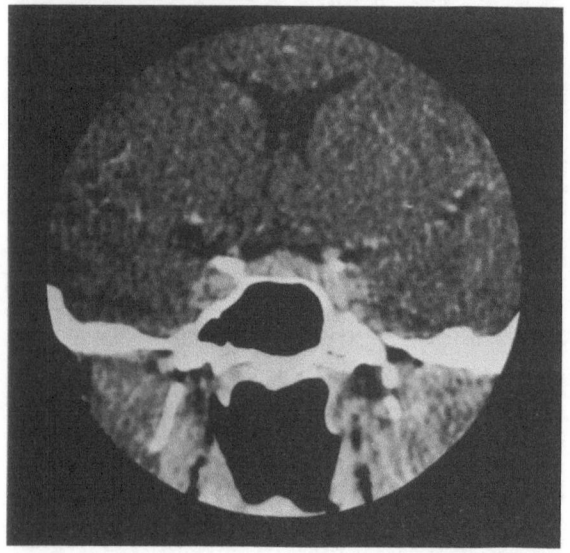

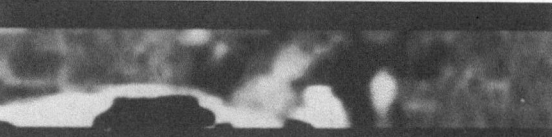

Fig. 18.5. Q: Is the convex pole of this pituitary gland suspect?

A: No, this pattern is normal when the sella turcica is flat. The volume of the gland is normal

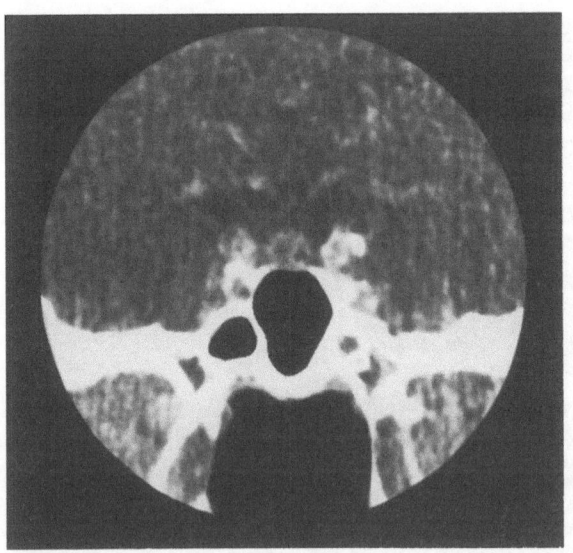 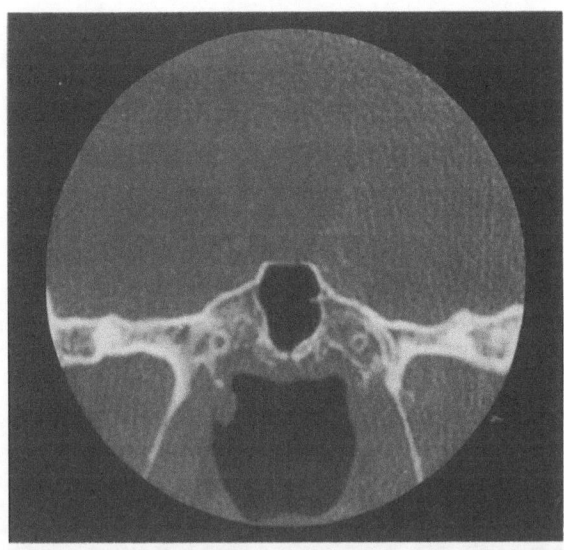

Fig. 18.6. Q: And what about this pituitary?

A: This is also normal; the sellar floor is here very short transversally. The volume of the pituitary is not modified

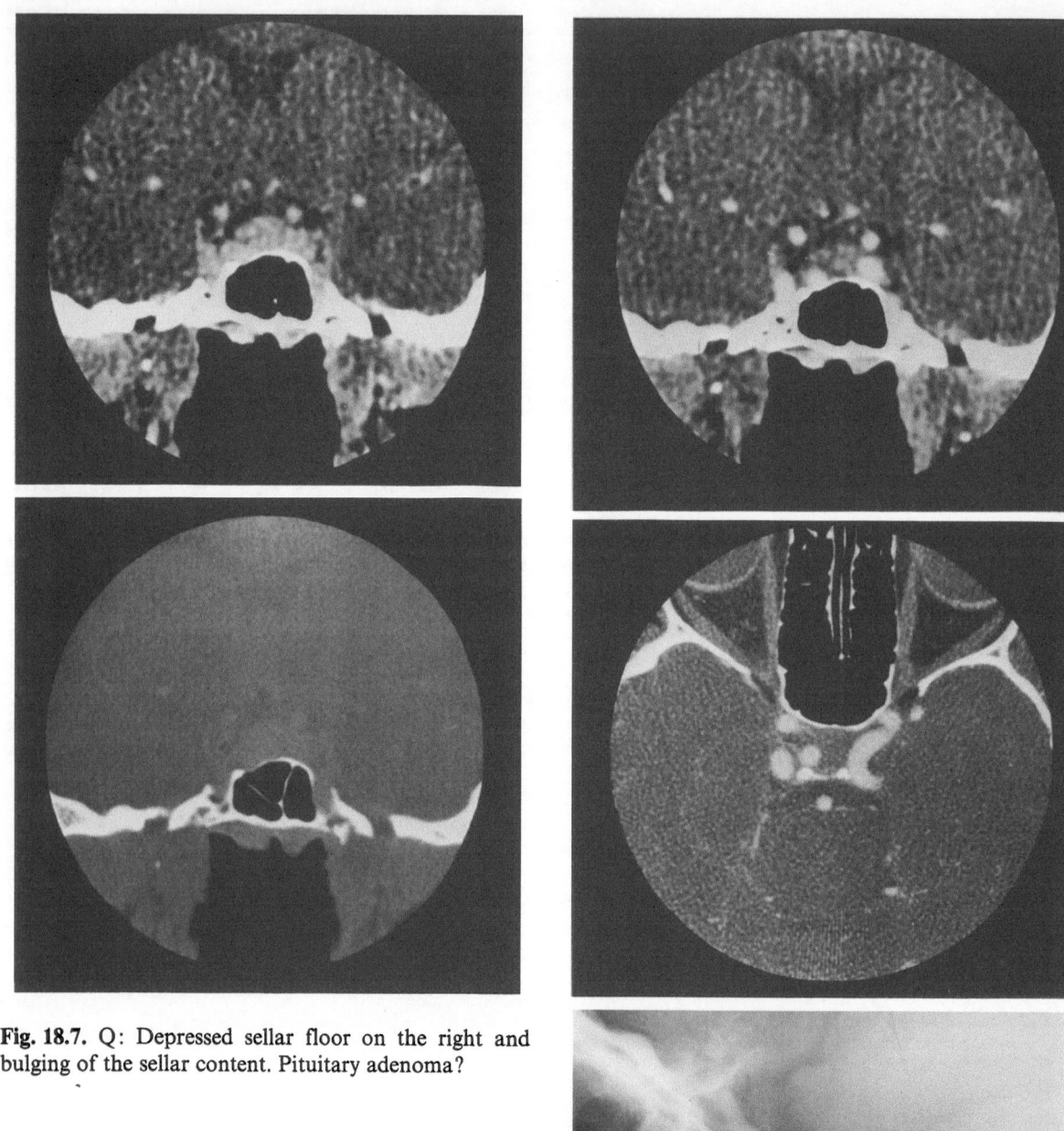

Fig. 18.7. Q: Depressed sellar floor on the right and bulging of the sellar content. Pituitary adenoma?

◄A: Coronal and axial dynamic CT scans reveal intrasellar carotid siphons. Normal pituitary gland consequently develops upwards. Once more the volume of the pituitary is normal. Now if you have a look on the lateral view of the sella, you could perhaps suspect the diagnosis from this X-ray film. The line doubling the anterior wall and the sellar floor (*white arrow*) looks like a carotid sulcus and a true double floor together: it divides into an upper line (*black arrowhead*) which joins the most posterior part of the sellar floor, corresponding to a depressed sellar floor and into a descending line (*white arrowhead*) which is the carotid artery imprint. Thus, you can conclude that the internal carotid artery lies on a depressed sellar floor, which is confirmed by dynamic CT scan

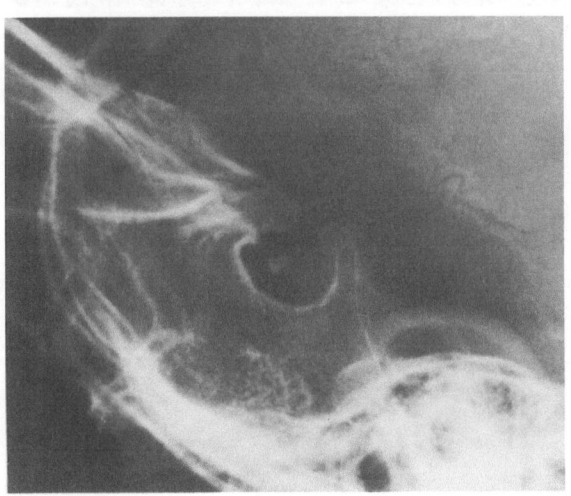

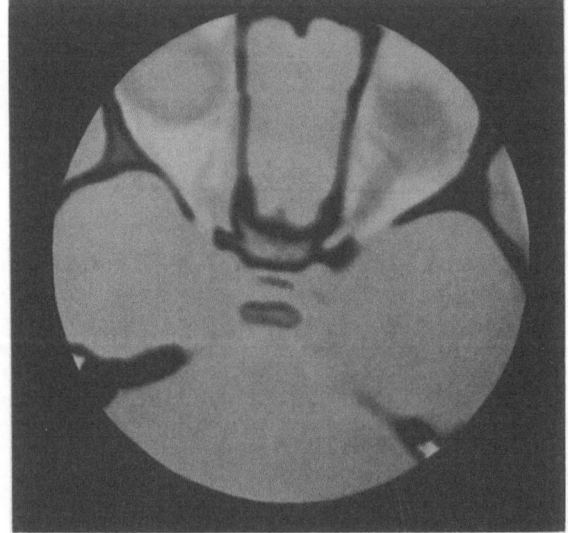

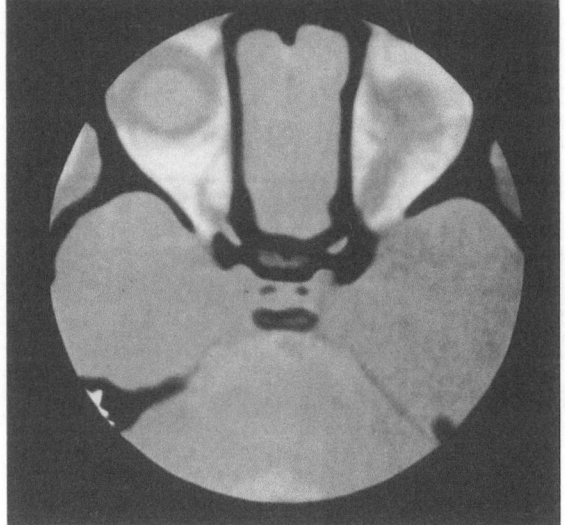

Fig. 18.8. Q: Discovery of an intrasellar "calcification" in this 4-year-old boy with delayed growth prompted a CT scan. Craniopharyngioma?

A: No. Axial CT scan with bone window shows a normal synostosis crest ending in partially calcified middle clinoid processes

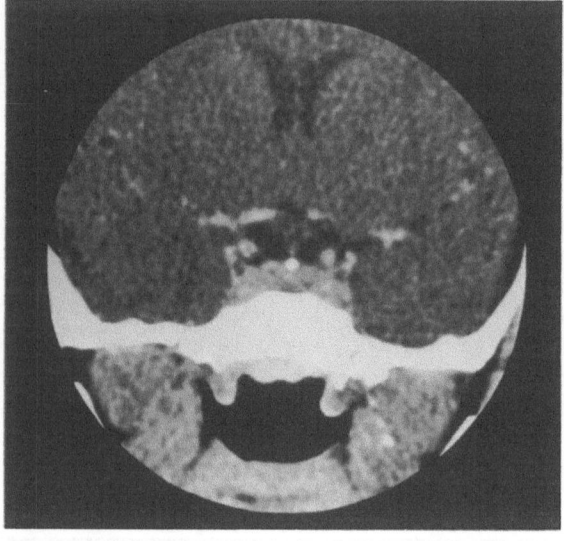

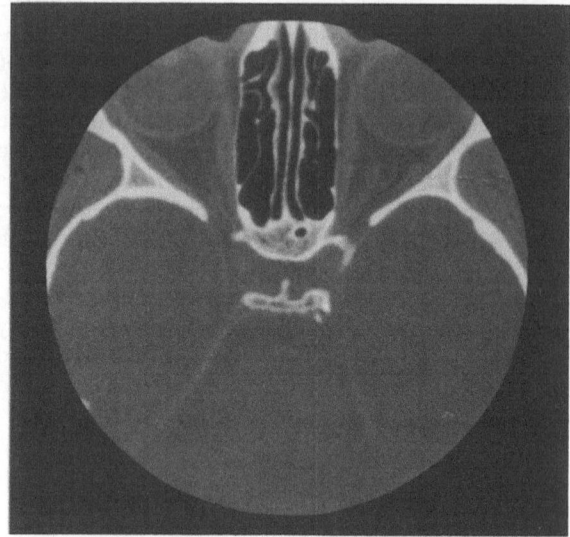

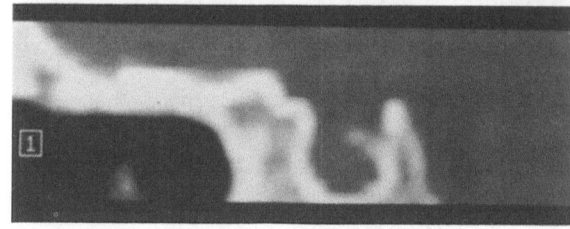

Fig. 18.9. Q: What about this dense point at the lower end of the pituitary stalk?

A: Axial and sagittal reformatted image demonstrate a sellar spine which is a rare normal variant

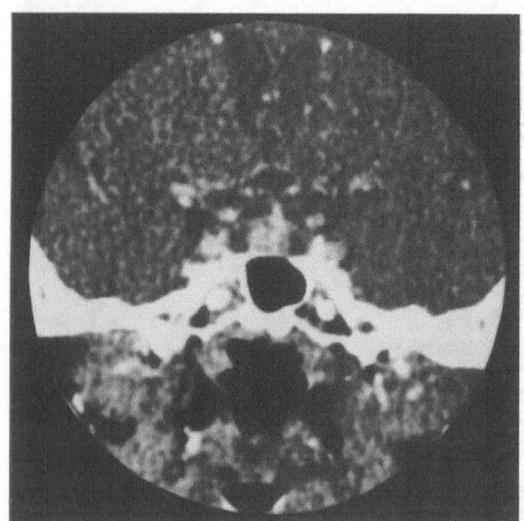

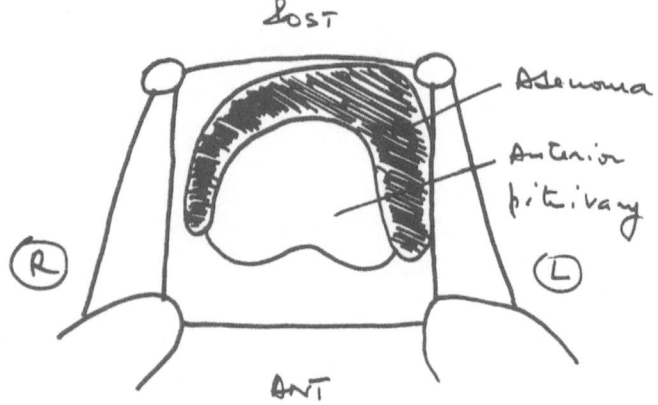

Fig. 18.10. Q: What about this "sandwich pituitary"?

A: At surgery, horse-shoe adenoma. See surgeon's drawing (G. Perrin, MD)

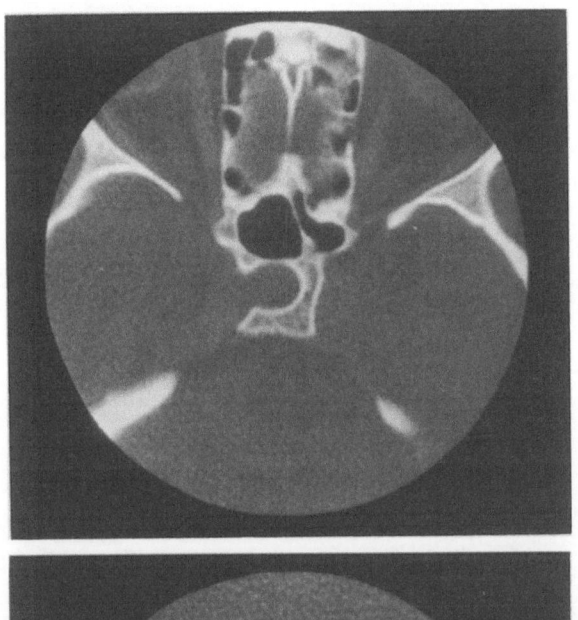
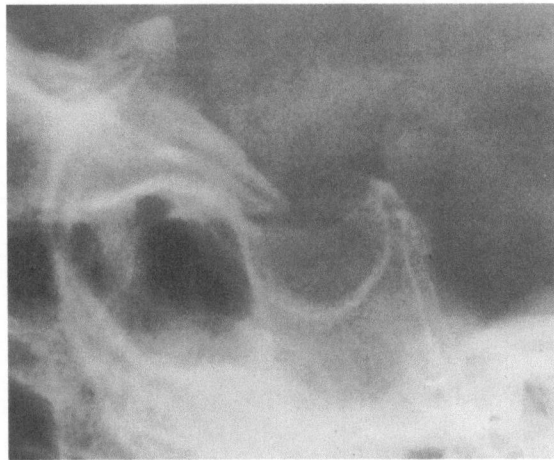
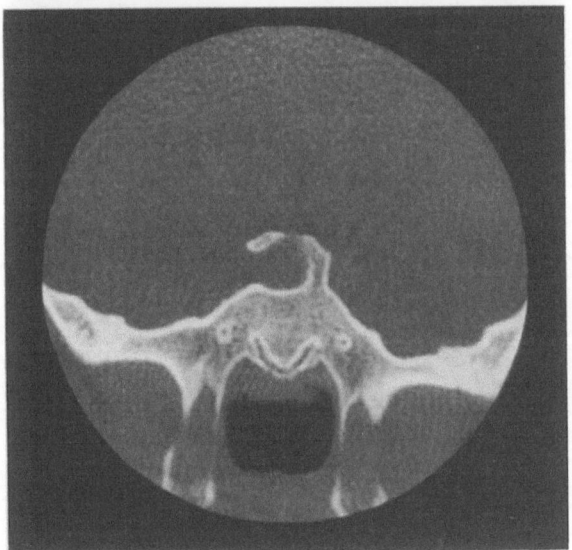

Fig. 18.11. Q: What about this unusual sella?

A: Ossification of a lateral wall of the sella is clearly shown on this lateral magnified view

Chapter 19

Magnetic Resonance Imaging of the Sellar and Juxtasellar Region

Michael Mu Huo Teng[1] and Klaus Sartor[2]

Introduction

Magnetic resonance imaging (MRI) is a tomographic method that utilizes the physical phenomenon of nuclear magnetic resonance (NMR). Most MR imagers operate at field strengths of between 0.15 and 1.5 Tesla (T), the superconducting magnets with field strengths above 0.3 T generally possessing better image quality. The imaging process is roughly as follows: first the patient is moved into the magnet. After a short tuning procedure bursts of radiofrequency (RF) waves, emitted from an RF coil inside the magnet, are radiated into the patient's body. The purpose of this is to excite the hydrogen nuclei or protons in the tissue. For this to be successful, the RF needs to match exactly the frequency with which the protons spin around the (z) axis of the magnetic field. Only then, at Larmor-frequency, resonance, i.e., energy transfer, occurs. By superimposing gradient fields over the main magnetic field and employing RF frequencies of defined band widths, selective excitation of predetermined body slices can be achieved. Following excitation the energy picked up by the protons is given off again, and a portion of it induces a voltage in the RF coil which now serves as antenna. The frequency of this signal encodes the origin in the body, while its amplitude represents the local proton density. The signal decays exponentially with a time constant T2.

For imaging purposes multiple RF bursts and signal recordings (pulse sequences) are necessary. In spin-echo imaging, currently the preferred technique, the time between repetitive cycles (TR) can be varied within a wide range. The same is true for the time between the excitation pulse and signal recording (TE). With each pulse sequence the net magnetization of the protons is tipped into the (transverse) xy-plane, thereby losing its component in the z-axis. Realignment with the z-axis (the axis of the main magnetic field) and restoration of its original magnitude again occurs exponentially, with a time constant T1. The T1 process is also called longitudinal or spin-lattice relaxation (of protons), whereas the T2 process mentioned above is called transverse or spin-spin relaxation. The relaxation times T1 and T2 are tissue dependent and therefore can be used to characterize tissue; T1 is usually considerably longer than T2.

In the spin-echo technique, the signal intensity is dependent, in addition to T1 and T2, on proton density or density of hydrogen nuclei H and an flow f (v). The relationship between these factors (given by the tissue), and TR and TE (controlled by the operator) is expressed in the following equation:

Signal intensity
$$I = H\,f(v)\,[\exp(-TE/T2)]\,[1 - \exp(-TR/T1]$$

The equation shows that shortening of TR increases the T1 influence on contrast while lengthening of TE increases the T2 influence. Therefore, pulse sequences with a short TR and a short TE, e.g., 300 and 30 ms, respectively, are T1 weighted. Conversely, pulse sequences with a relatively long TR and a long TE, e.g., 1500 and 120 ms, respectively, are T2 weighted.

---

[1] Department of Radiology, Veterans General Hospital, National Yang-Ming Medical College, and National Defense Medical Center, Taipei/Taiwan
Formerly Fellow in Neuroradiology and Instructor, Mallinckrodt Institute of Radiology, Washington University School of Medicine, St. Louis, Missouri/USA
[2] Associate Professor of Radiology/Neuroradiology, Mallinckrodt Institute of Radiology, Washington University School of Medicine, St. Louis, Missouri/USA

Pulse sequences with a relatively long TR and a short TE, e.g., 1500 and 30 ms, respectively, may be called "mixed." In the abscence of any flow it is essentially the proton density that influences the signal intensity.

The technology applied to transform all the signals recorded into a gray scale image is similar to the one used in CT. The first MR image, of thin-walled glass capillary tubes, was obtained by Lauterbur in 1973. In 1977 Damadian et al. reported the first MR image of a living human, and in 1979 Hawkes et al. were successful in generating the first image of a human brain. In their paper published in 1980, Hawkes et al. also demonstrated examples of a pituitary adenoma, a craniopharyngioma, and a partially thrombosed parasellar aneurysm of the internal carotid artery.

Normal MR Anatomy

For evaluation of the sellar and juxtasellar regions, two features of MRI are particularly advantageous. The first one is the multiplanar capability of the new method: axial, sagittal, and coronal sections are easily obtained without need for changing the patient's position; also, there is no difference in the resolution of these images. The second major advantage is the lack of bone-induced artifacts: Dense osseous structures have virtually no signal; hence, they do not interfere with the signal emitted from adjacent soft tissues.

On midsagittal cuts and on cuts through the sella (Fig. 19.1), we are able to distinguish the pituitary gland itself, the pituitary stalk, the optic pathways, the suprasellar cistern, and the third ventricle; neighboring soft tissue structures are also well appreciated. The signal intensity of the normal pituitary gland differs significantly from the signal intensity of the surrounding structures. The bony confines of the sella appear black on all pulse sequences, and so does the air-containing sphenoid sinus. The suprasellar cistern is dark on T1-weighted images, gray on T1/T2-mixed or proton density images, and bright on T2-weighted images, i.e., its signal intensity changes in the same way that that of most of the other CSF-containing spaces does.

Because of the high blood flow, the carotid artery lateral to the pituitary gland appears almost black. This contrasts sharply with the relatively bright appearance of the cavernous sinuses ("even echo-rephasing" secondary to slow laminar blood flow; compare with Fig. 19.9d). Therefore, the separation of pituitary tissue from the cavernous sinus may be difficult on axial or coronal cuts.

On T1-weighted spin-echo images, fat has the strongest signal intensity, followed by white matter, gray matter, musculature, CSF, and bone. On T2-weighted images, gray matter has a higher signal than white matter, and the strongest signal comes from CSF. The signal intensity of normal pituitary tissue remains close to that of brain.

Pituitary Adenomas

Because of its characteristic normal appearance, significant changes in size or signal intensity of the pituitary gland are easily recognized on MR scans. Extension of neoplastic tissue into the clivus, the sphenoid sinus, or the suprasellar space is particularly well appreciated (Fig. 19.2). Lateral extension is also readily seen; however, the distinction between cavernous sinus and tumor may be difficult at times. In our experience, the majority of pituitary adenomas have an abnormally high signal intensity on T2-weighted images (long T2). Occasionally the signal may be similar to that of CSF, necessitating additional T1-weighted or "mixed" images to rule out empty sella (which then would appear dark). T1-weighted images are also helpful in demonstrating previous intratumoral hemorrhage which causes the signal to be high (short T1 instead of the more frequent long T1 of pituitary tumors). In some cases, T1 and T2 values are not appreciably different from the respective values of the normal gland; this is reflected in Fig. 19.4. Generally, spin-echo imaging with two echos (one, for example, at 30 ms, the other one at 120 ms), plus a relatively long TR (1.5–2.0 s), and a slice thickness of 5–10 mm should be sufficient in most macroadenomas. Thinner slices are necessary for reliably diagnosing microadenomas, which is presently best achieved with magnets that have field strengths of 1.0 T and up. The same applies to the evaluation of the postoperative pituitary fossa, especially after transsphenoid surgery. Recognition of a primary empty sella is usually easy when T1-weighted as well as T2-weighted images are obtained.

Craniopharyngiomas

The vast majority of these tumors are suprasellar in location. Generally the pituitary gland remains normal in size, shape, and signal intensity. On T1-weighted images most of the lesion, or at least a large portion (which may or may not be cystic on CT), has a low signal intensity. Differentiation from CSF, which has an even lower signal intensity, and brain (higher signal) is usually possible. On T2-weighted images, most craniopharyngiomas have a very strong signal intensity, appearing considerably brighter than CSF and quite different from pituitary adenomas (Figs. 19.3 and 19.4).

Meningiomas

The MR diagnosis of meningiomas, in general, presents a challenge since the contrast between tumor and adjacent brain is poor on all spin-echo and inversion recovery pulse sequences. Juxtasellar meningiomas are no exception, and especially the small ones that have little effect on the suprasellar cistern and are unaccompanied by brain edema may be very difficult to detect. In fact, they may be missed altogether. This is partly explained by the calcium contained in many of these tumors, whose T1 and T2 values are often not too different from the ones of brain. Intravenous administration of gadolinium-DTPA, a paramagnetic agent, causes enhancement of meningiomas or T1-weighted and "mixed" images, which can be helpful in such diagnostic situations.

Other Tumors

Gliomas involving the suprasellar portion of the optic pathways (Fig. 19.5), hypothalamic gliomas, germinomas (Fig. 19.6), malignant lymphomas, chordomas (Fig. 19.7), and other tumors occurring in the sellar/juxtasellar region are often easily detected by MRI. They tend to have signal intensities similar to brain tissue on T1-weighted images, but relatively high intensities (however, usually not as high as craniopharyngiomas) on T2-weighted images. What remains difficult, though, is the differential diagnosis. For example, unless extension of the tumor onto the orbital portion of the optic nerve can be demonstrated, glioma of the chiasm may be indistinguishable from germinoma. Edema of the hypothalamus and adjacent portions of the basal ganglia seems to be more frequent in hypothalamic gliomas and malignant lymphomas. Chordomas have a fairly characteristic appearance (Fig. 19.7).

Sarcoidosis

Involvement of suprasellar structures by this disease is also relatively easily recognized, provided the granulomatous changes are advanced enough and not just microscopically small (Fig. 19.8). However, on the basis of MR features alone, it may be impossible to differentiate such changes from some of the neoplastic processes mentioned above. In the limited number of cases we have seen, clinical information as well as the results of other radiologic studies, e.g., chest films (showing bilateral adenopathy) were crucial in making the distinction.

Aneurysms

As a result of high blood flow nonthrombosed aneurysms appear black on MR scans, on T1-weighted as well as "mixed" and T2-weighted images (Fig. 19.9). Thrombosed portions tend to have a high signal on T1-weighted images. This is due to the absence of flow and a paramagnetic effect with shortening of the T1-relaxation time. Certainly very small aneurysms with diameters of only a few millimeters could be missed, but MRI is not meant to be the definitive test for aneurysms. Rather it should be used for further evaluating the topography of known lesions or assist in differential diagnosis. As a noninvasive procedure the new imaging technique can be utilized, for example, in the differentiation "one mass vs aneurysm." In the past arteriography and more recently digital venous angiography and dynamic CT have been used for this purpose. MRI would seem to be the only available test in patients with a history of severe allergy to contrast material.

Lesions with Sphenoid Origin

Normally, the cortex of the sphenoid bone and the (air-containing) sphenoid sinus appear black on MR images. The cancellous portions of the clivus give a relatively strong signal that is similar to the one of subcutaneous fat. This is mainly related to the constitution of the bone marrow which has a relatively short T1, thus appearing rather bright on T1-weighted images. On T2-weighted images, normal bone marrow is essentially isointense with brain. As a result neoplastic infiltration of the sphenoid bone and clivus is easily recognized, particularly on T1-weighted images. Similarly, replacement of air within the sinus by abnormal tissue or fluid should not be missed either. Metastatic disease, local extension of nasopharyngeal carcinoma, changes caused by chordomas (Fig. 19.7), chondromas and other aggressive tumors of this region, as well as inflammatory processes originating in the sinuses are optimally demonstrated, therefore, by MRI. An example of mucopyocele is shown in Fig. 19.10.

Summary and Outlook

The major advantages of MRI in the evaluation of the sellar and juxtasellar region are: (a) multiplanar capability, (b) superior soft-tissue discrimination, (c) absence of bone-induced artifacts, (d) flow sensitivity, and (e) lack of invasiveness and radiation hazards. The major drawbacks are: (a) volume averaging artifacts, (b)

inability to detect subtle calcification, and (c) artifacts and potential hazards from metallic surgical/dental materials and devices. Undoubtedly, some of these problems will be overcome soon. Specifically, thinner slices without sacrifice in signal-to-noise ratio should reduce the artifacts and diagnostic difficulties associated with the relatively thick slices presently available. In all probability, MRI will attain a prominent if not dominant role in the future.

References

Araki T, Inouye T, Suzuki H, Machida T, Iio M (1984) Magnetic resonance imaging of brain tumors: Measurement of T1. Radiology 150:95–98

Bilaniuk LT, Zimmermann RA, Wehrli FW, Snyder PJ et al. (1984) Magnetic resonance imaging of pituitary lesions using 1.0 to 1.5 T field strength. Radiology 153:415–418

Bradley WG Jr, Waluch V (1985) Blood flow: Magnetic resonance imaging. Radiology 154:443–450

Damadian R, Goldsmith M, Minkoff L (1977) NMR in cancer: XVI FONAR images of the live human body. Physiol Chem Phys 9:97–108

Davis PL, Kaufman L, Crooks LE (1983) Tissue characterization. In: Margulis A, Higgins CB, Kaufman L, Crooks LE (eds) Clinical magnetic resonance imaging. Radiology Research and Education Foundation. San Francisco

Fullerton GD (1982) Basic concepts for nuclear magnetic resonance imaging. Magn Reson Imaging 1:39–55

Han JS, Huss RG, Benson JE et al. (1984) MR imaging of the skull base. J Comp Ass Tomog 8:944–952

Hawkes RC, Holland GN, Moore WS, Worthington BS (1980) Nuclear magnetic resonance (NMR) tomography of the brain: A preliminary clinical assessment with demonstration of pathology. J Comp Ass Tomog 4:577–586

Hawkes RC, Holland GN, Moore WS, Corston R, Kean DM, Worthington BS (1983) Application of NMR imaging to the evaluation of pituitary and juxtasellar tumors. AJNR 4:221–222

Lauterbur PC (1973) Image formation by induced local interactions: Examples employing nuclear magnetic resonance. Nature 242:190–191

Mark L, Pech P, Daniels D, Charles C, Williams A, Haughton V (1984) Pituitary fossa: A correlative anatomic and MR study. Radiology 153:453–457

New PFJ, Rosen BR, Brady RJ, Buonanno FS et al. (1983) Potential hazards and artifacts of ferromagnetic and nonferromagnetic surgical and dental materials and devices in nuclear magnetic resonance imaging. Radiology 147:139–148

Oot R, New PJ, Buonanno FS, Pykett IL et al. (1984) MR imaging of pituitary adenomas using a prototype resistive magnet: Preliminary assessment. AJNR 5:131–137

Zimmermann RD, Fleming CA, Saint-Louis LA, Lee BCP, Manning JJ, Deck MDF (1985) Magnetic resonance imaging of meningiomas. AJNR 6:149–157

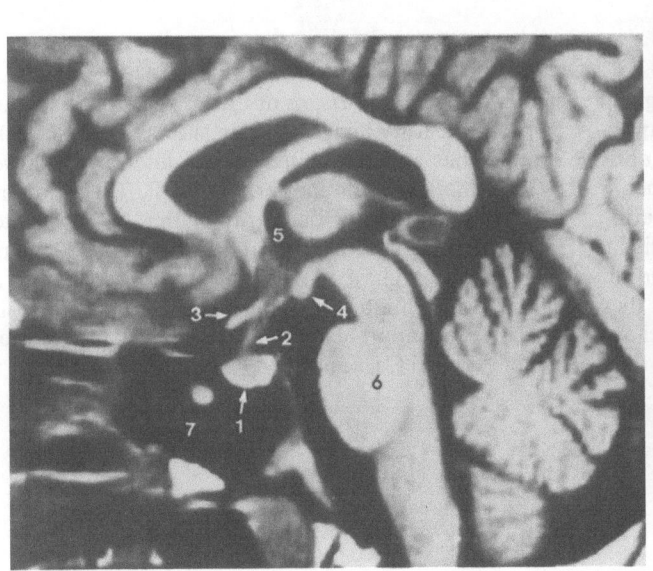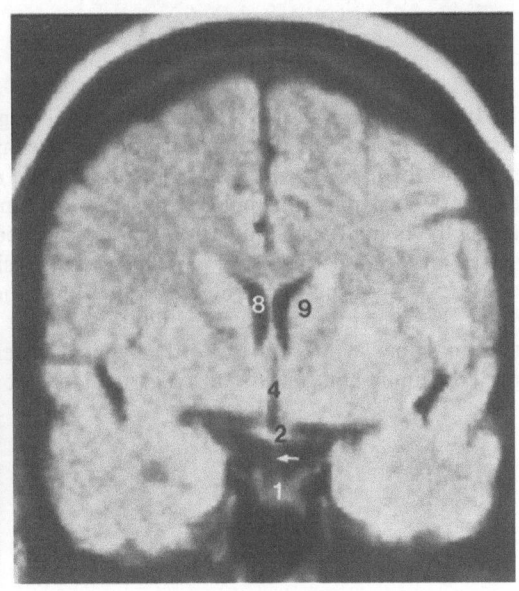

a   b

Fig. 19.1a, b. Normal appearance of the brain and sellar region. Midsagittal 3-mm thick partial saturation 1.5 T image with good demonstration of the pituitary gland (*1*), pituitary stalk (*2*), optic chiasm (*3*), mamillary body (*4*) and third ventricle (*5*) pons (*6*), shenoid sinus (*7*), frontal horn (*8*), caudate nucleus (*9*). (Reproduced with permission of the General Electric Co.)

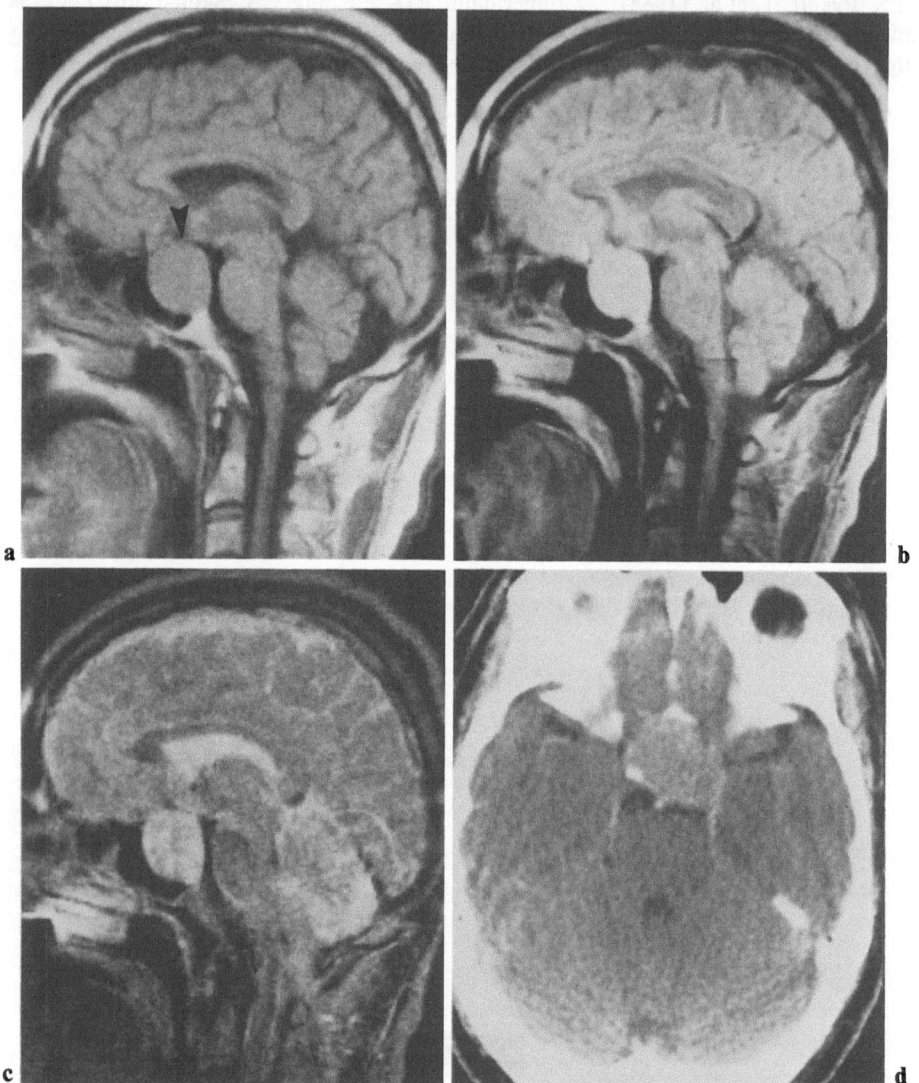

Fig. 19.2a–d. Large pituitary adenoma. **a** T1-weighted midsagittal image (SE 300/30) revealing suprasellar extension of the lesion; *arrowhead* points to elevated optic nerve and chiasm. In this pulse sequence subcutaneous fat has the highest signal intensity, therefore appearing *white*, while brain tissue, tumor and CSF have low signal intensities, appearing increasingly *darker*. Bone and air within the sphenoid sinus appear *black*, because they have no signal at all, or very little signal. Cancellous bone containing bone marrow also has a relatively bright signal, e.g., lower portions of the clivus. **b** T1/T2-mixed or proton density image of same slice as in **a** (SE 1500/30). Fat has still the highest signal intensity, followed by gray matter, white matter, and CSF. The lesion appears slightly brighter than gray matter. **c** T2-weighted image of same slice as in **a** and **b** (SE 1500/120). Now CSF has the highest signal intensity, followed by gray matter, white matter, and muscle. The signal intensity of the lesion approaches the one of CSF. **d** Contrast enhanced axial CT scan

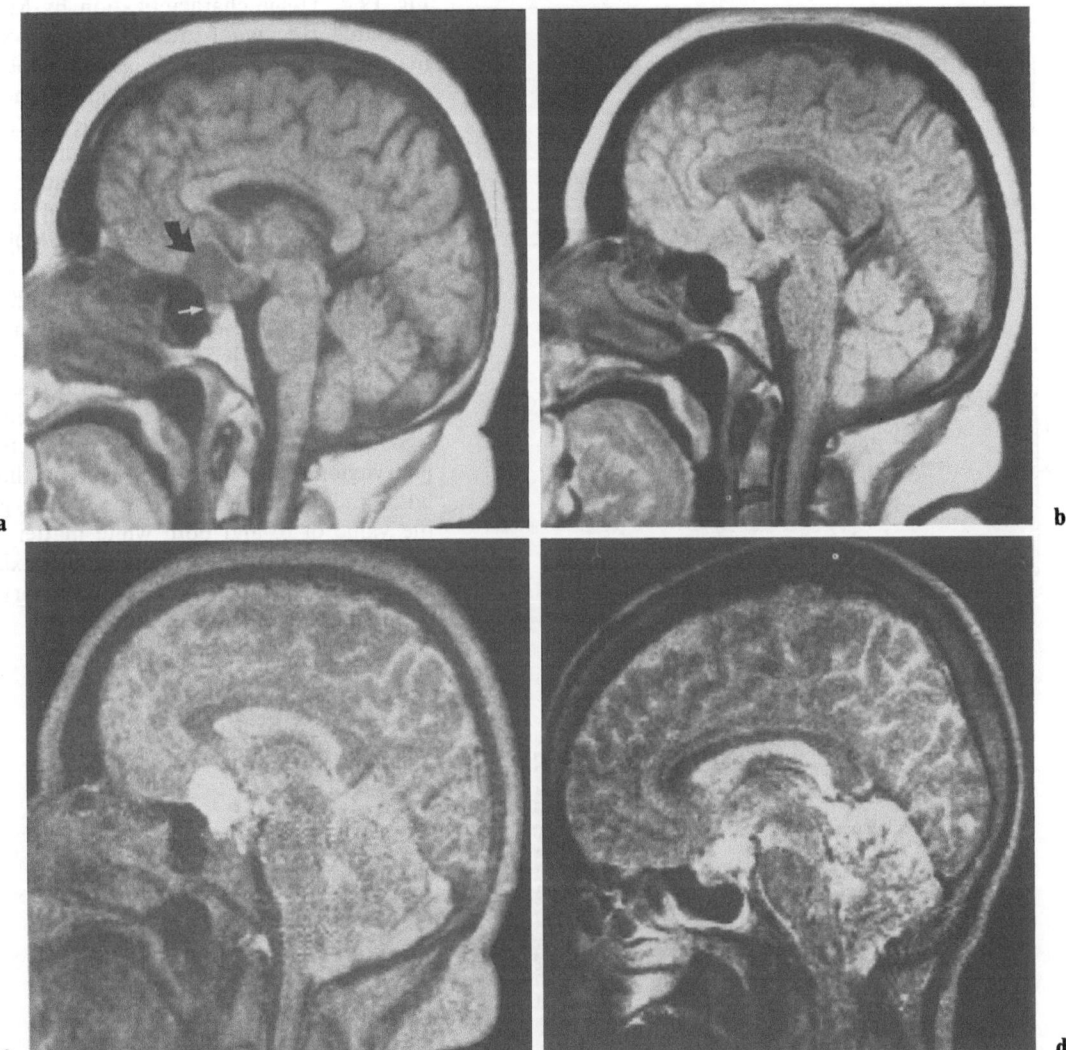

Fig. 19.3a–d. Craniopharyngioma. **a** T1-weighted mid-sagittal image showing the cystic component (*black arrow*) of the lesion anterior to the more solid component that bulges into the interpeduncular cistern. The tiny *white arrow* points to the normal pituitary gland below the (suprasellar) mass. **b** T1/T2-mixed image of the same slice as in **a**. The lesion is clearly separable from the pituitary gland, but barely from the surrounding brain. Only the posterior border contrasts with the relatively low signal intensity of CSF in the interpeduncular cistern. **c** T2-weighted image of the same slice as in **a** and **b** rendering the lesion, at least its cystic component, easily recognizable. **d** T2-weighted image of another case with craniopharyngioma

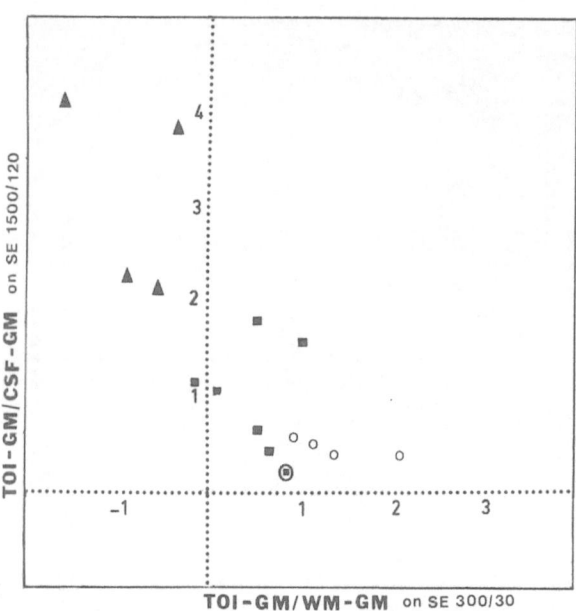

Fig. 19.4. Tissue characterization by MR imaging. We measured the signal intensities (pixel values) or gray matter, white matter, CSF, and normal pituitary gland in five cases, of pituitary adenoma tissue in seven cases, and of craniopharyngioma tissue in four cases. Then we calculated – using the pixel values on the SE 300/30 pulse sequence – the following ratio: [Tissue of interest (TOI) minus gray matter (GM)] divided by [white matter (WM) minus gray matter (GM)]. This provided coordinate values of the x-axis. Also, we calculated – using the pixel values on the SE 1500/120 pulse sequence – the ratio: [Tissue of interest (TOI) minus gray matter (GM)] divided by [CSF minus gray matter (GM)], thus obtaining corresponding coordinate values on the y-axis. As the figure shows, the resulting two-dimensional tissue signatures allow a clearer distinction between craniopharyngioma and pituitary adenoma than between the latter and normal gland. In this system, gray matter has the value of 0, and both white matter and CSF have the value of 1 on the x-axis and y-axis, respectively. o, normal pituitary tissue; ■, pituitary adenoma; ▲, craniopharyngioma

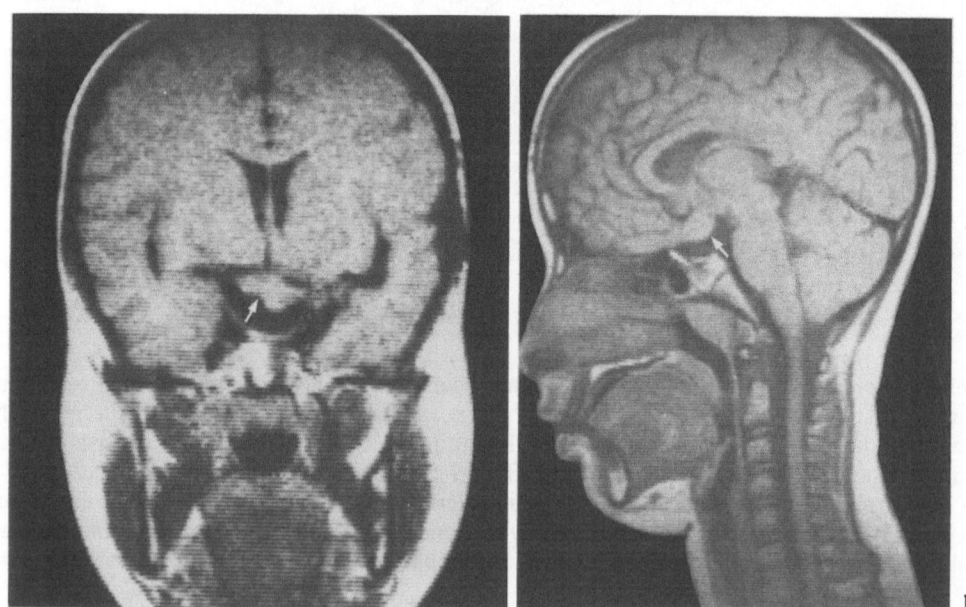

a   b

Fig. 19.5a, b. Optic glioma. **a** T1-weighted coronal image showing thickened and somewhat nodular optic chiasm (*arrow*). **b** T1-weighted midsagittal image with *arrow* pointing to enlarged optic chiasm and optic tract

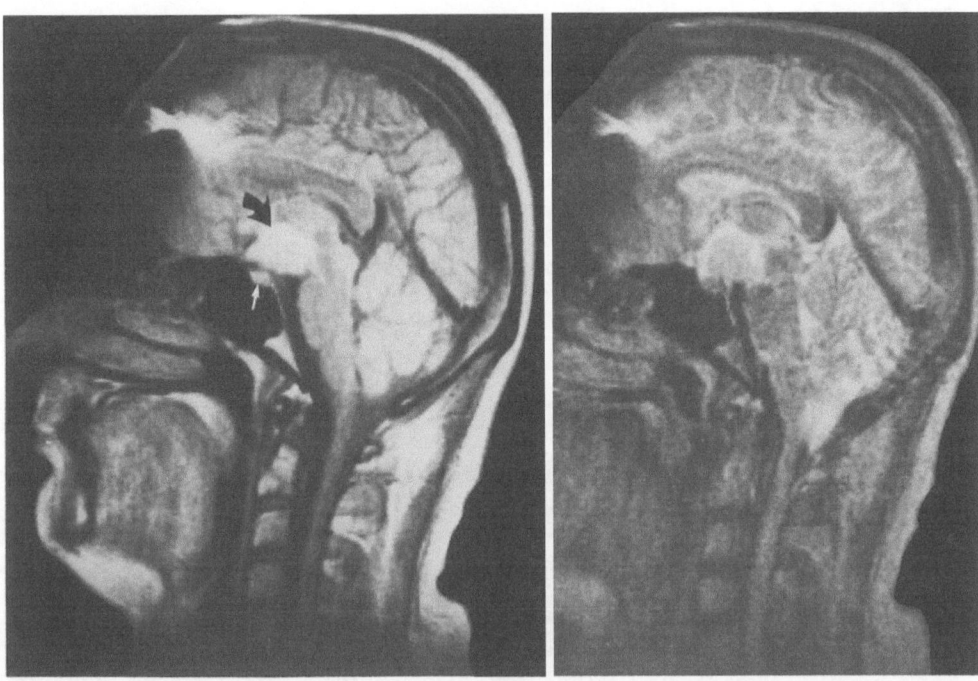

a   b

Fig. 19.6a, b. Germinoma ("ectopic pinealoma"). **a** T1/ T2-mixed or proton density midsagittal image. *Curved black arrow* marks the pear-shaped suprasellar mass which has a high signal intensity and extends into the interpeduncular cistern. Superiorly the lesion projects into the third ventricle, inferiorly it is clearly separated from the normal-appearing pituitary gland (*white arrow*). **b** T2-weighted image of same slice as in **a**. Artifact caused by bullet fragment secondary to self-inflicted gunshot wound

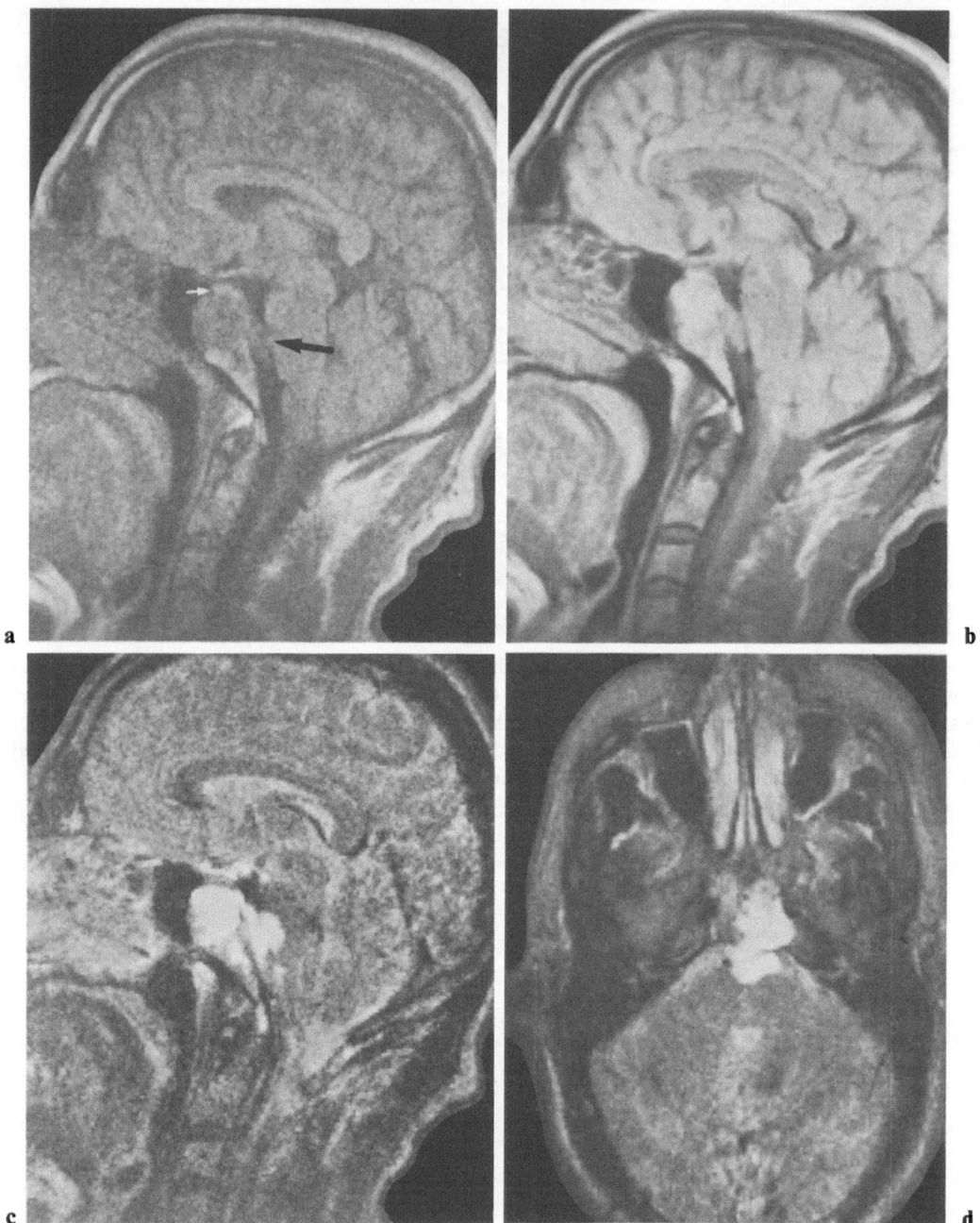

Fig. 19.7a–d. Chordoma. **a** T1-weighted midsagittal image revealing abnormal tissue between pons and back of clivus (*black arrow*) that extends into the clivus and bulges into the sphenoid sinus. The normal pituitary gland having a slightly higher signal intensity is seen on top of the lesion (*white arrow*). **b** T1/T2-mixed or proton density image of the same slice as in **a** shows the prepontine component of the mass somewhat better; the pons is flattened anteriorly and slightly pushed back-wards. **c** T2-weighted image of the same slice as in **a** and **b** demonstrates the tumor best; in this pulse sequence the pituitary gland – which has been flattened and is displaced superiorly – has a much lower signal intensity than the lesion. **d** T2-weighted axial image displaying the intraclival/infrasellar and the posterior fossa components particularly well; the basilar artery – appearing as a black dot – is pushed laterally

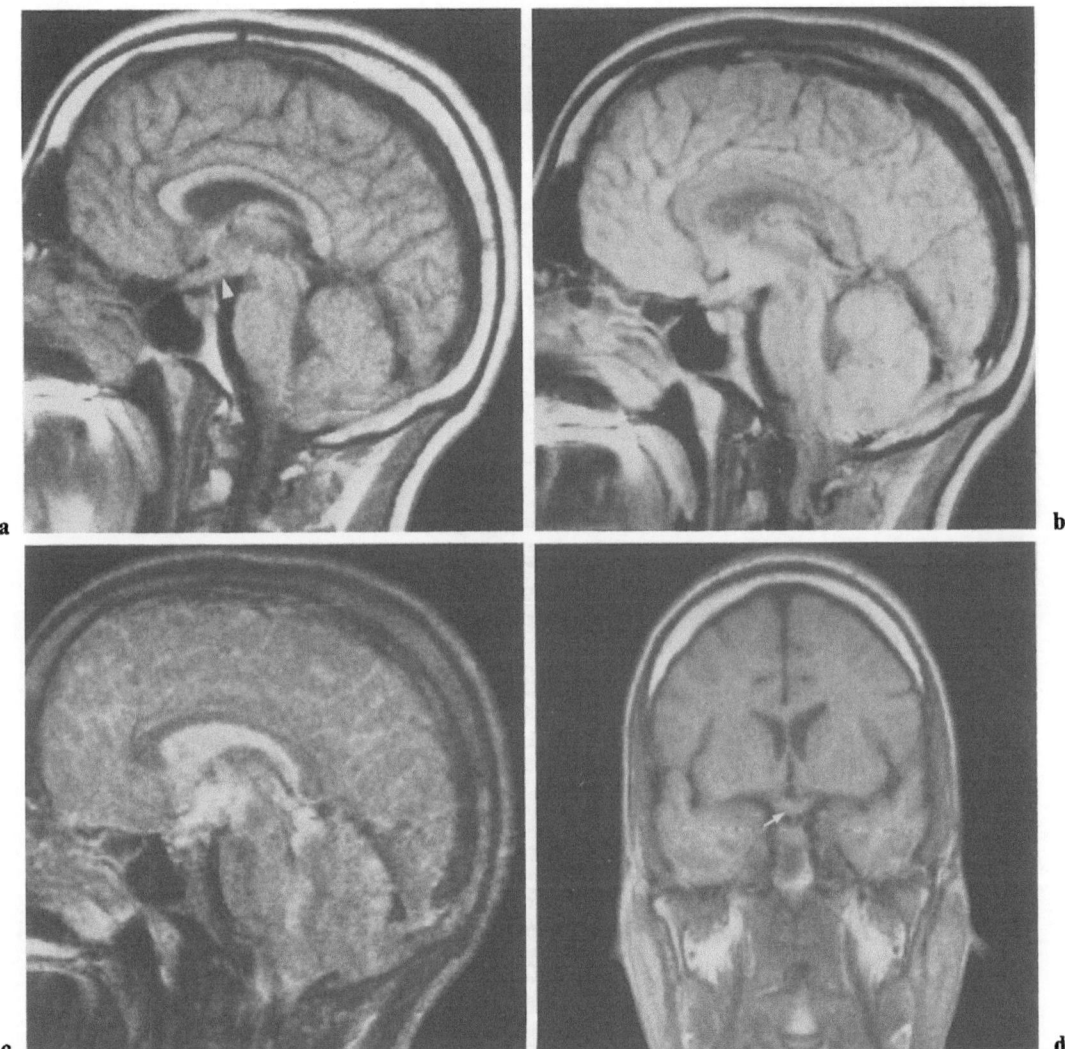

Fig. 19.8 a–d. Sarcoidosis. **a** T1-weighted midsagittal image. Optic nerve, optic chiasm, and optic tract (*arrowhead*) are markedly swollen; the lesion also appears to involve the hypothalamus. **b** T1/T2-mixed or proton density image of the same slice as in **a**. **c** T2-weighted image of the same slice as in **a** and **b**; the unusually high signal intensity in the region of the anterior third ventricle is consistent with hypothalamic disease. **d** T1-weighted coronal image showing the enlarged optic chiasm (*arrow*); note the thickened pituitary stalk below the chiasm

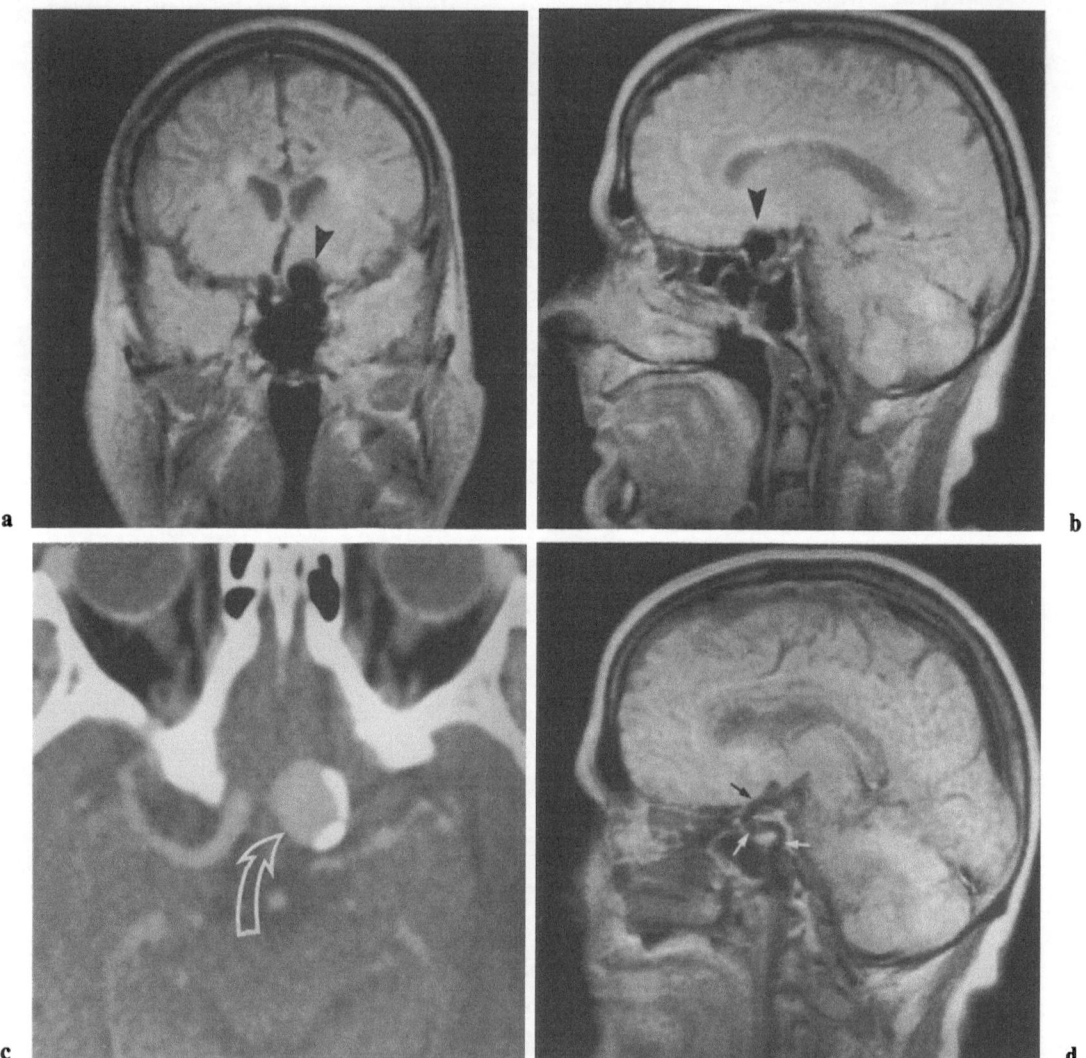

Fig. 19.9a–d. Carotid aneurysm. **a** T1/T2-mixed or proton density coronal image demonstrating the spherical aneurysmal sac (*arrowhead*). The lesion exhibits practically no signal because of high flow; note the lateral displacement of the third ventricle. **b** T1/T2-weighted parasagittal image; the aneurysm (*arrowhead*) arises from the supraclinoid portion of the internal carotid artery (compare with **d**). **c** Axial contrast CT scan showing the enhancing lesion (*arrow*); the lateral wall of the aneurysm is calcified. **d** T1/T2-weighted parasagittal image of normal side with the carotid siphon (*arrows*) well contrasted against the high signal background of the cavernous sinus. The high signal of the sinus is related to slow flow (paradoxical enhancement)

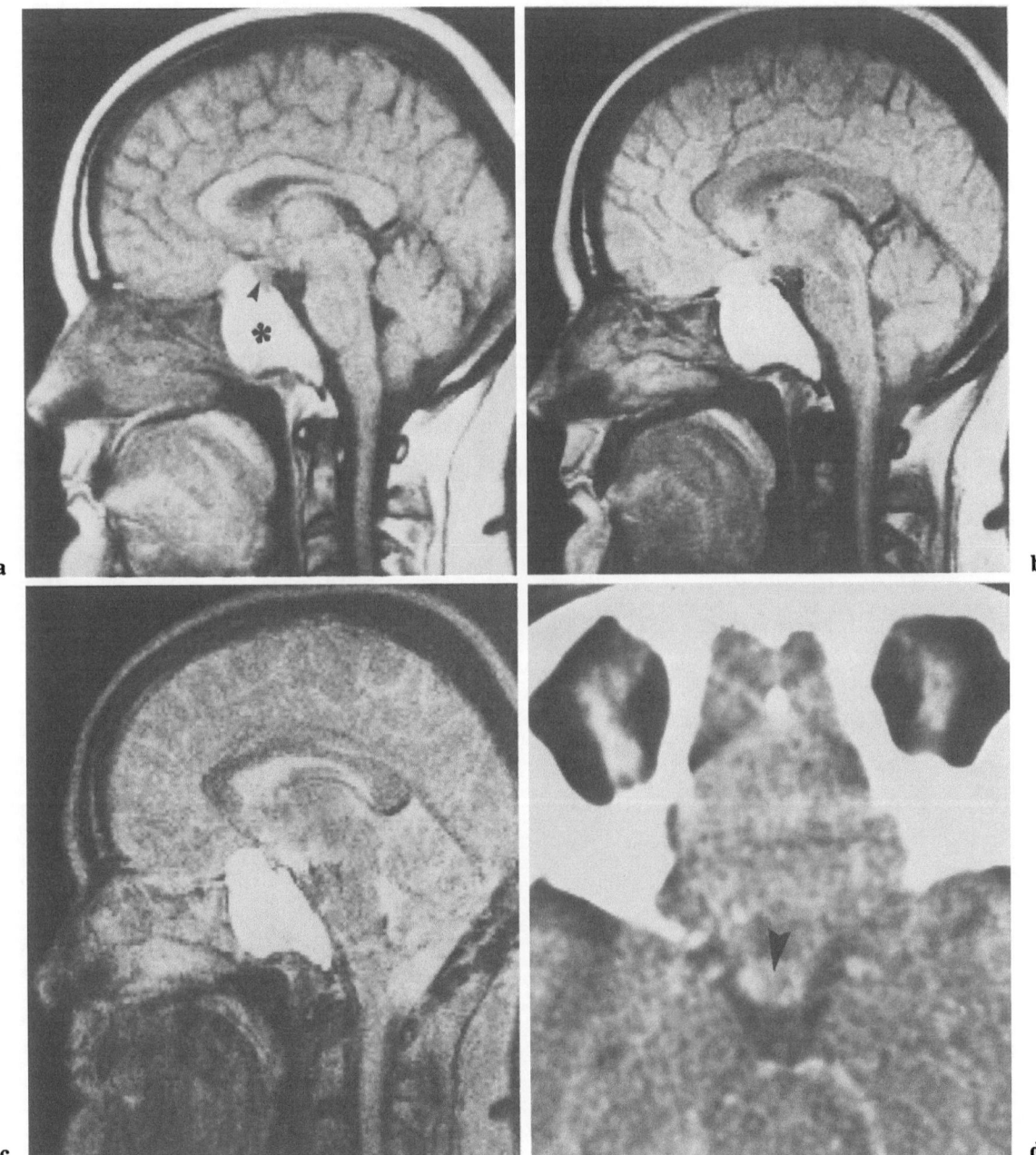

Fig. 19.10 a–d. Mucopyocele of the sphenoid sinus. **a** T1-weighted midsagittal image showing expansion of the entire sphenoid sinus. The sinus (*) has a high-signal intensity suggesting a fatty and/or hemorrhagic fluid filling it. The otherwise normal-appearing pituitary gland (*arrowhead*) has been displaced superiorly. **b** T1/T2-mixed or proton density image of the same slice as in **a**; the signal intensity becomes even higher reflecting, in addition to a short T1 value, a long T2. **c** T2-weighted image of the same slice as in **a** and **b** with a marked contrast between the lesion and the brain. **d** Axial contrast CT scan revealing a suprasellar position of the pituitary gland (*arrowhead*); under such circumstances the gland may be easily confused with a lesion of suprasellar origin

Subject Index

Italic page numbers refer to the figures